WERTHEIM PUBLICATIONS IN
INDUSTRIAL RELATIONS

BEYOND NATIONALIZATION

The Labor Problems of British Coal

GEORGE B. BALDWIN

32167

HARVARD UNIVERSITY PRESS

Cambridge, Massachusetts

1955

© Copyright, 1955, by the President and Fellows of Harvard College

Distributed in Great Britain by
Geoffrey Cumberlege, Oxford University Press, London

Library of Congress Catalog Card Number: 55–10966
Printed in the United States of America

A book such as this deserves a double dedication.

Privately, it is for HARRIET and ALAN.

Publicly, it is for those people in Britain, from all walks of life, who taught me so much about their coal industry and their country.

WERTHEIM PUBLICATIONS IN
INDUSTRIAL RELATIONS

George B. Baldwin, *Beyond Nationalization: The Labor Problems of British Coal,* 1955

Philip Taft, *The Structure and Government of Labor Unions,* 1954

Val R. Lorwin, *The French Labor Movement,* 1954

Donald J. White, *The New England Fishing Industry: A Study in Price and Wage Setting,* 1954

Theodore V. Purcell, S.J., *The Worker Speaks His Mind on Company and Union,* 1953

Lloyd H. Fisher, *The Harvest Labor Market in California,* 1953

Walter Galenson, *The Danish System of Labor Relations: A Study in Industrial Peace,* 1952

John T. Dunlop and Arthur D. Hill, *The Wage Adjustment Board: Wartime Stabilization in the Building and Construction Industry,* 1950

Ralph Altman, *Availability for Work: A Study in Unemployment Compensation,* 1950

Dorothea de Schweinitz, *Labor and Management in a Common Enterprise,* 1949

Walter Galenson, *Labor in Norway,* 1949

Leo C. Brown, S.J., *Union Policies in the Leather Industry,* 1947

Paul H. Norgren, *The Swedish Collective Bargaining System,* 1941

Johnson O'Connor, *Psychometrics,* 1934

William Haber, *Industrial Relations in the Building Industry,* 1930

Wertheim Lectures on Industrial Relations, 1929

J. D. Houser, *What the Employer Thinks,* 1927

ERRATA

Chart 5, page 150: *the key should read:*
    Heavy line = productivity   Light line = wages

Chart 8, page 215: *the key should read:*
    black-dot line = involuntary   white-dot line = voluntary

Figure 4, page 251: *the scale should read:*
    1 in. 10 mi.

FOREWORD

Professor Baldwin's study of the British coal industry is the fifth volume, out of a total of seventeen in the Wertheim series, which treats labor problems outside the United States. The comparative method is peculiarly well suited to the analysis of labor-management relations. The experience of a single country or an industry is seen in a full spectrum of developments, and attention is directed to both the common and the distinctive factors in the particular case. The British coal industry exhibits at once characteristics common to the British industrial-relations scene generally and other characteristics common to the coal industry around the world. The interplay of these forces, in the aftermath of nationalization, constitutes an exciting problem, and Professor Baldwin has provided a rewarding volume.

A great merit of the present study is that it considers the principal labor problems of the British coal industry without becoming entangled in the ideological issues of nationalization. Professor Baldwin invites us to examine what has happened in the last decade to collective bargaining, to the joint consultative committees at the pit level, to the level and structure of wages (including piece rates), to the recruitment, training and attendance of the labor force, and to the attitudes and policies toward technical change. These subjects occupy the principal chapters of the volume. The careful review of the record in these aspects is a prerequisite to any general judgment on nationalization in this industry.

In addition to substantive policies, Professor Baldwin has directed attention to important issues of organizational structure: for instance, the greater degree of centralization within the National Union of Mineworkers compared to that before nationalization and the relation of the Labor Division within the National Coal Board to line management. The organizational arrangements for decision making are seen to arise from real problems and in turn to influence subsequent decisions.

The study is an illustration of the significance of technical conditions of production in labor-management relations. The isolated and hazardous place of work, the changing conditions of work at the coal face, and the advancing technology of mining provide a distinctive flavor to the labor problems of the industry: recruitment, safety, and wage rate setting. Students of industrial relations would do well to recognize more clearly the challenge and the bent to labor-management relations provided by such technical conditions. A comparative international study of the longshore, maritime, building, railroad, air transport, textile, printing, or coal industry would provide significant evidence for this point of view. Professor Baldwin has rightfully insisted upon the importance of these technical conditions in the British coal industry for labor-management relations.

Although Professor Baldwin is most cautious about the future and attempts to provide no general theory of labor-management relations under nationalized industry, his analysis of particular problems provides much raw material for more general speculation. He notes a shift from "protest unionism" to "administrative unionism" (p. 51), despite a continuing bargaining role for the union. The proportion of coöperation and opposition in the union's policy toward the nationalized industry varies with the issue. The impact on the industry of the long-term policy of recruiting management and technicians from the sons of miners and from outside the industry may be expected to have long-term effects. The analysis of union policy and the behavior of individual miners toward technical change is likewise of wider significance. Any general appraisal of the impact of nationalization on labor-management relations will profit from this analysis of specific and stubborn problems.

JOHN T. DUNLOP

CONTENTS

TABLES

CHARTS

FIGURES

ILLUSTRATIONS
following page 6

AUTHOR'S PREFACE

This book concerns "the most difficult and the most urgent" set of problems of one of the world's great industries, an industry with a romantic and prosperous past that has latterly come on difficult times.

Unlike its American counterpart, the British coal industry is not suffering from the competition of rival fuels: Britain has no oil or water-power resources to speak of, no natural gas, and no known deposits of uranium (save for a few mocking traces found in some of her coal seams). The central problem of British coal today is not a problem of markets, which indeed was the great problem from the early twenties until 1941. The present problem is one of production, of adequate output. Ever since the early days of the Second World War, the British householder and British industry have lived in the shadow of a major fuel crisis. And in Britain a fuel crisis means a coal crisis. This crisis is not attributable to nationalization, though many offer it as a chilly commentary on this experiment.

From about 1550 until the First World War — for nearly four centuries — the British led the world in coal mining, and the phenomenal expansion of the industry between 1870 and 1913 had given it pride of place among Britain's basic trades. The Great War marked the end of this romantic era. The period from 1920 to 1940 saw a succession of market contractions and labor crises that demoralized alike the coalowners, pit managers, and numberless mineworkers and their families. Within the space of twenty years the industry cut back total output by 20 per cent, lost 40 per cent of its traditional export markets, permanently extinguished a third of its jobs at a time when no others were to be found, and cut average weekly wages by 50 per cent (in twenty years mining dropped from the first or second highest-paying industry in Britain to eighty-first in a ranking of one hundred). The average rate of return on capital fell from 9 or 10 per cent just before the First World War to 2 or 3 per cent twenty years later. In 1926, as part of this painful unraveling process, the industry experienced the longest industry-wide strike in its history, a strike which began as the only General Strike in British history. The strike was "general" nine days: the miners stayed out seven months, and lost. In 1951 some Yorkshire miners were still having money stopped off their wages in settlement of debts incurred in 1926.

The Second World War provided an opportunity for men to escape from the pits. Thousands did, with the result that as demand recovered during and after the war the mines could not produce nearly the tonnage the country wanted. Every ton of coal had suddenly become precious; every production problem urgent; every absentee a slacker.

In 1945, when the Labor Party won its decisive postwar victory, the industry was largely staffed by miners and managers who had lived through the hard years of the twenties and thirties. For years the miners' Union and the Labor Party had preached one prescription: nationalization. When every London clerk knew that the coal industry had been one of the country's stormiest and most unrelenting political and economic problems, it is little wonder there was no serious opposition to the Coal Industry Nationalization Act of 1946.

Coal nationalization has been with us now nearly a decade. Clearly but somewhat unexpectedly, nationalization has not been able to free Britain of continuing anxiety about its fuel supply; indeed, early in 1955 the British Government announced plans for the most ambitious program of investment in atomic energy yet planned by any government in the world. This will certainly not prevent coalmining from always being an enormously important industry in Great Britain; but whether underground output will ever expand from its present level around 210 million tons to the 230–250 millions planned for 1965 is uncertain.

It was the Reid Report, the authoritative technical survey of the industry submitted to Parliament in 1945, that called labor problems "the most difficult and the most urgent" of all those confronting the industry. No one seriously pretended that the industry could be revived by a skillful handling of labor problems alone; but without a marked improvement in this critical and complex area, no program of reform, whether public or private, was likely to be much of a success.

In addition to the decisive economic importance of labor problems in this labor-intensive industry, I have been particularly interested in the socialist dimensions of nationalization. Socialists had long claimed that they could make a unique contribution to the labor problems of modern industry and this claim had been advanced with special force in the case of British coal. A decade's experience has shown this claim to be much less valid than many once believed: instead of nationalization dominating the industry's labor problems, these problems have persisted and have dominated nationalization.

This judgment does not mean that nationalization has been unimportant or irrelevant. It simply means that people expected too much of nationalization and understood too little about the problems of the industry. Absenteeism, recruitment, wastage, methods of wage payment, unofficial strikes, the transfer of labor, rival unionism, joint consultation — problems such as these have an existence of their own which is largely independent of the form of industrial ownership. And the way such problems are handled has more to do with a miner's performance than the old issues of private ownership and the profit motive. These issues were important to the winning of nationalization at the political and ideological level. But it has been an historical weakness of socialist thought to assume that effective economic motivations for the workshop would be supplied by the ideological symbols of a political struggle. Nationalization was unquestionably a necessary reform, but it was by no means a sufficient one. To understand the labor problems of British coal and to deal with them effectively, we are led beyond nationalization.

Where does one go when he goes beyond nationalization? One is led to a more explicit and more detailed examination of specific problems than is usually given by persons primarily concerned with the political dimensions of nationalization. That at any rate was the progression of my own thought: I began by studying coal nationalization as one of the great social experiments of our time and found myself increasingly forced to study the specific problems of the industry. The main body of the book consists of case studies of some of these problems. I am convinced no one can really understand nationalization unless he takes the trouble to master this kind of detail. One must also try to approach these problems in a particular frame of mind — a problem-centered attitude that is as free as possible of emotional bias and political prejudgments. I hope this study may contribute something toward the development of this kind of attitude, outside the industry and within it.

The book is a condensation of a doctoral dissertation submitted to the Massachusetts Institute of Technology in 1952. An investigation of this kind would have been impossible without an extended period of field work in several of the British coalfields. I spent nearly sixteen months in them, from February 1950 through June 1951, a period divided almost equally between intensive field work in many districts and a period of continuous residence in Northumberland. Two fellowships made possible this work in Britain: for twelve months from September 1, 1949, I was a Research Training Fellow of the Social Science Research Council; during the 1950–1951 academic year I held a Fulbright Fellowship at King's College, Newcastle-Upon-Tyne, which is part of the University of Durham. Most of the published statistical material used has been brought down through 1953. From a continuing but distant study of developments in the industry through the spring of 1955, I am persuaded that most of the insights and judgments expressed in the text have not become dated since the original research was completed in 1951.

Individuals whose help deserves explicit mention are Sir Geoffrey Vickers, V.C., a member of the National Coal Board from 1948 until early 1955, who made possible my initial introductions in the industry; Miss Beatrice A. Rogers, whose versatile editorial talents have served many members of the M.I.T. economics faculty; Mrs. Ruth Whitman, whose professional editing was both sure and sympathetic; Mr. John

Davidson of Ashington, Northumberland, who read most of the manuscript and drew up several of the charts while a graduate student at M.I.T.; Mr. Gene S. Peterson, whose drafting skill produced the finished charts; Mr. Noel Gee, Chief Press Officer of the National Coal Board and Miss Margaret Ward, Photographic Librarian, to whose joint efforts I am indebted for most of the photographs used; and Professors Charles P. Kindleberger, Jr., and Charles A. Myers of the M.I.T. Department of Economics, who respectively stimulated and then generously encouraged and supported my interest in the subject.

For permission to use Figure 1 and all photographs except numbers 1, 6, and 7, I am indebted to the National Coal Board. Numbers 1 and 7 are reproduced through the courtesy of *Picture Post* magazine; for number 6 I have to thank the *Daily Herald*. Figure 2 is from the book *The Secrets of Other People's Jobs* and is reproduced through the kindness of Odhams Press. Chart 2, together with much textual material on the reorganization of the National Union of Mineworkers, is reproduced with permission of the *Quarterly Journal of Economics*. The maps in Figures 3 and 4 are from *A Guide to the Coalfields*, published by the Colliery Guardian Publishing Co., Ltd., who jointly with H. M. Ordnance Survey consented to their reproduction.

My wife prepared the organization charts of both the National Coal Board and the National Union of Mineworkers (Charts 1 and 2); she also compiled the index, although the responsibility for it is mine. But these are only the most visible of countless contributions during the years this book has been in the typewriter.

Cambridge
June 1955

GEORGE B. BALDWIN
Industrial Relations Section
Department of Economics
Massachusetts Institute of
Technology

BEYOND NATIONALIZATION
The Labor Problems of British Coal

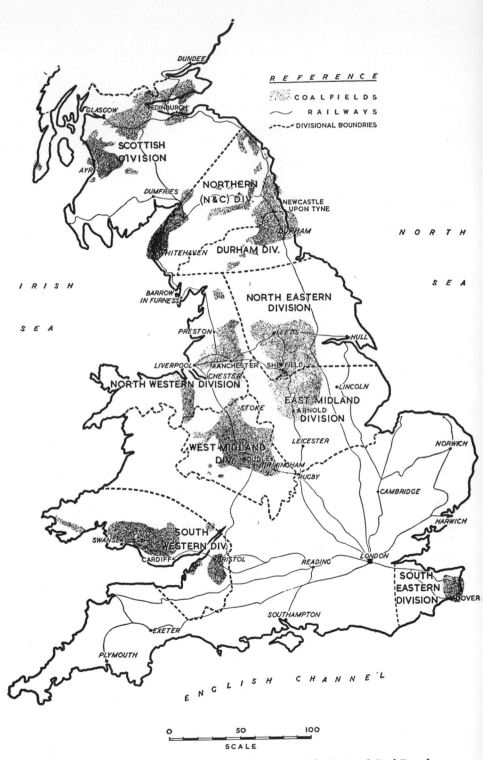

Fig. 1. Britain's Coalfields and the Nine Divisions of the National Coal Board.

CHAPTER ONE

THE TECHNOLOGICAL BASIS OF BRITISH MINING

An understanding of the production processes involved in British mining is indispensable to a comprehension of the industry's labor problems. These involve wage structure and its relation to production incentives; the way in which absenteeism affects output; the effect of the many technological changes now going on in the industry; the importance of the age distribution of the labor force and some of the wider aspects of recruitment and wastage; as well as the representation problems in the industry's joint consultative machinery. Some connection between an industry's technology and its labor problems is hardly unusual; but in coalmining the relationship is dramatic and direct.

Coalmining is a "dark, dirty, and dangerous" occupation. This common phrase sums up three important reasons why miners have a psychology peculiar to their trade and why nonmining people often regard miners with an ambiguous mixture of respect and rejection. There are, however, more fundamental differences between mining and other trades that underlie the darkness, the dirt, and the danger. These basic environmental aspects of mining are the dynamic character of the working environment, its unpredictability, and its lack of standardization.

The environment is characterized as dynamic simply because the walls, roof, and floor of the places where a large proportion of miners work are constantly in motion. True, this motion is normally so slow that it is not visible over short periods of time; but everyone connected with the pit is aware of the forces that are at work. The movement follows relentlessly from the fact that underground mining consists of making hollow spaces in solid matter, a process Nature resists and seeks to undo. As soon as extraction occurs, the balance of forces in the earth, sometimes for miles around, is disturbed, and pressures are set up which continue in operation until a new equilibrium is reached. These pressures express themselves underground as a crushing action, the roof tending to sink and the floor often tending to rise or "heave." The particular way in which these pressures manifest themselves will, of course, depend on the depth at which extraction takes place (the main determinant of pressure) and the nature of the geological strata above and below the seams being worked.

Extraction disturbs not only the condition of the earth underground but surface conditions as well. Indeed, surface subsidence is often plainly visible to the knowing eye of local residents and would be measured in terms of a few feet over an area where a thick seam of coal has been extracted. This settling process may take three or four years to work itself out into a new equilibrium. Thus owners of surface property (not only houses, stores, churches, or factories but owners of farmland as well) can be severely hurt by damage to their property. Mining can render buildings uninhabitable, can ruin a farmer's drainage and hence his fields, and can shear a town's sewage pipes — to suggest but a few of the considerations which mining engineers and surveyors must take into account. In a country as crowded as Britain, the problem of deciding whether or not it is safe to mine a particular area is often a major question of local politics.

The pressures produced by extraction are inexorable and overpowering — that is, they cannot be completely resisted, they can only be partially controlled. This means that the steps man can take to control the huge forces set in motion by mining operations are limited to delaying actions against the purely local manifestations of these forces. Pit-props do not serve to hold up the hundreds or thousands of feet of "overburden" that separate the working place from the surface of the earth; they serve the much more limited function of holding up, temporarily, the immediate roof, or stratum, which lies just above the coal seam. But sooner or later this roof must cave in, so great are the forces at work. The question of when and in what manner this roof caves is of the greatest importance; and roof control is much the most important geological aspect of environmental control exercised at the coalface, the place where there is the most dynamism. Roof control not only affects the safety of the working place; it also affects the ease of working the coal, for when the roof caves into the "waste" (or "goaf" or "gob," as the worked-out area is variously called) it relieves pressure on the coalface itself and renders the coal easier to work. Conversely, the regularity of the production cycle vitally affects roof control.

The gradual settling of the overburden does not affect the coalface alone — it also affects the roadways leading to the face. Once the ground has settled, roadways can be regarded as a stable environment; but before this settling has worked itself out, roadways become crushed down to half their size or less and must be periodically reconstructed by "ripping" (taking down material from the roof) or "brushing" or "dinting" (taking up the floor). Roadways in British mines are usually supported

by steel arches erected every couple of yards; these are twisted and
distorted into grotesque shapes, or are pushed deep into the floor (like
croquet wickets being forced into the ground) unless the roads are
periodically enlarged as subsidence occurs.

The unpredictability of the environment does not mean the spontaneous occurrence of falls of roof (the major cause of serious accidents
in both Britain and the United States); but rather that the conditions
under which routine mining operations must be carried on may alter
without warning — not necessarily quickly, though sometimes so. The
coal may become unexpectedly dirty (full of unsalable foreign matter);
humps and rolls may occur in the floor or roof strata, making work more
difficult; faults may be encountered where the coal is temporarily lost
because of unsuspected breaks attributable to a shearing of the strata,
ages back; the roof or floor may turn wet, so that work must be carried
on with clothes soaked through and mud underfoot — or perhaps a
particular face or district must be abandoned altogether; the nature of
the roof may alter, for example, turning in a few days from a hard,
dependable top into a soft, uncertain one.

All these and many other unpredictable changes in the natural environment affect the safety of the work place, the ease and productivity
of work, the degree of unpleasantness in the "working conditions," the
methods by which the coal can best be worked, the costs required to
get out coal from particular places and the wages which a given face
or workplace will yield. There is thus not only the obvious physical risk
involved in mining: there is an economic risk inherent in the environment which hangs over both the management and the wage earner. This
is more likely to be true of British than American mining, because British
mining must contend with natural conditions greatly inferior to those
of the United States and because the British wage system is more
committed to piecework than is the American (a fact mainly attributable
to differences in natural conditions). A controversial element in many
wage situations is the type and amount of "allowances" which shall be
negotiated to compensate wage earners for some of these unpredictable
factors that cannot be taken into account when price lists are first negotiated.

The dynamic and unpredictable character of the miners' environment implies that the working environment cannot be made as standardized or uniform as the working places of men in most occupations.
Thus, even where the factors listed above as unpredictable can sometimes be foreseen, little can be done to offset their effects. An 18-inch
seam of coal cannot be made into a 3-foot 6-inch seam; a wet roof cannot
be made into a dry one; a dipping seam cannot be made level; a bad
roof cannot be made "hard"; a dirt band in the coal cannot be avoided;
a gassy seam cannot be made free of gas; a soft or wet bottom cannot

be made firm. In these matters, Nature has the upper hand and no amount of effort can wrest it from her. Men must adapt themselves to the environment — as in seafaring, agriculture, aviation, and warfare.

There are three main stages involved in getting coal ready for sale: (1) the original "winning" or "getting" of the coal from the "face" of coal which extends along the working place; (2) the transport of the loosened coal from the coalface to the shaft bottom and thence up the shaft to the surface; and (3) the preparation of coal for market — the cleaning and sizing operations which are performed on the surface. This broad breakdown underlies the three main categories used in classifying pit manpower in Britain — namely, men at the coalface, elsewhere underground, and on the surface. Taking the industry as a whole, about 40 per cent of its employees work at the coalface, about 35 per cent elsewhere underground, and the remaining 25 per cent on the surface.

Coalface Operations

Eighty-five per cent of British coal is worked by longwall methods as distinguished from the room and pillar work which is almost universal in eastern United States. The main distinction between these two fundamental methods of working coal may be suggested as follows. Assume two similar areas of coal, 200 yards wide and a mile long. Longwall working will take a three- to five-foot strip of coal off the entire 200 yards every day, thus completely extracting the coal as the area or "panel" is worked; with all support removed, the roof must fall in and roadways serving the face must be built through the rubble produced by the falling in of the roof. In room and pillar working, on the other hand, extraction would proceed by making several "rooms" into the 200-yard panel, each room advancing daily as several three- to five-foot strips are taken from the advancing end of each room. As these parallel rooms advance, they are connected periodically, so that the walls of coal separating the rooms become, in effect, huge pillars. These pillars continue to hold up the roof, which does not cave in as the rooms advance into the panel. When the rooms have reached the end of the panel, the pillars themselves are extracted — beginning at the far end of the panel where the collapse of the roof will not interfere with future work "on the retreat." Generally speaking, the proportion of total coal in a given panel that is successfully extracted is lower with room and pillar working than with longwall; this is an important reason (but not the only one) for the much wider use of longwall working in Britain.

Room and pillar working originated in Britain and is still extensively used on the northeast coast (especially in County Durham). Natural conditions do not completely determine which method of work to employ — fashions exist even among mining engineers — but they are certainly very important; for example, the deeper one goes, the greater becomes the pressure and the more difficult it is to rely on pillars to support the roof; or the pressure may cause spontaneous combustion in the pillars. Several attempts have been made recently by prominent British and American engineers to encourage a return to more room and pillar mining in Britain, but the current feeling prevailing in Britain today is that such a return seems less feasible than was thought possible when a team of American engineers challenged British methods during World War II.

In longwall work, the cycle of operations at the face consists of five steps spread over three shifts — a "coaling" (or loading or winding shift), a "repair" shift, and a "preparation" shift. The five steps, in their normal sequence, are drilling, cutting, firing, "filling off," and moving up. We will indicate which steps belong on which shift after we see what each of them involves. Figure 2 helps to visualize the layout of a common type of longwall face.

DRILLING THE SHOT-HOLES

Coal is normally exploded or "shot" off the face through the use of explosives inserted into holes bored into the face to a depth of 3 to 5 feet. The traditional miner's pick is now largely obsolete in primary coal-winning, though there are a few important seams, such as the Barnsley in Yorkshire, that are still mainly hand-got. Pneumatic picks, in common use on the Continent, are widely used in those few British coalfields whose coal is suited to their use (notably South Wales and Durham).

Since the cutting operation which succeeds drilling involves piling coal-dust along the face in a way that makes facework awkward, drilling is usually done before cutting. The drilling is done by specialized men of whom no great skill is required. The drills themselves are power-operated rotary drills against which the operator exerts pressure with his shoulder, chest, or thigh (see Plate 1). A coalface of 150 yards can normally be drilled by a single driller during one shift, the inch-diameter holes being spaced every few yards at varying heights and at varying angles. The job is not a particularly dusty one, nor is it very noisy. In a pit which "turns" or winds coal on only one shift, the drilling would usually be done on the first or preparation shift, the shift before the main "coaling" shift.

CUTTING THE COAL

Shooting or firing the coal causes an explosive expansion inside the face; if there were no provision for absorbing this force, there would be danger of fracturing the roof and of failing to secure an even strip of dislodged coal all along the face. The cutting operation is performed to control the way in which the explosion takes place. The cut is most frequently made at the bottom of the seam (the coal is "undercut"); it may also be cut at the top or anywhere in between (see Plate 2).

The cutting machine is pictured in Figure 2. It will be seen that a coal-cutter operates on the principle of a band saw, the teeth of which are picks which can be periodically removed for sharpening. The length of the cutter-bar or "jib" may vary from about three to five feet, depending on the depth of cut. The cutter inches its way along the face slowly, pulling itself along a steel cable which is made fast at the end of the face. As the picks bite into the coal, a hollow space of three to six inches vertical depth and three to five feet horizontal depth is left in the coal. The extraction of this space means that a pile of powdered coal (the "duff" or "gummings") is left along the face just as dirt is piled up whenever a trench is dug on the surface; the cutter-bar flings this duff out in a loose fashion that causes a great deal of dust on the face. The cutting machine does not usually succeed in throwing all the dust clear of the cut and often the latter must be cleaned out in a separate hand operation. This job is done with a special long-handled shovel and may be performed either by an unskilled man (known as a "slatter," "gummer," or "duffer") or it may be considered part of the normal duties of the packers, pan-turners, or even of the cuttermen. Sometimes there is attached to the cutter a special "gumming bar" which performs the operation automatically.

Ordinarily the cutter is operated by a cutting team of two or three men, one of whom is in charge of the operation. Their work normally comes on the same shift as the drilling and just before the loading or coaling shift. While the cutters or machinemen may be working alone on a face, there may also be other men present on the face at work on the packs or engaged in moving over the conveyor (see below).

The use of coal-cutters has spread rapidly during the past twenty-five to thirty years, and this technological change has greatly increased the dust problem underground. Since a pit must be well ventilated to carry off gases, the air is constantly moving along the coalface — and in so doing it picks up and carries coal-dust. The very high incidence of lung diseases among miners arises from breathing this dust-laden air. The reason the two lung conditions, pneumoconiosis and silicosis, are worse in South Wales than in any other district is that the coal there (especially in the anthracite area on the western end of the South Wales coalfield) is hard and the coal-dust is unusually fine and capable of penetrating the lungs more completely.[1]

---

[1] In silicosis, these particles harden the lungs, as if they were filling up with cement; in pneumoconiosis, the particles surround the lung's cell tissues, depriving them of air, but they do not solidify. The characteristic involuntary, heavy coughing of many miners has the effect of breaking the enveloping particles surrounding the lung cells, allowing the man with pneumoconiosis to use these cells again in breathing.

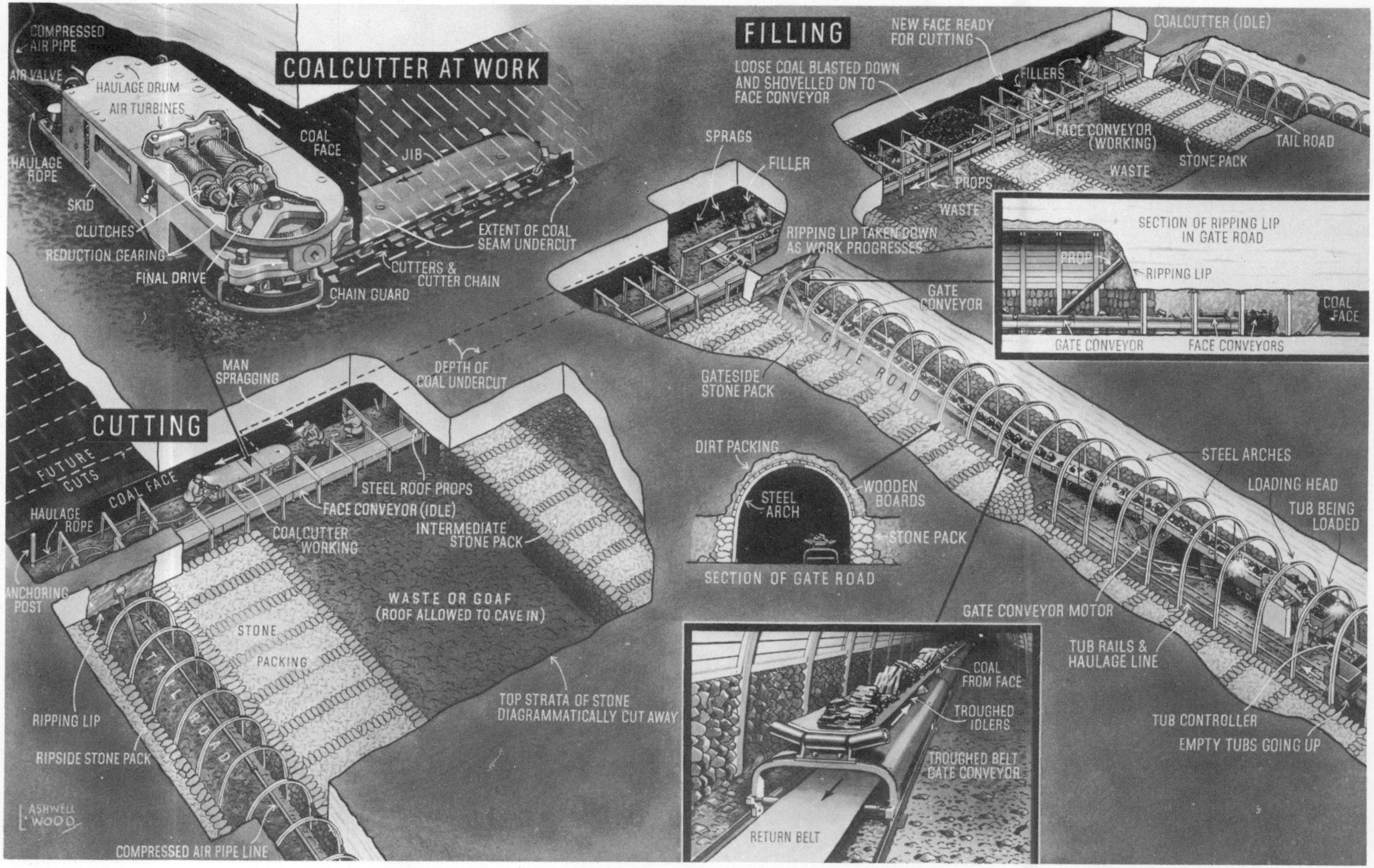

Fig. 2. Layout of a common type of longwall face (double unit).

From *The Secrets of Other People's Jobs*.
© Odhams Press

1. Drilling team preparing shot-holes with electric drill.

2. Face being undercut by machinemen with coal-cutter. Note undercut to right of machine operator's boots.

3. Strippers filling off a coalface onto a face conveyor. Note hydraulic props, steel straps, and men's lamps in the distance.

4. Hand loading into tubs at East Walbottle Colliery, Northumberland.

5. Packers building a pack on gob side of face conveyor. These will help roof break clean as it settles.

© National Coal Board

6. Rippers with pneumatic pick removing stone overburden. Stone will be taken out of pit in tubs

© Daily Herald

7. Pit ponies are getting fewer, but will remain useful in parts of some pits for years to come.

8. Typical trainload of tubs clipped to the "rope" on main haulageway. Empties being returned inbye.

9. A battery locomotive hauling large mine cars. Most British locomotives are diesel-powered.

10. Roadway showing long-run effects of pressure on arch supports. Eventually old arches will be taken down, the road enlarged ("ripped"), and new supports installed.

© National Coal Board

11. Full tubs at the pit bottom waiting to be loaded into cage, just arriving. Onsetter controls loading and winding by bell signals to engineman on the surface.

© National Coal Board

12. A picking table on surface where boys are removing stone from coal. Many modern washeries are being bu
that eliminate this operation.

SHOT-FIRING

After the coal has been drilled and cut, it is ready to be fired. This operation, rigidly controlled by law, must be performed by specially trained men who do nothing else (the shot-firers). Ordinarily, the coal is shot down soon after the cutters have finished their operation, at the end of the cutting shift and just before the "fillers" or "strippers" come onto the face to get the coal on its way. However, there is likely to be some firing going on in any pit at almost every hour of the day or night, as secondary shots are sometimes required during the coaling shift in order to ease the fillers' work. A pair of firemen (or only one) would suffice to serve a single coalface.

"FILLING OFF" OR "STRIPPING"

The big shift at any pit is the one on which the fillers or strippers load off the loosened coal onto the face conveyors which carry the coal along the face to a central point from which it is taken off the face (often by another conveyor) by passing out a roadway which goes off at right angles; this secondary roadway or "gate" leads out to the main haulage road, a larger, better-maintained road than those leading in to the faces (see Plate 3).

The number of fillers on a single coalface depends on the depth of the cut, the height of the seam, the length of the face, and the customs of the pit. There would be ten to twenty fillers for every 100 yards of coalface, each man usually working at the same place along the face every day and each responsible for loading the loosened coal onto the conveyors and for erecting props and "straps" in his working place (straps are bars of steel or wood three to four inches wide and four to six feet long which lie against the roof on top of the props; see Plate 3). A filler may or may not help out the men working on each side of him if he happens to finish his customary stint before they do — these work relationships depend on the personal relationships of the men involved and on the type of wage contract under which they work.

The fillers' work is not very highly skilled (though it is certainly not unskilled); it is, however, hard work, and face workers generally pass their peak efficiency in their middle thirties. The growth of machine mining during the past generation has deskilled faceworkers to some extent. The predecessor of the modern "filler" was the "collier," an all-round pick-and-shovel miner who needed more "know-how" (and hence longer training) than the modern filler. The term "collier" is still used in Durham and Yorkshire but it is usually reserved for the diminishing group of men who still work on hand-got faces.

The atmosphere is often very dusty and hot (in some pits the men work almost naked) and sometimes wet (no protective clothing can keep a man dry in a wet seam). It is very rare that men can work standing up — they normally work from their knees, and in very thin seams (under 24 inches) they must work lying down. The normal working posture is exceedingly hard on a man's knees and elbows and the diseases of these joints, "beat knee" and "beat elbow," are common among facemen.

MOVING UP

After the coaling shift is completed, the face has been advanced three to five feet and is ready for new drilling, cutting, and filling off. Before the cycle can begin again, however, four additional operations must be performed which serve to move up the working space and equipment in step with the advance of the coal itself — as a worm brings up its tail after pushing forward its head. These four moving up operations are (1) shifting the face conveyor to a new position nearer the advanced coalface; (2) the construction of a new line of "packs" (shock absorbers which cushion the fall of the roof when it caves); (3) the drawing off of the waste; and (4) the ripping of the "lips" at the junction of the gates or roadways leading to the coalface from the main haulage road. Each of these operations is performed by separate teams.

Conveyor Shifting. This operation, sometimes called "pan-turning" or "flitting," involves the dismantling of the conveyor belt (or metal pans) so that the conveyor parts can be moved between one row of props, where they are reassembled in the new "track." The "driving end" of the conveyor and the "tension end" must be pulled over to their new positions (done with a chain-lever tool called a "Sylvester"), the belt rolled up, moved over, then unrolled and rethreaded through the two ends (much as movie film is inserted into a projector) and the unfastened conveyor ends must then be joined together in a new, stapled joint. This operation, which often requires the full shift, is done by a team of 4 to 6 men.

Packing. The position of packs along the face may be seen in the diagram, Figure 2, and the building of a single pack is illustrated in Plate 5. These are built by men known as "packers" who, while not highly skilled, must nevertheless have a certain amount of "know-how" — there are good packs and bad ones. The number and size of packs on a face will be set by the manager and the number of packs a single packer will be responsible for during his shift will be set by negotiation. There would be about 4 to 6 packers on a 100-yard face and they would do their work on either the repair or preparation shifts.

Packs are permanent constructions; on most faces, but not all, temporary constructions called "chocks" are sometimes built in order to control better the way in which the roof caves in. Once the roof has caved, the chocks (built of criss-crossed timbers like railroad ties built into a square "log cabin") are dismantled and the timbers can be reused. Where chocks are used, they are in addition to the packs and are built by the packers or "coggers."

Drawing Off the Waste. When the packs are built and the time has come to cave in the roof, this is done by pulling out the row of props that lies on the gob side of the newly built packs. These props are pulled free with the aid of a "Sylvester," which usually is able to salvage the props for reuse. Where chocks are being used, these would then be released (they contain a wedge-shaped device at one point which can be knocked out, thus removing the support and allowing the roof to settle, crack, and come to rest on the packs).

Face Ripping. The position of the "lip" can be seen at the junction of the gate roads and the coalface in Figure 2. In order that the road may keep up with the advancing face, this lip must be ripped down during each cycle or turnover of the face. This operation, performed by ripping teams of about six men, involves moving the face end of the gate conveyor out of the way temporarily drilling into the rock roof from the road end, blasting the lip loose, erecting new roadway girders, clearing away the dirt or rock resulting from the blast, and finally replacing the gate conveyor — not in its old position but now extended 3 to 5 feet to keep up with the advance. The handling of the gate conveyor is a responsibility of the pan-turners, who also handle the face conveyor; the ripping itself is done by a separate team.

Ripping is as skilled as any of the coalface operations — it requires considerable skill as to the best tactics to use in getting down the roof. It is also a relatively dangerous operation, as the ripping lip lies just forward of the line along which the roof breaks at the edge of the waste — in other words, the gates are advanced in disturbed ground that, along the rest of the face, is being allowed to collapse.

The three shifts which include these operations together constitute one complete cycle of operations. The most productive shift and the one on which the most men are employed is the filling shift; none of the other face operations can usually be performed while the fillers are loading off the day's output. The second or repair shift includes the packing, conveyor-shifting, the drawing off of the wastes, the ripping of the gate lips and the advancing of the gate conveyor. The third or preparation shift involves the drilling of the face, the cutting of the coal and shot-firing; we also noted that the packers may be found at work on this preparation shift as well as on the preceding repair shift.

These remarks on the distribution of the separate operations over the three shifts apply only to a single coalface and not necessarily to the whole pit, as there may be half a dozen or more widely scattered faces in a single pit (often working different seams at different levels) and these will not usually all be planned to operate on the same phase of the cycle at the same time of day. If coal comes off too many faces at the same time, bottlenecks quickly develop on the main haulage roads or at the shaft bottom. These bottlenecks are not uncommon in British pits because of inefficient haulage and winding arrangements (see below). As Figure 2 shows, the separate units on a single double-unit face may be worked on different phases to produce balance and flexibility in the pit's operation. The price that must be paid for this flexibility and balance (which allows many pits to wind coal on two shifts, making them "double-turning" pits) is the necessity of maintaining in working order a sufficient number of faces — a task that requires more development and maintenance costs and the use of more capital equipment than if only a small amount of faceroom is scheduled.

It is extremely important to complete each phase of the cycle within each twenty-four-hour period. If this rhythm is interrupted on any coalface, production is disorganized, men must be temporarily transferred to unfamiliar working places, and proper roof control may be temporarily lost. Much the most common cause of failure to complete the cycle is the failure of the fillers to fill off their face during their normal shift. Such failure may arise from any number of reasons — shortage of supplies at the face, a breakdown of the face conveyors, a breakdown of transport on the main haulage roads (which backs up onto the faces and causes waiting time), the occurrence of accidents, falls of roof, and — especially important — the failure of the normal complement of men to show up for work on their scheduled shift (especially on Saturdays and Mondays). This means that when the pan-pullers come on duty on the following shift, they must themselves clear away whatever coal remains on the face before they can turn over the conveyor; alternatively, the fillers themselves must work overtime to finish their work — which naturally involves overtime costs and the willingness of fillers to work overtime.

Anything which overcomes the rigidity of the foregoing characteristic cycle of operations on a given face represents progress, and there are several emerging developments in British face practice which promise to give relief from this problem and to increase the efficiency of work at the face. The most important of these technological changes is the development of machines which both cut and load the coal in one operation — "cutter-loaders" as they are called. There are at least seven such machines which have reached the testing stage since the war. In addition, there is another machine (the Meco-Moore made by Anderson, Boyes and Company) already well proved, whose use is gradually being extended in and beyond the two main districts, Nottingham and North Derbyshire, where it has had considerable success for the past fifteen years. These cutter-loaders all combine the two operations of loosening the coal from the face without the need for shot-firing and loading it onto the face conveyor or into some other means of transport, such as shuttle cars. Drilling, shot-firing, and filling are thus eliminated and it often becomes possible to load coal off the same face on two of the three shifts.

None of the foregoing machines are appropriate for all conditions nor are any of them suitable for some conditions; but it seems highly probable that great room will be found for the rapid and extensive spread of their use in the country as a whole, once the "pilot plant" phase has been completed. It is probably fair to state that British mining practices and equipment at the coalface are at least as good as, if not better than in any of the continental coalfields. It is in the transporting of the coal from the faces to the surface, and its handling on the surface, where the great backwardness of the British industry lies and where the greatest opportunities exist for capital development and manpower savings.[2] The next section is devoted to the other underground operations, the chief of which is transport or haulage.

Underground Operations Not at the Coalface

There are three broad categories of work to be performed underground that are not classified as work at the coalface. These categories are (1) direct work in operating the haulage system that takes the coal to the shaft bottom and thence to the surface and which carries supplies "inbye" (or towards the coalface); (2) service work required to maintain and extend the roadways, compressed air pipes and electricity cables; the erection, dismantling, and servicing of conveyors; carpentry work (building doors for ventilation control, fitting out the humble underground offices near the shaft bottom where records are kept and officials meet between shifts); bricklaying, when support-walls are required or permanent ventilation barriers are desired; the straightening of steel props and bars for reuse on the faces; stabling and tending the pit ponies (not, of course, at all pits); moving equipment into the pit and from place to place; splicing broken cables and extending and maintaining in good working order the rope haulages; maintaining haulage locomotives in good operating condition at the relatively few pits where such locomotives have been introduced; installing and maintaining pipes and pumps to carry water out of the pit; and (3) development work, consisting mainly of driving new roadways, or "drifting."

Most of these men are on day-wage rates (though as we shall see in

[2] The best summary of the industry's technological problems is the *Reid Report* (1945), whose main findings will be found in Appendix I.

Chapter Five on wages, there are some jobs off the face which are also paid on a piecework or "contract" basis) and these classes of men are usually referred to loosely as "datal" or "oncost" workers.

Except over very short distances, coal is transported mechanically and men are needed chiefly at points of interruption in the transportation system, where tubs are brought off a face by hand or pony "putting" (the word means pulling when done by ponies and pushing when done by men) and must subsequently be clipped on to a wire rope which hauls a trainload to the pit bottom; again, coal which is carried off a face by gate conveyor is often transferred into tubs at the junction of the gate and a main haulage road. At such junctions, called loading points, men are required to shunt the tubs into position, to regulate their filling, and to clip them onto the main haulage rope (see Plate 8). Most loading points require at least two men and many require four or more.

Before following the coal "outbye" we had better draw more explicit attention to the class of "putters" who are often a sizable group in pits where haulage has not been mechanized (sometimes for good reasons). The "putters" man the transport system between the coalface and the main haulage roads, pushing the empty tubs to the loading point at the face end and then pushing them (usually only a few yards) to the haulage road. Where the distances are longer, or the gradients too steep for hand putting, ponies are often used: their drivers are not called "putters" but simply pony-drivers (Plate 7).

The characteristic method of pulling tubs on the main haulage roads of British pits is by means of a wire rope. At each point where tubs are collected for assembling into trainloads, the tubs must be clipped on to the haulage rope and signals must be sent to the driver of the haulage engine indicating that the rope should be started or stopped. Between these loading points (relatively near the faces) and the pit bottom, there are usually long stretches of roadway (from a few hundred yards to over a mile, perhaps) where no routine jobs are required. When the loads have made their trip to the pit bottom, they come to an enlarged space that functions as a marshaling yard for transferring the tubs into the cage in which they are wound to the surface. This enlarged area is known as the "pit bottom" (Plate 11). In it there may work anywhere from three to ten men, all of whom are directly or indirectly occupied in putting full tubs onto the cage or sending empties back inbye. At the junction of the pit bottom with the main haulage road, one or two men will see to the unclipping of the tubs from the rope; another will put the tubs toward the shaft bottom; and at the shaft bottom itself is the man who is in charge of the loading of the cage — called, appropriately, the "on-setter." When the cage is in position, this man opens the safety barrier and, usually with the help of an assistant, pushes the full tubs into the cage. The act of pushing in a full tub is usually sufficient to push out an empty on the far side of the shaft, though an attendant on that side is often required to pull it out. These empties often run by gravity (or are mechanically carried) along a semicircular roadway which carries the empties back to the junction of the pit bottom and the main haulage road, where they are once again clipped to the rope and sent inbye. Once the cage is loaded, the onsetter sends a signal to the surface and the coal disappears on its speedy ascent.

This description is characteristic of pit-bottom layout and winding methods at most British pits today. The most important change being introduced at this point is the substitution of "skip winding" for the winding of individual tubs in a cage. A "skip" is a large bucket or bunker attached directly to the winding rope; it is loaded from a storage bunker built in the pit bottom. Not only does this layout require fewer men at the pit bottom but, more important, it permits a more continuous flow of coal-winding, as storage permits evening out the irregularities in the flow of coal to the pit bottom which inevitably occur because of unevenness in the rate at which loaded tubs arrive at the pit bottom (there is a normal peak on the filling shift about one or two hours after the fillers get to their faces). This system also greatly speeds up the turnaround time of each tub (since they do not go to the surface) and there is less likelihood that certain faces will have to mark time while waiting for empties to reach them. Also, skip-winding permits faster winding speeds and hence more winds can be secured per shift; it thus offers an important potential increase in shaft-handling capacity — a serious bottleneck in many pits and one likely to increase as face mechanization progresses. By the summer of 1952, eighteen pits either had skip-winding or were in process of being converted to it.

Two of the many other jobs in underground operations require special mention — ripping and drifting. We have already described the daily ripping of the face lips. A similar task must be regularly performed on many of a pit's roadways as part of their normal maintenance (since the roof, sides, and floor of many roads are periodically disturbed by geological pressures). "Back ripping," or ripping that is performed back from the faces, consists of enlarging or repairing roadways by taking down twisted and distorted arches, firing or pulling down some new stone with picks, and reërecting new arches to support the enlarged roadway.

"Drifting" is the name given to the process of driving new roadways through stone (or sometimes through the coal itself). It is usually associated with the development work (that is, the preparation of new faces and roadways) that is continuously going on in any well-run pit — today, of course, there is much "drifting" in connection with reorganization schemes. "Drifters" or "stonemen" usually work in teams of four to six men who drill holes for explosives, shoot down the rock, erect arches for roof supports, and load the rock into tubs to be taken out of the pit. This is slow work and a roadway in stone will often not advance more than a yard or so every twenty-four hours; thus it may take one, two, or three years to build a major new haulage road. The Coal Board is making special efforts to increase the speed of drifting.

There are four customary grades of officials responsible for the underground operations: shot-firers, deputies, overmen, and undermanagers. As described above, shot-firers have no supervisory duties except to prepare shots to be fired in accordance with legal requirements and to police the area. Deputies are the real first line of supervision on any coalface and there would normally be one deputy to each face or half of a double-unit face; the deputies' main responsibilities relate to enforcing safety regulations, securing aid in the event of mechanical breakdowns, and seeing that facemen are adequately provided with supplies of pit-props, straps, packing and chock material, and so on. Deputies, although they correspond roughly to the foreman level in American manufacturing, have much less responsibility for personnel handling than would be true of an industrial foreman: they cannot hire or refuse to hire labor, they are not usually given cost reports for their faces, they rarely if ever discipline a man except for safety breaches, and they are themselves members of a trade union, though not the rank-and-file mineworkers' union in most districts.

The overman ("oversman" in Scotland) is responsible for a number of faces comprising usually one district in one seam of a pit. He is normally underground all his shift, travels the pit frequently, and is in close touch with his deputies, receiving reports from them and issuing instructions to them.

The undermanager is not so called because he is the next man under the pit manager but because he is the principal management official underground. He is mainly responsible for all underground operations and spends much more time underground than does the pit manager. Generally there is one undermanager for each pit, but where the pit is large and more than one seam is being worked, two, three, or four undermanagers may be responsible for particular areas or seams. In some pits, there is an assistant undermanager. The undermanager would of course be directly responsible to the pit manager.

Surface Operations

All but a few British pits are shaft mines, not drifts, of which there are proportionately many more in the United States. South Wales and Scotland each have a large number of drift mines, but not any other districts. The operation of the engines which wind the cages (or "chairs") up and down the shafts — and every pit is required to have two shafts as a safety precaution — is in the hands of a very small craft called winding enginemen, of whom we shall hear more in Chapter Two. There is usually only one winder on duty in the engine-house for each shift, though there is often an oiler on duty as well. His controls are levers operated from a control-box to which bell signals are sent by the "banksman," who is responsible for loading and unloading the cage at the pit top. The banksman thus corresponds to the onsetter at the pit bottom. As there are people, coal, or supplies entering and leaving the pit on all three shifts, and over the weekends, there must be a winding engineman on duty at all times.

There is almost invariably a colliery operated steam-raising plant that either provides steam for the winding engines or is used to drive generators for electric winders and for general use on the surface and underground. A crew of three to eight men (chiefly stokers) is required to operate the boilers.

As the tubs are taken off the cage by the banksman and his assistant (called in some districts a "puller-off"), they are "putted" along a track to a weighing machine where the full tub is weighed so that the weight and the face (or even the individual) from which it came can be recorded for purposes of wage payment. After being weighed, the tubs move along a short distance in the wooden superstructure that characteristically surrounds the pit top to a point where they are dumped, either mechanically or by hand. The loose coal then passes along screens, which are moving metal tables (which often shake the coal along without themselves moving, except backward and forward) that carry the coal and all impurities loaded with it past teams of men whose job it is to pick out such things as stones and pieces of wood which are not salable (see Plate 12). The men (in some districts, such as parts of Cannock Chase and in Lancashire, women are employed on this work) working on the screens or "picking belts" or "picking tables" are almost always either lads too "green" to be sent underground or men too old to continue on underground work, or men at intermediate ages who have been forced out of the pit by injury. The screens, therefore, provide a class of jobs for men who have served their time, so to speak. But under modern wage scales, hand-picking is an inefficient and costly method of cleaning coal, and future plans call for mechanizing coal preparation on an extensive scale.

Since a modern coal-cleaning plant requires much less labor than the traditional picking belts, modernization projects involve a sizable redundancy problem involving an age group difficult to transfer or to retrain. Whenever a pit is closed, the lowest percentage of transfers is invariably found among the men who work on the screens.

A pit will usually have a fitting shop (machine shop), an electricians' shop, a pipefitters' shop, a blacksmith's forge, a carpenters' shop, and similar service facilities requiring men with well-developed manual skills. In a small pit, these craftsmen would, of course, work out of a central shop. The men who man these and other departments are now formally known as "craftsmen." Except at two or three large and progressive companies, colliery craftsmen had never been apprenticed to their work, as they are in most other industries. This reflects the somewhat lower status of colliery craftsmen as compared with their counterparts in other industries, the fact that the regular craft unions have never received wide recognition in the industry, and the lack of concern for systematic training programs on the part of colliery firms in the past.

Current wage agreements recognize nineteen separate Grade I and seven Grade II craftsmen's positions (the Grade II jobs are not different from the Grade I jobs; their holders are simply less skilled men). The work of some craftsmen is confined exclusively to the surface, but many men will divide their work between the surface and underground; "fitters" (or mechanics), electricians, and pipefitters, especially, must go down the pit frequently. The mechanical trades are usually in charge of the colliery enginewright; the other crafts likewise have their separate foremen and the main surface operations are usually in charge of one or two surface foremen who report directly to the pit manager.

At large pits there is often a considerable amount of railway trackage and one or two shunting engines to move full and empty railway wagons (cars); the men who operate and maintain this equipment are usually colliery employees, not railroad employees.

At most sizable pits in the country there will be found pithead baths — a combination locker-room-showers, usually with a canteen and dining hall built into one end. A bath serving one thousand men will require about four bath attendants to keep the baths clean, sell soap and towels, tend the boiler, and so on; these will be under a "baths superintendent." The canteen is staffed by half a dozen or so local women. Today there are baths at just over half the Board's 930 pits; these serve about three-quarters of the industry's workers. The extension of baths has held a high priority since nationalization and the opening of a new bath in a colliery village is an event of great importance. Apart from their convenience to the miners themselves, baths mean that a town's buses will be cleaner, that less dirt will be brought into homes, and that miners will leave their work looking no different from men in other manual occupations.

The white-collar jobs on the surface are not numerous — a few clerks for making bookkeeping entries for stores, cost records, and wage payments, and a typist or two to handle correspondence and to get out the many reports which pit managers must submit to higher officials. Colliery clerks are as often men as women — most managers have a male chief clerk of their own, not a secretary. In general, most of the clerical work and accounting is performed at the Area offices, not at the pits.

Every pit will have a surveying office, manned by two to six (or more) surveyors whose job it is to maintain the pit's working plans and whose duties involve frequent regular underground trips to survey the progress of the work. Since nationalization, separate safety and training officers have been appointed at all pits of any size and these individuals usually have a room in the pit's offices; they do not normally have any clerical staff of their own.

All these jobs would have to be performed in British pits whether the industry were nationalized or not. The key jobs, as we have seen, are those performed at the face. It is the conditions under which the facemen work, the arduous nature of their jobs, and the importance of their work which underlie many of the industry's labor problems. The special characteristics of facework determine the industry's method of wage payment, its wage structure, and the status system that dominates trade union affairs and the social life of colliery villages. Within the broad group of facemen, the specialized jobs give rise to distinctions of pay and prestige within the group itself.

Pit managers are under continual pressure to get as many men as possible onto face jobs and to minimize the number of men on "oncost" or "unproductive" jobs (including, of course, surface jobs). Thus the weekly figures on manpower at the face have been of as much public concern during recent years as the trend of total manpower in the industry. But after we have once fixed in mind the two key figures for underground jobs and face jobs as a proportion of all jobs performed at a pit (about 75 per cent and 40 per cent, respectively), we can then go on to emphasize the huge number of individual job titles which have taken root in the industry.

This multiplicity of names for jobs has also posed great problems for experienced union and management officials. In 1950, negotiators began to wrestle with the complex, delicate problem of straightening out the industry's wage structure. For a start, the Board and the Union drew up a list of all the job titles that could be found in the coalfields. This list contained more than 6000 job names. When job descriptions were used to eliminate different names for the same jobs, there still remained some 750 names. By March 1955 the parties had completed the most far-reaching rationalization of the wage structure ever undertaken in the industry: in place of the 6000 job titles that had existed five years previously, only 361 were to be recognized in the future and these were to be grouped into thirteen industry-wide job classifications (eight underground, five on the surface).

In succeeding chapters it will help to keep in mind the basic work processes outlined above and the special nature of the miners' environment.

CHAPTER TWO

THE COAL BOARD AND THE
MINERS' UNION

The mining industry is dominated by two great and complex organizations, the National Coal Board and the National Union of Mineworkers. The influence that each of these has on labor relations in the industry is determined partly by its own internal characteristics and partly by the quality of its relationships with its opposite number. This chapter is chiefly concerned with certain aspects of each institution's internal structure and problems.

I. The Coal Board

Ever since nationalization there has been a running public debate as to whether or not the Coal Board is soundly organized.[1] This is a many-sided argument and from it we shall select only those issues which concern the Board's handling of labor relations. The complaints leveled against the Board on this score fall generally into the following four categories:

[1] This chapter was written before the appointment of the five-man Advisory Committee on Organization which the Coal Board established in December 1953 "To consider the organisation of the National Coal Board and to make recommendations to the Board." The 105-page *Report* of this committee, headed by Dr. Alexander Fleck, Chairman of Imperial Chemical Industries, Ltd., was published in February 1955 (NCB, *Report of the Advisory Committee on Organisation*, 2s. 6d.). Its findings and recommendations were unanimous (with the exception of the Chairman's dissent on one point) and met with widespread approval in nearly all sections of the press.

Two earlier studies of Coal Board organization share the Fleck Report's conclusion that the fundamental organizational problems of the industry, and in particular the problem of decentralization, are not likely to be solved by the proposals for structural reorganization that dominated public debate from 1947 to 1955. See W. H. Haynes, *Nationalization in Practice: the British Coal Industry* (Harvard Business School, 1953) and the Acton Society Trust's pamphlet, *The Extent of Centralisation: a Discussion Based on a Case Study in the Coal Industry* (1951). It remains to be seen whether or not the Fleck Report will command sufficient authority to remove the issue of Coal Board organization from politics for the next few years. My own views of Coal Board organization parallel broadly those of the Fleck Report.

1. Overcentralization. The Board is so organized functionally that too many decisions capable of prompt and local determination are instead subject to delayed and absentee decision through constant reference to higher levels.

2. Bureaucracy. The Board has created too many new jobs, thus loading the industry with unproductive functionaries who raise costs and lower morale.

3. Salaries and perquisites. The Board's salaries are so high and its perquisites so unfair that the Board cannot win the confidence of the rank and file.

4. Unfair bias. The individuals appointed to most of the high positions on the Board have come from the old colliery companies and, since leopards do not change their spots, the industry is still being run by persons whose outlook is not likely to give nationalization a fair test.

Left-wing critics usually endorse all four complaints; conservative critics endorse the first pair but deny the second. The net effect of these criticisms is said to explain the alleged failure of the Board to win the miners' confidence, a failure evidenced by continuing strikes, persistent absenteeism, disappointing output, high labor turnover, and similar symptoms.

THE BOARD'S STRUCTURE

Chart 1 outlines the Board's formal structure. The fifty-odd Area organizations are the critical ones from an operating point of view, and the principal positions found in a typical Area are listed on pages 21 and 22 below.

The Nationalization Act provided only for the establishment of the National Coal Board and said nothing about the further organization of the managerial function; this was left up to the Board itself,[2] subject to the approval of the Minister of Fuel and Power, who appoints the Board and to whom it is immediately responsible. With some exceptions, the Divisions correspond quite closely to the natural discontinuities in the location of the industry; they show a similar correspondence to the regional organization of the Miners' Union — a factor of very considerable

---

[2] In practice, the scheme of Divisional and Area organizations was worked out primarily by Sir Arthur Street, who was Deputy Chairman of the Board from its appointment in July 1946 until his sudden death in February 1951. The structure adopted was very similar to that worked out during the war when the Ministry of Fuel and Power was established. Sir Arthur had been a lifelong civil servant (until he joined the Board), with a prewar career in the Ministry of Agriculture and a wartime career in the Ministry of Aircraft Production. He was a prodigious worker and won for himself widespread respect throughout his adopted industry.

A detailed description of the wartime organization will be found in Professor W. H. B. Court's definitive study, Coal (His Majesty's Stationery Office and Longmans, Green, 1951); see part IV, "The Central Development of the Control."

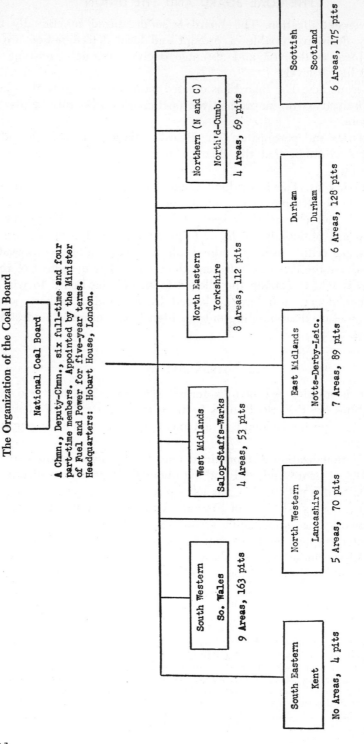

Chart 1

The Organization of the Coal Board

National Coal Board

A Chmn., Deputy-Chmn., six full-time and four part-time members. Appointed by the Minister of Fuel and Power for five-year terms. Headquarters: Hobart House, London.

South Eastern
Kent

No Areas, 4 pits

South Western
So. Wales

9 Areas, 163 pits

North Western
Lancashire

5 Areas, 70 pits

West Midlands
Salop-Staffs-Warks

4 Areas, 53 pits

East Midlands
Notts-Derby-Leic.

7 Areas, 89 pits

North Eastern
Yorkshire

3 Areas, 112 pits

Durham
Durham

6 Areas, 128 pits

Northern (N and C)
North'd-Cumb.

4 Areas, 69 pits

Scottish
Scotland

6 Areas, 175 pits

Total Areas: 51 (1954)   Total pits: 863 (Sept. 1954)

Note: Divisions are run by Boards appointed by, and collectively responsible to, the National Board. Areas are run by General Managers, individually responsible to their Divisional Board but appointed by the National Board.

[20]

MAIN FUNCTIONAL RESPONSIBILITIES OF EACH LEVEL AND
REPRESENTATIVE SALARIES AS OF 1947

I. *The National Board:* Fixes the level and structure of prices. Conducts collective bargaining for all classes of employees on all major terms of the employment relationship. Plans long-range investment objectives and allocates capital among the Divisions. Directs the conduct of scientific research. Negotiates many supply contracts. Holds to account the Divisional Boards. Borrows money from the Government. Conducts legal proceedings before the High Court.
Salaries: Chmn. £7500; full-time members £5000; part-time members £500.

II. *The Divisional Boards:* Chmn., Deputy-Chmn., and four Directors (Production, Marketing, Finance, and Labor). Coördinates marketing on a regional basis. Responsible for appraising the performance of Area General Managers. May authorize capital expenditures up to £250,000 per project. Bargains collectively with district unions on a limited range of subjects. Provides numerous staff services to Areas (e.g., engineering, auditing, legal services, coördination of welfare activities, recruitment and training of new entrants, architectural design).
Salaries: No fixed scale; individual full-time members receive from £2000 to 5000 "in the light of qualifications and experience"; part-time Divisional Board members receive £500. No other senior officials in Divisions receive more than £5000.

III. *The Area Offices:* Area General Manager responsible for overall performance. Positions normally found in Area organization are listed below. Plans and designs much of the industry's investment; may authorize capital investment up to £100,000 per project. Appraises the performance of individual colliery managers, whose costs and output are closely watched. Often involved in the negotiation of pieceworkers' price lists and the settlement of strikes.
Salaries: See below.

IV. *The Collieries:* The Pit Manager has statutory responsibility for the safe operation of his pit (periodic safety inspections are conducted by the Mines Inspectorate, which reports directly to the Minister of Fuel and Power). Manager is individually responsible to the AGM for performance results. Much time spent negotiating pieceworkers' price lists and settling grievances.
Salaries: Maximum for Mgrs. is £1650 plus concessionary coal but minus the value of most other "perqs." Minimum salaries vary by Divisions, ranging from £1000 in Scotland and South Wales to £1100 in Yorkshire and the East Midlands (as per BACM-NCB Agreement of July 19, 1951).

Principal Positions in a Typical Area
(1947 salary ranges in parentheses.)

Area General Manager (up to £4000)
Area Production Manager (up to £3000)
Deputy Area Production Manager
Area Finance Officer (£1200–1950)
Area Marketing Officer (£850–£2300)
Area Labour Director (£600–£950)
Area Welfare Officer (£600–£950)

Area Secretary
Area Estates (Housing) Officer
Area Planning Officer
Area Safety Officer
Area Statistical Officer
Area Chief Accountant (£ 1200– £ 1950)
Area Chief Scientist (£ 700– £ 1300)
Area Supplies Manager (£ 750– £ 1250)
Area Electrical Engineer
Area Mechanical Engineer
                                    (£ 900– £ 1750)
Area Coal Preparation Officer (£ 500– £ 1000)
Area Surveyor (up to £ 1200)
Area Mechanization Officer (£ 750– £ 1250)
Area Cashier
Area Cost Accountant

Note: Most Areas are subdivided into "Sub-Areas" which are in charge of Sub-Area Agents, known simply as "Agents." These men are line officials who link small groups of pits to the Area office. An Agent normally has a small staff of technical specialists (e.g., Sub-Area Electrician, Mechanical Engineer, Surveyor, etc.).

Source: Appendix V, NCB's 1947 *Annual Report.* Salaries for all above positions not available. Position in rate range depends on size-class of particular Areas and length of service. Most of the above salary levels have been increased somewhat since 1947.

---

advantage in the field of labor relations and one often neglected by persons critical of the Board's structure.

The financial requirements of the Nationalization Act, like the organizational provisions, touch only the National Board, not the Divisions, Areas, or individual pits.[3] The National Board is required to see that the industry pays its way over an undefined span of "good years and bad"; there is no requirement that each Division, Area, or pit be self-supporting.

The Issue of Overcentralization. The charge that the Board's structure is overcentralized, with the result that decisions which are delayed too long and are wrong ones when finally made, is more convenient than compelling. The real difficulty — and a real difficulty does exist — is administrative rather than structural overcentralization. This means that the depressing effects of overcentralization can be relieved by closer attention to a more precise definition of executive responsibilities at the various existing levels without the necessity for structural decentraliza-

[3] For an analysis of the ways in which the Coal Industry Nationalization Act differs from subsequent acts nationalizing other British industries, see D. N. Chester, *The Nationalised Industries: A Statutory Analysis,* Institute of Public Administration, July 1948, 48 pp., and the revised edition of 1951.

tion; indeed, structural decentralization without administrative reform would leave the root problem largely untouched.[4]

The critical level at which the depressing effects of overcentralization on labor relations are felt is the colliery and one of the best "bill of particulars" of the problems felt by pit managers is that outlined below, supplied by a manager in the Midlands. Few of these managerial grievances bear directly on labor problems; but in their effect on management's morale they are bound to affect the attitude of managers toward their jobs and toward the men through whom alone they can achieve results.

1. Managers do not know to whom to go when they have a problem or a personal grievance.

2. There are inequities in pay and perquisites from one pit to another.

3. Managers cannot make raises (for example, regrade men) without first securing permission from the Sub-Area Agent (who asks the Area Mining Engineer, who then asks the Area General Manager). Yet "who knows the merits of individual cases better than the pit manager?"

4. Too many reports are required and managers are allowed insufficient help in their preparation. When this problem was put to the Divisional Chairman and a "report on the number of reports" was submitted to document the manager's complaint, nothing further was heard on the matter and no reforms were initiated.

5. A manager does not have enough "thinking time" — he is considered inattentive to duty if he is not constantly down the pit.

6. There is no objective basis for making appointments to managerial positions. Too often decisions seem to be either arbitrary or biased.

7. Too many AGM's refuse to let managers have detailed financial information about their own pits because they do not want such information to "get about."

8. There is no orderly career ladder within managerial positions.

9. There has been a loss of the personal touch. For example, in the old days a manager who had a good year with his pit was often given a year-end bonus. These have been eliminated, resulting in diminished incentive.

10. The Board has been stingy with respect to salaries — when it fills positions which have been vacated, it has tended to pay the new appointee less than it paid the former incumbent. The Board has claimed that it pays

---

[4] This is substantially the same conclusion as that reached by the Fleck Committee. After pronouncing as "sound" the "main structure of the organization — Headquarters, Divisions, Areas and Collieries," the Committee went on to analyze the factors that determine performance within this structure. Among the many recommendations made the most important were steps to improve the quality of management at all levels in the organization — a complex reform ranging from higher salaries for top management to broader training for pit managers.

The best statement of the case for structural reorganization is contained in the pamphlet, *The Structure and Control of the Coal Industry,* written by Col. C. G. Lancaster, M.P., in association with Sir Charles Reid and Sir Eric Young (published by the Conservative Political Center, June 1951). The very similar proposals advanced by the Conservative Party itself are analyzed explicitly in the Acton Society Trust monograph, *The Extent of Centralisation.*

according to experience and ability, "but the tendency has been too marked for this to be true."

The significant fact about these representative grievances is that they all involve problems of administrative policy and technique and personal relationships within the managerial hierarchy; they are not the kind of problems that would disappear with structural reorganization. Indeed, the Conservative Party, in more than three years of rule, has not seen fit to make any of the changes in Coal Board organization which some of its members had argued were indispensable. The party's reluctance to alter the Board's structure reflects doubt as to the probable effectiveness of the decentralization remedy and an unwillingness to incur the open opposition of the NUM, which, while not uncritical of the Board's administration, is nevertheless confident that no Tory reform would be a good one.

THE STAFFING OF THE BOARD

Three aspects of the staffing of the Board have involved problems of labor morale and the acceptance of managerial leadership. These problems are: (1) How many positions, and what types, are necessary in view of the functions to be performed? (2) What individuals are to be selected to fill these positions? (3) How much remuneration ought these positions to carry? Satisfactory answers to these questions are never easy and scarcely become more so when an effort must be made to square them with the moral and political perspectives of socialism.

Coal Board Jobs. Ironically, the Coal Board has not spawned a burdensome bureaucracy, though it has not been able to avoid a widespread feeling among the mineworkers that it has. The Board has made a convincing case for the 40 per cent increase in nonindustrial staff which has occurred since 1947, a case which rests on the large increase in the number of functions now charged to the industry which formerly were not so.[5]

The growth in nonindustrial staff (from about 30,000 in early 1947 to 43,000 at the end of 1953) has been at the colliery and area levels and not at the divisional and national levels against which the "bureaucracy" charge is usually made. Indeed, the costs of running the nine divisional and the national headquarters have remained steady at $\frac{7}{10}$ of 1 per cent of the industry's total costs ever since 1947, and the total salary bill for all persons at both these levels has remained less than the

---

[5] All the Board's Annual Reports contain information on this point; see esp. pp. 32–34 of the 1947 *Report* and p. 59 of the 1950 *Report*. The Fleck Committee referred to the "ill-informed criticism, both inside and outside the industry," on the staff problem and expressed the view that "The proportion of staff . . . is certainly not too high. Indeed, we believe that the efficient management of the industry requires a higher ratio, as well as a higher average quality." (*Fleck Report,* para. 113).

annual cost of prenationalization directors' fees alone (that is, omitting the salaries which some directors earned).[6] It is the transfer from private to public knowledge of this kind of information about personal remuneration that makes it so difficult to judge the significance of staff growth under nationalization.

Despite the reasonable defense which can be made for the increase in the number of salaried personnel in the industry, it remains true that the Board has not been able to cultivate any popular understanding of the necessity for these new appointments. Although the top leadership of the National Union of Mineworkers took an admirably cautious and responsible attitude toward rank-and-file grumbling about "Coal Board bureaucracy," it took the union machinery four hesitant years to produce an official union view on the bureaucracy issue.

At every annual conference of the NUM since nationalization there has been at least one resolution censuring the Coal Board's organization, and in 1949 there were more resolutions on this topic than on any other. That year the conference directed the National Executive Committee to establish a special committee to investigate NCB administration. This task dragged on for more than four years, chiefly because top officials of the Union could not be spared from other duties to get on with the work. In 1953 the Executive Committee came to the conclusion that "on the evidence we have before us there is no reasonable cause for dissatisfaction with the establishment of the National Coal Board." Conference delegates accepted this verdict from their Executive Committee, though the latter returned responsibility to the districts by saying that it would stand ready to investigate any specific complaints referred to it by the areas.[7]

Although this judgment is undoubtedly an honest and courageous one on the part of the Union's top officials, it would be surprising if it put an end to grumbling at lower levels. Nevertheless, the Union's leadership has acted as a buffer between the Board and rank-and-file criticism of it and, without any clear views of their own on this inherently difficult judgment, they have defended the Board against vague and unsupported popular criticism. The Union has not, however, been able to make the positive contribution which it might be hoped a mature union could make to the administration of a nationalized industry: it has not mobilized and presented effectively to the Board the many thoughtful and well-informed suggestions about the Board's organization which can be found among its local officers. The Union will probably remain unable to perform such tasks until it becomes more adequately staffed or commissions some independent body to conduct a professional audit on its behalf.

[6] Harold Wilson, *New Deal for Coal* (1945), p. 207.
[7] See *NUM Conference Report*, 1953, p. 42.

For nearly a generation coal nationalization had held the twin objectives of technical reorganization and an extension of industrial democracy. In the short run, these goals were irreconcilable, for the first required the retention of the managerial personnel who had run the industry in the "bad old days" while the second seemed to require that they be replaced with new technicians who would operate within some context of "workers' control." In practice, the alternative of replacing experienced managers with new ones not discredited by history was not a practical option, in part because of the statutory requirements governing managers' qualifications, requirements that labor itself had long supported in the interest of safety. Consequently, at the time of nationalization there could be no argument against the impossibility of finding socialist-minded men who could qualify for the key managerial posts at the operating level (the collieries and Areas). Perhaps at some vague future date the Labor Party or the trade union could put forward "men of their own" who would be technically trained. But the immediate question in late 1946 was not whether to retain most of the existing management personnel but rather how much influence they would enjoy in the new organization and how they would use this influence. The impossibility of producing any visible change in the men who ran the pits had the disadvantage of making scapegoats of the existing managers for those elements who wanted to grumble about the progress of nationalization. But this difficulty was simply the price that had to be paid for trying to work a revolution without having one: such "revolutions" take much longer, they lack decisiveness, and they proceed with more ambiguity.

The above argument did not apply only to technical production jobs. It proved equally impossible to find men who knew anything about finance, marketing, and the law except among men who had spent their lives in such work, something very few socialists or union officers had done. Consequently the Board membership, at national and divisional levels, had to be drawn largely from men available from these professions. The same situation prevailed in the Labor Department, which was staffed almost exclusively by men who had been officials of the NUM.

The only position that was free from this functional determinism was the chairmanship of the national and divisional boards. As a matter of policy, each of the original eight divisional chairmen was chosen from outside the industry, though the deputy chairmen were chosen from men with production experience in the industry. This policy on chairmen was adopted partly because it was hoped that a man with no past in the industry might have a better future in it and partly because the divisions were much bigger administrative units than anyone in the industry had been accustomed to manage. However, this policy failed to yield any decisive advantages and in all four of the vacancies which have opened in divisional chairmanships since the original appointments, the new chairmen have all been men with long experience in the industry. This shift has come about because of a growing recognition that a man's origin and background are not as important as his personal qualifications and a knowledge of the industry.[8]

Apart from abstract discussions about former army generals, trade-union officers or coalowners as desirable Board appointees, one might assume that the NUM (or the Labor Party, under a Labor Government) would try to secure some degree of control over the selection of specific individuals for Board appointments. This has not, however, been the case — certainly not in any formal sense. The Board does not clear appointments with the NUM before making them final, and informal consultations seem to have been rare, both under Laborites and Conservatives. Among NUM leaders, however, there has been considerable feeling that the Board ought to consult union officials before making appointments (at least to the key production and labor relations jobs, though not to such jobs as marketing, finance, law, purchasing, and so on). There have been cases where union officials have energetically protested appointments *ex post,* though these have had little or no effect.[9]

Closely related to the problem of filling management jobs is that of remuneration. Nationalization is one aspect of a socialism whose concern for social justice implies less inequality in the remuneration enjoyed by people in different walks of life. Fortunately, this ideological objective was being well served by Britain's income-tax policy, and the Labor Government did not have to make any sharp break with the salary levels to which people in the industry had become accustomed. Indeed, it would have been impossible to impose a socialist salary structure on the industry without a wholesale exodus of management personnel or a major revolt among them if they remained. The principal change has been the establishment of a salary structure which contains fewer variations around the average than is normally found in most industries. All

[8] The only former "miners' leader" to head a division is James Bowman, a former vice-president of the NUM and for many years general secretary of the miners' district union in Northumberland. He was made chairman of the Northers (Northumberland and Cumberland) division when this was established at the beginning of 1950. When the NCB was reorganized early in 1955, Mr. Bowman was promoted to the post of Deputy Chairman — not because of his trade union background but because he was widely recognized to be an unusually able administrator.

[9] For example, when the Deputy Chairman of the West Midlands Divisional Board was appointed in October 1952, a mining engineer was selected for the post (another mining engineer is chairman). The new mining engineer replaced a former NUM official, who had retired. The Area Union of the NUM protested vigorously that the job ought to have been filled by another ex-union man in order to "maintain the balance" on the Divisional Board. The Area Union's protest was carried to Sir Hubert Houldsworth, the pro-Labor Chairman of the National Board; he agreed that there ought to be a "balance" between the technical and ex-union men in the two top jobs in the divisional boards, but he took the view that this balance ought to be calculated on a national basis and need not be maintained within each division.

formal bonus plans (which were rare) and informal Christmas bonuses for a good year (which were fairly common) have been abolished, and salaries and "perqs" [10] are now determined by collective bargaining between the Board and the new unions that have been formed to represent the management grades. Problems of individual grading seem to have caused more grumbling among management personnel than the absolute salary levels which define the limits of the grading process.

The issue of equitable salaries in a socialized industry is not, of course, completely solved by the existence of a steeply progressive income tax. There are many socialists who still ask, "What kind of socialism is this that pays £3000 or £5000 to a former colliery director — worse still, to a former union official?" The only answer is that it is the kind of socialism which preserves historical continuity and the only kind possible without a violent social upheaval on the Russian model. But the price of this continuity has been to carry along from the past much of the traditional grumbling about managerial salaries. He would be a rash man who could say what level of salaries would be low enough to remove this question from the shadowy background of labor relations. One lesson the study of modern labor relations suggests is that there are many more hopeful ways of raising rank-and-file morale than by lowering management salaries.

There is a widespread but unclear belief in union circles that "there should be more of our people on the Board." This does not represent any weakening of respect for the necessity of technical qualifications for appointment to the key production jobs. Instead, this belief implies that technically trained managers should either (1) originate from the working class, or (2) be known to be sympathetic to nationalization (and should probably be members of the Labor Party). But experience makes it clear that the objective symbols of a manager's class origin or political affiliation have little or nothing to do with the degree of respect he commands from the men under him or opposite him in the union. Given the nature of a coal mine, qualities of personality are probably far

---

[10] A "perq" is a form of real income which is provided either in kind or at the expense of the firm instead of out of an individual's income. It is thus a form of petty privilege and might include any or all of the following: free house; free coal; free gas, electricity, and water; free house repairs and maintenance; the use of a gardener; or, less commonly, the use of a company car or even a chauffeur. Before nationalization these were enjoyed, in varying combinations, by most senior production officials of a colliery (the manager, the agent, and the managing director). In many districts, however, everyone in the industry received some perqs, by far the most important of which was "miners' coal," which was part of the wage structure in every district except Lancashire and Cumberland. In the latter districts this privilege was won in 1950 after a major unofficial strike in 1949. In Northumberland and Durham, all miners have traditionally received either a free house or a cash rent allowance in lieu thereof.

Since nationalization the Board has tried to convert most managerial perqs into cash additions to salaries; however, this has not gone down easily, as cash income is taxable while perqs are not.

more important. But the important point, so far as progress toward better labor relations in the industry is concerned, is that the functional nature of management has much more to do with how managers behave than their social origins or political affiliations. It is wishful thinking for union leaders to believe (and of course not all of them do) that managers will some day be much easier to get along with when the present hangovers from the old régime have been replaced by "our people." It is in the nature of things, whether in America, Britain, or Russia, that operating officials with responsibility for costs are difficult people for union leaders to get on with. The problem is slightly different in the case of higher nonoperating officials at the policy-making levels at the divisional and national levels; of this, more will be said shortly.

Vague grumbling about the necessity for the capture of management by pro-union individuals is at bottom a muted British version of the class war. The strength of this protest is likely to diminish if the Coal Board is successful in opening up careers to talented individuals irrespective of class origin. Such a development is well under way and consists mainly of the generous provision of educational opportunities wholly paid for by the Coal Board.[11] In a very real sense, nationalization is creating a much larger degree of social mobility within the mining community than has been true in the past.

The Labor Department. Of the six to eight members of the National and Divisional Boards, there is invariably one member (an ex-union officer) who is in charge of the Labor Department.[12] A divisional Labor Department consists of about ten men, predominantly ex-union officers

---

[11] The Board has offered 100 full university scholarships annually for boys leaving secondary schools who are willing to take mining degrees. However, the Board has not been able to find enough qualified applicants to take up all the scholarships offered: from 1948 to 1951 only about 300 awards were made out of the 400 scholarships available. Two-thirds of these 300 scholarships went to boys from "within the industry," the other third representing "outsiders" who had decided to try mining as a career. Despite the inability to award all the available scholarships, mine management is definitely attracting young men: the mining departments of British universities have had many more students since nationalization than before. Indeed, the fear is already present that there will soon be more mining engineers than the industry can usefully employ. This is not yet a serious problem, but a healthy competition for managerial jobs seems not far off. It thus becomes important that the Board develop effective procedures for appraising the performance of future managers and for assuring that they get the right kind of training. Many thoughtful critics of the Board would like to see young managers and technicians given more training in modern management and less training in traditional mining techniques than is now being done.

[12] Throughout the Board there is a distinction between "labor relations" and "welfare," though at all levels except the national, welfare activities are usually conducted by a special Welfare Officer within the Labor Department. "Welfare" does not include subjects involving union-management relations, though some of the Welfare Officers do have union backgrounds. We shall not concern ourselves with the welfare side of the Labor Department's work, though this is a large and important subject of foremost significance in providing leisure-time facilities in mining communities.

but including some former colliery officials. Each Area has its own Labor Officer, who normally has an assistant or two, and these Area Labor Officers likewise are almost always former union officers. There are no Labor Department representatives at the pits themselves. The only positions at the pits which have any explicit personnel functions are the training officers, whose duties concern almost exclusively the carrying out of the Ministry's training regulations; the latter concern chiefly juveniles and foreign labor.

The Board's Labor Department occupies a position both unique and ambiguous. The miners' union, unlike its counterparts in the French and German coal industries, chose not to seek formal representation on the Coal Board for fear of losing its independence of action if it acquired managerial responsibility. At the same time, the Union wanted to have "friends" within the Board and the obvious place for these was in the Labor Department (the department is in fact largely staffed with former NUM officials). This means that the NUM and the Coal Board have very different views about the purposes that the Board's Labor Department ought to serve. The Board recognizes that there is need for the introduction of modern personnel and labor relations practices into the industry, and consequently it feels that its Labor Department should be a staff department to help production officials. Union officials, on the other hand, tend to regard the Board's Labor Department as the defender of labor's interests within the Board — interests which are thought to be under inevitable attack from either "production people" or "finance people" — rather than as the arm of management concerned with labor and personnel problems. That such an ambiguity in the position of the Labor Department should arise is not surprising. While it seems likely to persist, it is unlikely to cripple the department's effectiveness except in a few Areas.

As noted, the National, Divisional, and Area members of the Labor Department are, with few exceptions, men with trade-union experience and, usually, with considerable experience as working miners. This fact has more advantages than disadvantages: the labor officers' lack of knowledge about the modern personnel attitudes and practices which the Coal Board rightly feels the industry needs is outweighed by their intimate knowledge of the industry — a factor of special importance in mining, which has a tradition, a language, a wage system, and a psychology more peculiar to itself than most industries. It is nevertheless true that in quite a few Areas both production and labor department officers feel that eventually some sort of personnel representatives may be required at the pit level; but such a development seems a long way off.

The union background of most labor officers does not mean that these men joined the Board directly from union positions. In a great many cases — perhaps half or so — they came through the Ministry of

Fuel and Power which was set up in 1942. The ministry established the positions of labor officer and investigation officer (the latter being assigned chiefly to the investigation of absenteeism), and these jobs were filled largely from the miners' unions. When the war ended, these individuals either had to return to union jobs (which by then had often been filled by others) or to manual jobs underground or on the surface. It may be taken as axiomatic that once an individual — particularly an individual in middle life — has left a manual job in mining for a white-collar job, it is a very difficult physical and psychological adjustment for him to return to underground work. It is as much a cut in status as in pay. It is not surprising, therefore, that when the Nationalization Act was passed in 1946 and the Coal Board began to assemble a staff, there was considerable pressure to find jobs within the Board for those former NUM officials and branch secretaries who had accepted jobs with the Ministry for the duration of the war. Inevitably and sensibly, there occurred a wholesale transfer of such individuals to the Labor Department jobs in the new Coal Board.[13] This transfer not only solved the personal problems of the individuals concerned; it also made available to the Board a group of men who possessed both long experience as miners and union officers as well as some administrative experience.

It might be thought that because the NUM has more of a proprietary interest in the Labor Department than in other Board posts it would exercise more control in appointments to this department than it does throughout the Board generally. However, this does not seem to be so, either formally by having a representative on the final selection committee or informally by making private recommendations on a personal basis between local NUM and Board officials.[14] The nearest the labor movement gets to having any say in the Board's appointments would seem to be that in many cases the Divisional Committee which interviews applicants who end up on its "short list" includes a representative of the Divisional Labor Department — normally the Labor Director, who is a former NUM official. Needless to say, there are people in the

[13] Mr. Tom Smith, the late Board member in charge of labor relations for the Northeastern Division and a wartime parliamentary secretary to the Minister of Fuel and Power as M.P. for Normanton, West Yorkshire, was largely responsible for making possible the transfer of the ministry's labor officers to the Coal Board. This meant that the original appointments were, in many cases, not publicly advertised but were filled privately. In subsequent vacancies, however, jobs have been advertised throughout the Board's structure as well as throughout the NUM and, less frequently, publicly.

[14] This statement refers to appointments to vacancies which have occurred since the original staffing of the Board. In some, possibly in many, cases NUM officials were asked in 1946 and 1947 who they thought would make suitable appointees — which meant that the union could select one or more nominees. The line officials responsible for making final decisions in appointments often welcomed this help. Private and off-the-record consultations obviously depend on personal relationships, which differ greatly from one administrative unit to another.

NUM who wish the Union had a much greater voice in determining
Board appointments.

The Board's labor officers have two difficult problems. One is their
relationship with the Board's production officials; the other is with the
labor movement from which most of them have come. Although the
labor officers are frequently regarded by the miners' union as the spokesmen for the men's interest within the Board, their lives are surprisingly
independent of the union movement. Labor officers do not attend meetings of their old unions except on rare occasions when invited in their
official capacities. The NUM has even gone to the extreme of denying
these individuals the right to retain honorary membership in the NUM.
This formal independence of the labor officers from the labor movement
is offset somewhat by the fact that few of the Area Labor Officers (and
by no means all of the Divisional Labor Department staff members)
have changed their residences since joining the Board; there has frequently been no sharp break with former friends and leisure time
activities.

There is even some feeling in the Union that the labor officers were
left too independent by the NUM at the time the Coal Board was set up,
so that they have in fact become too dependent upon "standing in well"
with the Board's line officials simply to protect their own jobs. Labor
officers were not given any assurances that they could return to their
former jobs within the union movement if they fell out with their new
employer.[15] This bridge-burning has left the labor officers in the position that if they do not get on well in their Coal Board jobs, they must
either compete for union jobs "from scratch," return to manual jobs, or
leave the industry. As a practical matter, it is not easy for one who has
left a union job to return to it, since it will already have been filled and
the individual will have been long absent from his potential constituents.
Nor do labor officers have a professional skill they can easily sell outside
the coal industry; their position is very different from that of a personnel
officer in private industry.

This is not to suggest that the Board's Labor Department is full of
dissatisfied ex-trade union officers who long to return to the union movement but are debarred from doing so. Quite the contrary: most of the
labor officers have adjusted tolerably well to their jobs — which often
represent a considerable career achievement for men who followed their
fathers into the pits at age thirteen. Nor has there developed as much
friction between the Board's production officials and the labor officers

---

[15] For example, both Abe Moffat and James Hammond of Scotland and Lancashire, respectively, referred to what they called the "onerous terms laid down" by
the union when the labor officers were picked. "The onerous terms were that they
had to be cut completely adrift." (See *1948 NUM Conference Report*, pp. 159 and
169.)

(many of whom used to do almost daily battle with the present production officials before nationalization) as might have been expected.

Nevertheless, there is a fairly widespread feeling in the miners' union, and even among some of the labor officers themselves, that they are not playing as influential a role in the Board's affairs as they and the Union had originally hoped. This disillusionment undoubtedly stems from an exaggerated and misplaced conception of what that role could or ought to be; but it is undeniably present in some quarters. There is thus some feeling that, if the Board's labor officers enjoyed the security of being able to return to white-collar jobs in the union, they would be much less likely to defer to the points of view of those with whom they might find themselves in disagreement (that is, the production and financial officials). As it is, this view holds, the ALO's are too independent of the Union to behave independently on the Coal Board.[16]

The chief criticism of the present ambiguous status of the labor officers is not the logical indefensibility of their position (the British are rarely embarrassed by logic), but the failure of the present arrangement to secure for the industry the type of effective personnel work which it could well use. The present situation subordinates the functional logic of the line and staff principle to the political reality of union insistence on symbolic control of the Board's Labor Department. While the Board may be able to train the ALO's so that they can perform personnel functions effectively and thus leave undisturbed the present unorthodox arrangements, it seems more realistic to hope that the drawbacks of the present arrangement can be bypassed through the introduction of more highly trained assistants in the Labor Department, men who will often not have union backgrounds and whose presence and function will be wholly neutral politically. But it will not be easy to find good men who will be willing to play a rather quiet second fiddle in a department which, it appears, must continue to be dominated by ex-union officials.

Very few of the issues I have discussed in connection with the organization of the Labor Department are treated directly in the Fleck Report. However, that Report does have some important things to say about the Board's organization for the handling of its labor problems. In brief, the Fleck Committee recommended that the Board establish a new Industrial Relations Department (to combine several functions heretofore handled by separate departments) and that in making senior appointments to this Department the Board rely less on the appointment of men with trade

---

[16] An isolated case may be cited where an "ALO" felt so out of step with Board policy as to resign (in Area No. 1 of the Northeastern Division, the Doncaster Area). In June 1950, James Kane announced that he was resigning out of disgust with the Board's capitalistic outlook. Kane was reliably reported to be a member of the Communist Party (there are certainly very, very few in the Board's Labor Department); he eventually returned to an underground job.

union backgrounds. The Committee also noted that only a very few pits in the country presently have personnel officers but that all large pits ought to have them to relieve pit managers of much of the pressure of personnel questions that claim so much of their attention. Only time can tell whether or not the NUM would accept the decrease in its influence within the Labor Department which the Fleck Committee's recommendation implies and whether or not the mineworkers, branch officials, and pit managers themselves would accept the introduction of personnel officers.

With respect to the labor officers' relationship to the production officials who necessarily dominate Area and Divisional affairs, it ought to be said first that in many cases satisfactory line-and-staff relationships have been established. The existence of friction between the Board's Production and Labor Departments, especially during the first two or three years of the Board's life, is widely known. Such friction would seem to have been inevitable as a major transition problem. Both individually and as representatives of the pro-nationalization forces, Labor Department people were men "on the make," while Production Department people represented groups that had been chased into the "doghouse." Both groups will have to alter their traditional ways of regarding each other — a requirement on which there has already been progress. But the domination of the British Association of Colliery Management by production officials led a majority of Labor Department members to withdraw from BACM, whose right to represent these grades was canceled by the Board, and to form an independent Labour Staff Association which received recognition in 1952.

Still, in a minority of Areas and perhaps in some Divisions, union leaders and even the Board's labor officers themselves continue to complain that the Labor Department "ought to have more control over board policy." It is not quite clear what this would mean. Sometimes it seems to mean that labor officers ought to have line authority over many aspects of labor policy. At other times the individuals who carry this complaint say that there ought to be more labor representatives on the Divisional or National boards, so that one Labor Director would not be a lone sheep among five or six wolves. But any effective change would require that "labor" secure a majority of the directorships. This may be the hope of a small minority of people in the Labor Party and, more importantly, in the NUM; but as previously noted, the official policy of both the NUM and the Labor Party is against this kind of "workers' control." Until the line and staff concept (which is new to the industry) is much more widely understood, there will be individuals and groups who will resist the essentially pro-management role which it implies for the Labor Department.

II. The National Union of Mineworkers

The NUM has always been the largest union of mineworkers in the world, and, with over 600,000 members, it is today the fourth largest trade union in Great Britain. Although it is not the only union active among Coal Board employees, it is the only union which represents employees at pit level, where more than 95 per cent of the industry's labor force work. There is a handful of pits where the men are not union members (for example, a few at the eastern end of the Leicester-shire coalfield), but they certainly do not constitute a "nonunionized" sector of the industry that presents any threat to the union. The NUM is about as secure from external threats as any union can be; and, while the union undeniably has internal problems, this new-found security represents a considerable change from the prewar period when non-unionism, rival unionism, and "poaching" by outside unions presented external threats to its prestige and effectiveness.

It goes without saying that the Union is now operating in an institutional environment vastly different from that in which the Union's traditional policies and procedures were formed, and it is a much-asked question whether or not the Union has adapted well to the new tasks that now seem required of it. On the other hand, there can be little doubt of the Union's quite remarkable success in solving the classical problem of trade-union administration, the reconciliation of democracy and efficiency.[17] The Union does not pay its officers the high salaries that compromise the moral position of officers in several American unions;[18] and the custom of giving many key officers life tenure is prevented from having mischievous consequences by reserving vital powers to local representative assemblies.

THE UNION'S STRUCTURE

The least self-evident aspect of the Union's structure is undoubtedly the difference between the Area unions and the Constituent Associations

[17] See Beatrice and Sidney Webb, *Industrial Democracy* (1902), chap. ii, "Representative Institutions."

[18] In 1951, when entry wages in the industry were £6 10s. and facemen were earning around £10–12, the president, vice-president, and secretary of the NUM were being paid about £12, £10, and £19 a week, respectively. In addition, these national officials were entitled to £2 a week as a tax-free "personal allowance"; they also received perquisites such as subsidized housing, light, and water, and generous but not unreasonable expense accounts. Area officers are paid on a national salary scale; in 1951 secretaries and presidents of area unions earned about £14–15 a week and miners' agents £13–14. Some of the larger districts pay their permanent officials around £2 a week over the national scale out of the district's own funds. Area officials, including the agents, usually receive some perqs in addition to their cash salaries.

Chart 2

Structure of the National Union of Mineworkers

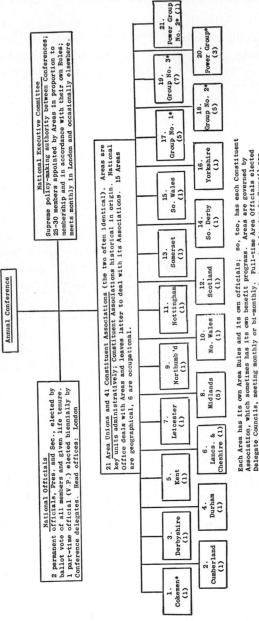

Annual Conference

National Executive Committee
Supreme policy-making authority between Conferences; 25-30 members appointed by Areas in proportion to membership and in accordance with their own Rules; meets monthly in London and occasionally elsewhere.

National Officials
2 permanent officials, Pres. and Sec., elected by ballot vote of all members and given life tenure. 1 part-time official (V.P.) elected biennially by Conference delegates. Head offices: London

1. Cokemen* (1)
2. Cumberland (1)
3. Derbyshire (1)
4. Durham (1)
5. Kent (1)
6. Lancs. & Cheshire (1)
7. Leicester (1)
8. Midlands (5)
9. Northumb'd (1)
10. No. Wales (1)
11. Nottingham (1)
12. Scotland (1)
13. Somerset (1)
14. So. Derby (1)
15. So. Wales (1)
16. Yorkshire (1)
17. Group No. 1* (5)
18. Group No. 2* (5)
19. Group No. 3* (7)
20. Power Group* (3)
21. Power Group No. 2* (1)

21 Area Unions and 41 Constituent Associations (the two often identical). Areas are key units administratively; Constituent Associations historical in origin. National Office deals with Areas and leaves latter to deal with its Associations. 15 Areas are geographical, 6 are occupational.

Each Area has its own Area Rules and its own officials; so, too, has each Constituent Association, which sometimes has its own benefit programs. Areas are governed by Delegate Councils, meeting monthly or bi-monthly. Full-time Area Officials elected locally for life (but are paid by National Union); branch delegates to Council are elected every year or two by branch members at special meeting.

Over 800 Branches or Lodges
Each governed by the Rules of its Constituent Association. Main officers are: Secretary, Delegate, President, and Treasurer; also Lodge Committee of officers plus 6-15 others. All elected every year or two at branch meetings.

Notes:
  Numbers in parentheses under Area Unions refer to number of Constituent Associations in that Area.
* Occupational Areas representing members in several different coalfields; all other Areas geographical.

(see Chart 2). As the distinction between the two entities is only to be understood historically, we must review the evolution of the present structure.

An extensive constitutional reorganization of the Union became effective as of January 1, 1945. On that date the old Mineworkers' Federation of Great Britain became the National Union of Mineworkers, a change in name which symbolized a shift from a federated basis of organization, adopted when the MFGB was originally formed by the joining of many autonomous county associations in 1888, to a much more centralized basis of organization in which the national union now has considerably more authority over its component parts than was formerly the case (the NUM is still much less highly centralized, however, than most American unions; a large amount of district autonomy still survives). The nearly autonomous bodies which were affiliated as the Mineworkers' Federation are, with one or two recent additions, the present Constituent Associations.

The "Area Unions" are new entities, created at the time of the 1945 reorganization. They are the administrative subdivisions of the new national union. Broadly speaking, therefore, what the 1945 constitution gave the Union was a new legal and financial relationship between the affiliated unions and the central organization — it did not interfere with the number or internal structure of the component parts. The old "district unions" have thus become administrative areas of a single National Union of Mineworkers, and the salaries of Area officials and staffs are now paid from bank accounts controlled by the national union (except in those few cases where an Area is willing to pay out of its own funds for additional personnel whose employment the National Executive Committee is unwilling to authorize).

Nevertheless, with that instinct for illogical and untidy but highly effective compromise that is so distinctively British, the new Area unions exhibit a very large measure of autonomy: they elect not only their own Area officials, for example, but also the Agents (who are thus not to be regarded as representatives of the national union stationed in the districts); the Area unions even retain the right to register themselves as separate organizations with the government's Registrar of Friendly Societies (an optional procedure that confers certain rights and immunities on a union); and they retain the right to set their own dues (above the national minimum) and to operate their own benefit programs. But, despite this continuing fact of a large measure of *de facto* district autonomy, the new rules do give many new powers to the central union, powers which the latter can exercise only with great caution if it does not wish to violate local loyalties and interests. One is not long in the coalfields before one learns that the NUM is still very much a highly decentralized federation of district unions.

The National Executive Committee corresponds to the International Executive Board of American unions. However, the Executive Committee of the NUM, as in most British unions, has more authority over, and is more independent of, the union's full-time national officers than is customary in the United States. The miners' Executive meets more frequently, and is more exclusively responsible to local interests than is true in most American unions. This correctly suggests that the full-time officials of the NUM are more concerned with executing than with framing policy — a situation not generally true of the national officers of American unions. The Executive Committee, to which representatives are appointed for two-year terms, sometimes includes one or two "rank and file" members; but it is the permanent officials of the Area unions who are normally put on the Executive by the Area unions, and it is this group which controls the Committee. The national officials of the NUM, who also sit on the Executive, have no authority over the appointment or removal of any Executive Committee member; neither the president nor the secretary of the union has any vote on the Executive, except in case of ties, when the president may vote. All the "emergency powers" over the affairs of district and local unions which are so often vested in the national officers of American unions are vested in either the Executive Committee or Conference of the NUM.

The power structure, administration, and some of the problems of the Union can best be approached by reviewing the difficulties which it faced under the old system of "district autonomy" and the movement for constitutional reorganization which gathered momentum in the 1930's.[19]

When the MFGB was in process of becoming the NUM in 1943–44, the national union had one full-time official, three staff members, and an annual income about one-tenth the £140,000 taken in by the Durham Miners' Association. These few facts suggest the extent to which the locus of power lay in the twenty-two autonomous districts and not at the Federation's London headquarters.

Such "district autonomy" meant that no effective central leadership could develop on the trade-union side of the industry. All the big jobs were in the district unions, not in London; the big funds were owned by the larger district unions, not by the MFGB; the salaries of all influential people in the Federation, except the full-time secretary, were paid out of the funds owned by the large districts, of which they were almost invariably officers.

Until 1945 (with a brief exception during the presidency of Robert Smillie in 1919–20), the positions of president and vice-president were

[19] See G. B. Baldwin, "Structural Reform in the British Miners' Union," *Quarterly Journal of Economics*, Nov. 1953, pp. 576–597.

part-time posts and carried nominal salaries of £125 and £60 (1939). The president, Will Lawther, was general secretary of the Durham Miners; the vice-president, James Bowman, was general secretary of the Northumberland Miners. Both men received most of their incomes from their respective district unions. Both maintained their residences in their districts, not in London.

The overwhelming bulk of union property was owned by the districts, not by the Federation; all union members were members of their own district union, not of the MFGB (whose "members" consisted only of the district organizations); and all collective bargaining was done on a district or pit basis, save from 1921 to 1926. The MFGB did not participate in the district negotiations and was not a party to the resulting agreements, copies of which it often did not receive.

Not only were there no representatives of the national organization in any of the districts, but no such representative was even supposed to come into a district to make investigations or to give official advice unless specifically invited to do so. The Federation had no full-time solicitor, no research or statistical department, no national officials resident in the districts corresponding to the organizers or international representatives of American unions, and no educational director. The Union published no newspaper, except for the four years 1927 to 1930, when *The Miner* was discontinued for reasons of economy.[20] Finally, there was no uniform scale either of contributions (dues) or of friendly benefits: the amount of each varied from district to district — the lowest dues in the 1930's (for employed members) being 6d. per week; the highest being about 1s. 3d. Many of the district unions had developed traditions and funds around particular benefit schemes — death benefits, pensions, out-of-work pay, accident payments, "shifting money" to help men move from one village to another, and so on.

It would be quite wrong to conclude from the foregoing account that the Miners' Federation was an impotent and uninfluential appendage to the trade union side of the industry. The Federation was probably the most powerful and most influential trade union in the industrial life of the country and certainly so in its political life. No other union in Britain could consistently send twenty-five to forty political representatives to Parliament; and in few industries did the workingmen enjoy the degree of legal protection which the "Miners' M.P.'s" had won for their constituents — in the fields of safety regulations, control over the length of the working day, the right to elect workmen's "auditors" (the checkweighmen), workmen's compensation, and the representation

[20] The MFGB did publish a *Monthly Bulletin of Statistics* from 1928 until early in World War II, but this appears to have been limited to the reprinting of statistics gathered by organizations other than the union.

of miners' interests before numerous government commissions. Almost all the propaganda emanating from the trade union side of the industry was prepared and issued by the Federation.

This political influence persists, of course, to the present day, and nomination to the "panel" of the Divisional Labor Party (from which final nominations are made for specific elections) constitutes an important form of recognition for faithful union service. The "Miners' M.P.'s" are organized into a Miners' Parliamentary Group that is formal enough to have a chairman, vice-chairman, treasurer, and secretary. The chairman and secretary regularly attend all meetings of the NUM Executive Committee. The MFGB was also the body which represented the miners in the Trades Union Congress and in the Labor Party — the separate district organizations were not officially represented on these bodies, although the MFGB delegations were comprised of men nominated by the district unions. In industrial questions, the Federation — through the monthly meetings of its National Executive Committee and through the Annual Conference — could exercise leadership in the formulation of policies; and it provided the one industry-wide forum where such policies could be discussed outside the district organizations. But — and it was a big "but" — the Federation was powerless to compel any affiliated organization to respect any policy it might recommend. The inevitable result was that districts often acted in their own interests in a way that could not always be reconciled with the wider interests of the Federation as a whole.

The industrial defeats in the 1920's and the disorganization and demoralization which ensued generated pressure within the Union for a reorganization of its structure. This movement began to receive serious consideration early in the 1930's.

Broadly speaking, the proposals discussed during the 1930's began with a theoretically ideal form of union organization but ended up with a set of compromises (and very good ones) made necessary by specific vested interests within the Union. This discussion — as revealed in debates at the Federation's Annual Conferences — began from the widely held feeling that there were "too many unions in the industry," both inside and outside the Federation. The internal remedy, then, was some kind of reduction in the number of miners' unions in the industry and the great debate was whether the reduction ought to be partial, through the amalgamation of certain neighboring county associations, or complete, through the merging of all the twenty-two county unions into one organization — a step which would involve the liquidation of all the county unions. Neither of these steps was eventually taken, and the reason they proved impracticable was the existence of the vested interests of the separate district unions, their officers and agents, and their members. The specific interests that continually frustrated the reform movement were these:

1. Jobs. The permanent officials of the district unions — district officers and Miners' Agents — understandably did not come round to support the centralization proposals until they were framed in such a way as not to constitute a threat to their personal security. Such a guarantee was eventually made.[21]

2. Real Property. This consisted of union offices, houses for union officials, houses for retired miners, some automobiles and rest homes (usually at one of the more popular seaside resort towns) for sick or injured members.

3. Funds. Districts which had often worked for many years to build up a structure of benefits were naturally jealous of the control of these funds. The variation in the types of benefit programs from district to district (some districts, such as South Wales, had no such schemes) had produced a structure of dues which varied considerably. A major problem in working toward greater uniformity and greater centralization was how to secure a uniform contribution from all mineworkers throughout Britain without disturbing the benefit levels to which the districts had grown accustomed (and which were often matters of great local pride).

In addition to these substantive interests, there were many subtler psychological threats implicit in the amalgamation movement. Any decrease in "district autonomy" implied a reduction in the status of district officials. Further, the "district" had always been, and still remains, an effective focus of group and individual loyalties; amalgamation threatened to dilute the sense of indentity which bound together many miners' leaders and their members. Finally, the drive toward unification sometimes appeared to certain of the high-wage districts as a threat to the district's wage levels.[22]

The reorganization movement could make no progress until ways could be found to reconcile the vested interests of the existing district unions with the needs of a more centralized national union. Credit for

[21] This problem was very similar to that confronting the Coal Board when it inherited many ex-management officials whose services were not essential to the reorganized industry. According to the nationalization statute, these men either had to be retained in service or paid compensation for loss of office. The difference between the Coal Board's position and the NUM's was that in the latter case the guarantee of personal security of the individuals concerned was a condition of "nationalizing the union" while it was not a condition of nationalizing the industry — it was simply an expression of the British sense of "fair play."

[22] This was especially true of the Nottinghamshire coalfield, which today pays the highest wages and has the highest output per manshift of all British coalfields. It has always been a high-wage, high-productivity field. After the 1926 strike, in the late 1920's, the Notts miners were organized into a Miners' Industrial Union by George A. Spencer, who felt that his men could do better for themselves outside the Miners' Federation. Although he was eventually persuaded to rejoin the Federation in 1937, he remained a vocal spokesman for those high-wage Midlands districts which feared that a more centralized union would threaten their very favorable wage standards.

the solution of this problem goes largely to Arthur Horner and Sam Watson who, in 1937, were given the responsibility of formulating a new set of specific proposals. These men approached their assignment from a new quarter: instead of formulating an ideal union structure as though they were building *de novo*, they set out to identify the way in which the various district organizations did in fact operate, and how the establishment of a unified organization might affect the district unions; they then framed realistic compromise proposals which they hoped would protect the vital vested interests of the districts without frustrating the shift of other essential powers and functions to the national union.

Horner and Watson drew up a set of proposals which were submitted to and accepted by the National Executive Committee, which recommended them to the union's 1938 Conference. These 1938 proposals provided the general formula through which unification was finally achieved in 1944. The core proposals were these:

1. All present full-time officials and staffs of the district unions were to be kept on the new union's payroll. They were later assured that no one was to suffer any loss of salary, forfeiture of pension rights, or loss of such perquisites as subsidized housing.

2. The existing system of branch responsibility to Area Executive Committees was to be left substantially untouched. Only eight separate Areas were proposed, and this implied that several existing district unions would be amalgamated into the new Areas. As Chart 2 shows, it eventually proved necessary to provide for 21 Areas and to retain the identity of some 41 separate unions as Constituent Associations, grouped under the 21 Areas. Not a single district union was in fact dissolved — a fate that several districts would have faced under the 1938 proposals.

3. Each Area was to be responsible for collecting 6*d*. per week for each of the members in the Area, the full 6*d*. to go to the National Union. It later proved necessary to reduce the payment to the National Union to 4½*d*. per week and to assure that 20 per cent of all monies collected from any branch would be returned to that branch for expenditure as it saw fit. Any Area which wished to collect more than 6*d*. per week from its members could do so and could retain the full amount to use as it saw fit.

4. The benefit funds which several important district unions had built up were to be left intact for the use of the districts which had accumulated them. The 1944 rules eventually required each Area to pay a one-pound capitation fee for each member; presumably this fee had to come out of general funds, not out of special funds earmarked for friendly benefits.

5. Control over district (and of course national) strikes was to pass from district to national hands. Since one of the main purposes of "nationalizing" the union was to secure coördination in bargaining strategy and policy among the districts, this change was accepted without dispute, though no tight control over district bargaining or national-union participation in the signing of agreements was contained either in the 1938 proposals or in the final 1944 Rules.

The Horner-Watson proposals proved too far ahead of the cautious reorganization sentiment in 1938, and they had to be temporarily dropped. The Second World War, however, inevitably brought a radical shift away from the old district basis of handling the industry's problems to an industry-wide or national basis: wages, prices, and production control all came under national regulation, and these developments greatly strengthened the hand of those who were convinced of the necessity for structural reform. When the Union's 1943 Conference opened in July, the Executive Committee had ready a new set of proposals, proposals which were, in fact, very similar to the Horner-Watson recommendations of 1938. Conference accepted the proposals, sent them back to the constituent unions for detailed criticism, and arranged for a Special Conference in August 1944, for their final consideration. The Special Conference at Nottingham resolved all outstanding issues and sent the new Rules to the membership for a ballot vote, which went 390,964 to 39,666 for acceptance.

A summary of the way in which the major issues were resolved at Nottingham can tell us much about the political pressures, and the administrative arrangements that exist in the Union. Solution of the following six issues dominated the Nottingham Conference:

1. Amalgamations: The power of the Executive Committee was limited to "recommending" to a Conference any amalgamations of adjoining Areas; only a Conference may force amalgamations.

2. Appointment of Officers and Executive Committee: The presidency was added to the secretaryship as a full-time office, both of which were to be filled by ballot vote of the members; the vice-presidency was left a part-time office, to be filled biennially by Conference vote. The Executive Committee had recommended that its own members be appointed for two-year terms, with the method of appointment to be left up to each Area, as had been customary. South Wales successfully secured an amendment which requires each Area Union to submit its nominees to a ballot vote of the Area members.

3. Terms of Office: The Executive Committee recommended that the union's full-time officers, once elected, should serve "at the pleasure of the union" (which meant, in effect, for life). Several Areas tried unsuccessfully to secure fixed terms. The life-tenure rule had long been customary in the miners' unions (at district as well as national level, though not at the local level). Life tenure is common, though by no means universal, among other British unions.

4. Retirement Age: Since 1933, national officers had been compelled to retire at 70. The Nottingham Conference accepted the proposal to reduce this figure to 65 and to forbid anyone over 60 from accepting a full-time office. Conference made the Executive Committee promise to institute a pension plan and to take over all the contracts of service which the district officials held with their district unions (since these officials would now become, technically, officials of the national union).

5. Salaries and the Capitation Fee: The Executive Committee proposed that each Area Union pay a one-pound capitation fee for each member, a device that would effectively shift the great bulk of district funds into the treasury of the national union. District officials, especially those from the larger districts, wanted to know what plans the national union had with respect to paying their officials before they would approve the release of

their funds. Conference gave these officials a moral commitment that no official would suffer any loss of income or perquisites as a result of the change-over.

6. Amount of the Per Capita Tax and Union Dues: The problem was to arrive at a figure which would secure enough revenue to the national union to enable it to perform all the industrial functions of the union, while leaving enough money in the districts so that the latter could still finance their benefit funds. This figure was set at 4½d., a half-penny less than the Executive had recommended. Constituent unions had to collect a minimum of 6d. per week from their members, but were left free to set higher dues if needed (as some did in order to finance their traditionally more elaborate benefits).

We should now have a general notion of the national union's organization and how this developed during the reorganization movement that began after 1926. The chief locus of power and activity of the NUM, however, still lies with the district unions, and it is their government which shall occupy us next.

The Area Unions. Every NUM Area has a small set of permanent officials (anywhere from two to five, excluding the Agents), a larger Area Executive Committee (the Area's permanent officials plus, say, from six to twelve elected additional members), and an Area Council (composed of one "delegate" from each branch or lodge in the Area, plus the Area's permanent officials). The delegate or council meetings which each Area holds every month or two constitute one of the most effective communications channels in this or any other union. These meetings hammer out decisions on every conceivable aspect of trade-union policy — though since the postwar shift of power to the national union, these policy decisions are often advisory rather than controlling. Since the meetings are attended by lodge delegates on the one hand and by the Area officials on the other (and among the latter are normally found the Area's representatives on the National Executive Committee), they constitute an extremely effective channel for (1) making clear to higher officers the attitudes of lodge representatives on specific questions and (2) explaining to lodge officials the points of view held by higher officials or reporting progress reached in negotiations.

One of the year's council meetings in every Area must be designated as the Annual Conference of that Area, and at this Conference the Area's policy is set on questions of both national and district interest. For example, delegates to an Area Conference may "mandate" the Area delegates who will attend the union's national Conference on certain of the resolutions which it is known will arise at that Conference. Delegate meetings also constitute the channel through which resolutions proposed by the branches must first be considered. If, for example, an Annual Conference of the Scottish Area of the NUM decides to turn down a resolution sent to it by one of its branches, that resolution cannot go to national Conference from the branch. But if the Area Conference approves a branch resolution, the latter may then be forwarded for consideration at National Conference where it will be moved in the name of the Scottish Area.

The frequency with which council meetings are held in an Area has much to do with how well-informed officials at any one level are with respect to what is going on above or below them. Originally, the NUM Rules provided for bimonthly meetings — that is, the NUM would bear the expenses of only one Area meeting every eight weeks, though Areas could request permission of the NEC for special interim meetings (they could hold more frequent meetings at their own expense). Indeed, before the "nationalization of the union," many districts had been accustomed to holding council meetings every month and at the NUM's 1947 Conference three large Areas successfully proposed (after an earlier defeat in 1945) that the NUM Rules be altered to allow (but not to require) monthly meetings.

Scotland and Northumberland were two large districts which voted against monthly meetings. Their arguments were that, since the formation of the NUM, the Area councils had been changed from policy-making to administrative bodies (a doubtful assertion, though certainly their policy pronouncements now carry less authority than under autonomy); also, some speakers felt that it was hard enough to cultivate sufficient interest in the lodges for bimonthly meetings, let alone monthly ones. They also pointed out that few if any Areas had bothered to ask the Executive Committees permission for special meetings and that in no case had such permission been denied.

The effectiveness of communications between the rank and file, or the branch officers and the Area Unions is not, of course, solely dependent upon the frequency and quality of delegate meetings, though these are extremely important. Of equal importance is the way in which the Areas are organized for administrative purposes. An Area's organization has much to do with the frequency with which Area officials get to the individual branches, the ease with which members and lodge officers can secure access to their Area officials, and hence the accuracy of information held by Area officials concerning particular problems at particular pits. These questions obviously affect the number of unofficial strikes in a district, the effectiveness with which strike situations are handled, the ability of union leaders to secure prompt redress of grievances or to bring about changes in working customs which they may feel justified (an especially important point since nationalization).

There are two characteristic administrative patterns in the district unions: "centralized" and "decentralized." Among the large districts, Durham, Northumberland, and Yorkshire represent the centralized form of organization; Scotland and South Wales are decentralized. Thus all five of the Durham Agents live in union-owned houses in Durham City and work out of the impressive miners' offices on Red Hill in that city; the Northumberland Agents work out of Burt Hall in Newcastle-on-Tyne. Yorkshire's five permanent officials (officers and Agents) all work out of the Barnsley office — though not all five live in Barnsley. In any of these coalfields, one can reach the farthest pit from the central office in about an hour by automobile. The Scottish Area, on the other hand, has resident Agents in five or six towns scattered throughout the Scottish coalfield (the Area headquarters were moved to Edinburgh from Glasgow shortly after the administrative unification of the Area in 1945). South Wales, more cut up geographically and with over two hundred lodges, has also kept Agents at key points away from the union's center in Cardiff. The Areas that remain decentralized today are generally those that were themselves organized as federations of semi-autonomous local associations as recently as 1934 (South Wales), 1944 (Scotland), and the Midland Federation (1947). Geographical inconvenience explains the origin of the decentralized organization.

The smaller Area unions are usually small enough so that they are not confronted with a decision as to whether they ought to centralize or decentralize — their two or three officials work out of a single office. This is true, for example, of Leicestershire, South Derbyshire, North Derbyshire, and Nottingham — the four Area unions which operate within the East Midlands Division of the Coal Board. There has been half-hearted talk of a merger of some or all of these four unions. Political differences and the all-important problem of redundancies have so far prevented any amalgamations. However, these four unions have set up a Divisional Standing Joint Committee, made up of representatives from each union, which meets monthly at Sherwin Lodge, the Coal Board's Divisional Headquarters near Nottingham.

A crucial question in all the big centralized districts is whether or not the officials maintain a sufficiently close touch with the rank and file and with branch officers. This is not an easy question on which to pass judgment. For example, both Yorkshire and Durham have over 100,000 members; Yorkshire, where the industry is organized on a somewhat larger scale, has about 115 branches while Durham — where branches are "lodges" — has about 170. Yet despite the comparative similarity of organization in these two huge districts, Yorkshire has a much poorer record on unofficial strikes and on local political eruptions that challenge the central leadership at Barnsley. There is, both within Yorkshire itself and in other districts, a considerable feeling that the Yorkshire Area's structure "wants looking at."

In August 1952, the Yorkshire Area's Council voted against a proposal to appoint eight full-time Agents, one for each of the Coal Board Areas in the Northeastern Division. The proposal would have removed some of the burden of the president's work by giving locally based Agents more responsibility for settling disputes that arose in their Areas. The vote was reported as "close," which means that while there was wide sentiment in favor of some administrative reorganization, there was even wider sentiment against the increase in operating costs which this would have involved. The Area's permanent officials may also have been reluctant to let go of their present authority.

There is very little such feeling, if any, in or with respect to the Durham Area. Part of the contrast between these Areas may also be explained by personalities. In addition, the Durham leaders do not believe in the wisdom of full-time union officials at the pit level — a practice that is fairly common in Yorkshire, particularly South Yorkshire. Such officers, it is felt, sometimes grow out of touch with conditions down their pit; also, their existence constitutes a temptation (both to the officer and to his constituents) to hunt out and magnify issues that would not be considered genuine if the aggrieved individuals had to seek out a part-time branch secretary whom they knew was intimately familiar with the condition of the pit.

The Local Unions. The miners' lodges constitute the "local unions" of the NUM. They enjoy positions of great formal authority (since they retain autonomy in many crucial matters) and of even greater power. The leading officers of the lodges are people of key importance in their village, their role being something like a poor-man's "squire." Socially and politically, the branch secretary is usually a far more important person than the local pit manager, whose authority usually stops at the pit. Quite frequently, lodge officials are actively interested in local government, either through membership on the urban or rural district council or on the county council (membership on a county council carries about as much status in Britain as does membership in a state legislature in the United States).

The miners' lodges are organized on a basis different from that generally true of British unions: the latter are organized geographically, whereas the miners' lodges are organized by place of work. Thus in a borough such as South Shields (County Durham), there would typically be only a single branch of the Amalgamated Engineering Union (AEU) or the Electrical Trades Union (ETU), though there would be several plants from which the members of each local union were drawn; but there would be one miners' lodge for every pit in County Durham. There have been a few localities in which the lodges had been organized on a village basis where the basis has only recently been shifted to the pit. Some of the lodges at the Walsall end of Cannock Chase shifted over from a village to a pit basis as recently as 1947; the same thing occurred at about the same time in parts of Fifeshire. Village organization requires fewer branches than does pit organization, but the latter affords members better service, since their officers are usually much more familiar with conditions at their particular pit; under village organization, of course, the branch officers must serve members working at several local pits.

The typical branch has four principal officers: the president, the secretary, the treasurer or financial secretary, and the delegate. These four are subject to the general direction of the lodge's executive committee and, like the latter, are subject to periodic election, usually by the members who turn up at a special meeting (not by a "ballot vote"). Branch elections are held every one or two years, depending on the particular coalfield. Most people in the industry — union and management alike — would probably agree that annual elections are too frequent, since it takes a new branch secretary about a year to "get to know his job." But, of course, at many pits with annual elections the incumbent secretary is reëlected year after year.

Thus the custom of electing "permanent officials" which is characteristic of the Area and national unions does not exist at the local level. One result is that political forces play a larger role in branch life than is true at Area or national level.[23]

Of the four usual branch officials, the lodge secretary is almost always the most important, though in Scotland the delegate is considered the chief local officer. The secretary is the chief executive officer of the lodge. It is he who has the most frequent contact with the pit manager, handles most of the correspondence between the branch and the Area union, receives most of the grievances and compensation problems of the membership, and who leads discussion at branch meetings. The president's position is often largely honorary — a post in which up-and-coming younger men are sometimes found and, more frequently, ex-secretaries who have been honorably retired from active direction of the lodge but whose experience is retained in the officer group. Ordinarily, the lodge president is chairman of lodge meetings and of meetings of its executive committee. The lodge delegate is the chief link between the lodge and the Area Union — the chief link is this personal one. It is he who attends the monthly or bimonthly council meetings at the Area union's central headquarters, carrying to it the views (and often the "mandate" or committed vote) of local members; it is also he

---

[23] This single fact has accounted for frequent frustration of some of the key labor agreements negotiated between the NUM and the Coal Board "in London." For example, the 1947 Five-Day Week Agreement, the 1947 Agreement on Foreign Labor, and the 1950 November Wage Increase Agreement all contained important "conditions" wanted by the Board and to which the NUM agreed. These "conditions" involved changes in local arrangements that were bound to be unpopular (e.g., increasing work-tasks, introducing "foreigners," and abolishing a system of "bidding" for work places); branch officials simply refused to take energetic action for these reforms — or, where they did, the turnover of branch officials was sometimes very high, as in Yorkshire in 1948 following union officers' efforts in 1947 to persuade facemen to revise their stints upwards.

who reports back to every lodge meeting what took place at the latest council meeting (his remarks are usually supplementary to the excellent printed reports of council meetings which are distributed to every lodge by most of the Area unions).

Most lodges hold open meetings every month, with the lodge committee often meeting once between the open meetings. At some of the big pits an individual lodge will have an office or meeting-hall of its own, but in more cases than not the lodge has no permanent office: mail goes to the house of the secretary or delegate and meetings are held in a rented church, Miners' Welfare Hall, Coöp rooms, or similar local facility. As for the meetings themselves, they are usually very poorly attended, though this is not nearly as telling a criticism as many people believe. The number of members who attend the monthly meeting naturally skyrockets when an important wage question is at stake or when a strike threatens; but these occasions are very exceptional and on the whole it is uncommon to have more than a handful of members present (very often the lodge committee constitutes a majority of a branch's open meetings).

Generally lodge officials are not full-time union officials but work at the pit. The lodge secretary and delegate must often take days off to attend to union business; hence many pit managers see that they are kept on jobs where their absence will not seriously affect production.

Only at the largest pits (say those with over 1500 to 2000 members — in 1948 there were only about 50 pits with more than 2000 men employed) would there be any need for a full-time official; and probably no pit has more than one full-time official, the lodge secretary. The coalfield having the largest number of full-time lodge secretaries is Yorkshire; a full-time secretary is probably the rule among the big pits of South Yorkshire. A district such as Durham, which is centrally organized like Yorkshire, has not more than one or two full-time lodge secretaries — at large pits on the east coast; in Durham, lodges are discouraged from having such officials. Few if any full-time lodge secretaries would be found in any of the smaller districts. Scotland and South Wales, with their decentralized organization, have very few full-time secretaries (South Wales has a handful at its largest pits).

There are no union representatives or "stewards" at the place of work beyond the members of the lodge committee and the lodge officers, who are distributed at random throughout the pit. There is no union representative on each coalface, for example, except by coincidence. This means that when grievances develop on a particular face, or from any particular section of the pit, the group involved will approach the lodge secretary (usually after work — on the pit top, if the secretary can be found, or later at his home or in the Workingmen's Club) and ask him to persuade the pit manager to receive a "deputation." This deputation will normally include informally selected representatives from the group involved, plus the lodge secretary and perhaps some other members of the lodge committee. Thus the lodge's structure in the pit is not built up on a "representative" basis, as is the typical American union in a manufacturing industry. The local unions in the American coal industry, however, do not have a "representative" basis, either, perhaps because the physical conditions of mining do not lend themselves to the familiar representative local union structure normally found in manufacturing industries.

There is a great gap between the large role which, one is told, the miners' lodges play in pit villages and the visible evidence of this role. Very few branches have their own meeting-halls or offices, conduct any educational work, sponsor any dances or choirs, maintain any cricket grounds or bowling greens or snooker tables or organize any coöp stores. Most of these activities will, indeed, be found in any mining village. But the coöp store is organized and run by the coöperative movement; beer-drinking (a British social and recreational custom of far greater importance than in America) is done at the local "pub" (or "local"), at the Workingmen's Club, or at the Miners' Welfare Institute (the latter provides an extremely wide range of social activities for men and women, boys and girls, and is normally the social center for most mining communities — the "miners' country-club"). Snooker, bowls, boxing, skittles, tennis, cricket, football, beer, lectures, dances, choral concerts, dramatic presentations, darts, bicycle racing, dominoes, boy's clubs — all these vital activities lie outside union sponsorship. Mining towns are pluralistic in their institutional structure, not monolithic: the strength of the union does not produce "union towns" in the sense that "company towns" are company towns. "The lodge" appears as little more than a set of officers who open their mail at homes where telephones seldom ring. How, then, can we talk about the key importance of the lodge in the social and intellectual life of the town?

Naturally, the branch's importance will vary greatly from one mining village to the next. Nevertheless, it is very often the case that the particular individuals in the village who are leaders in the branch will often be found as members of the committees which regulate the affairs of the coöp, the Club, the Welfare Institute, the local Labor Party or the local council. It is this informal network of "interlocking directorates" that gives the miners' branch so much power — much more power than authority.

The division of labor involved in manning the many positions in the separate (but fairly standardized) organizations found in pit villages must often operate as a sort of informal patronage system operated by a small group of key officials of the local branch — where these men are strong individuals. And often their power is exercised not by taking overt leadership in community projects (though this they do too) but more subtly, less obtrusively, less consciously. Indeed, a large part of their power can be defined negatively — they do not take initiatives so much as they constitute a sieve through which anyone else's initiative must pass.

MAJOR PROBLEMS OF THE NATIONAL UNION

The picture of the NUM which emerges from the foregoing description is that of a union which had from its start hit upon an organizational structure remarkably successful in conducting the business of a trade union. The federated form of organization, while it posed serious internal problems so far as negotiating strategy was concerned, assured the growth of strong local traditions and interests with which the reorganization movement of the 1930's finally had to come to terms. The union's present structure thus still constitutes one of the most effective forces for "decentralization" in the industry, while at the same time local interests, since the reorganization, are no longer able to frustrate national policy. Indeed, the union's reorganization in 1945 had the happy effect of preserving most of the virtues of "district unionism" while throwing off most of its weaknesses, admirably preparing the union for the needs of nationalization.

But the constitutional machinery of government is not the whole of any union's life, and, sensible and workable as this machinery is, the NUM is not without its problems. There are, it is true, some potentially serious problems that have to do with the union's government, and in their dealings with such important fringe groups as the winders, the "power group," the cokemen, and the clerical grades. But the most serious problems have to do with the new functions which many thoughtful people, both within and outside the NUM, feel the union ought to be performing as part of its role under nationalization. These new functions involve issues of policy and of leadership, not of constitutional arrangements.

With nationalization, the NUM has shifted, psychologically, from "protest unionism" to "administrative unionism," from a distant position whence it stood as a challenger of the system of ownership and operation to a more intimate position where it stands as a sponsor and friend of the new order which is in large part its own child. Despite the undeniable vigor with which the traditional bargaining role of a union is still fulfilled, union officials today must spend a large amount of their time and energy serving with management on joint committees (for example, on consultative committees, on the administration of the new pension and Social Welfare schemes, on ad hoc "production drives" to boost output) and concerning themselves with the problems of the industry, the same problems that keep management awake at night.

This shift in the union's role raises obvious questions about the ability of union officials to avoid becoming mere apologists for management and to retain enough independence so that they can still act as independent critics of managerial policies and performance and yet be a useful communications channel between management and their own members.

Another set of new functions which the union has assumed since 1945 is the representation of its members' interest in a liberal and efficient administration of the welfare benefits that have become part of the welfare state. At the local level, this means in large part giving members competent representation before the medical tribunals which adjudicate accident claims under the Industrial Injuries Act which has now replaced the old Workmen's Compensation Acts. At national level, it means the provision of an effective lobbying service, chiefly by the NUM lawyer and by Miners' M.P.'s, to see that the administrative policies and interpretations of the Ministry of Health and National Insurance meet the needs of the mining population. Vis-à-vis its own members, the union has performed a vital service in making them aware of the benefits and procedures provided by the postwar welfare legislation. In short, one emerging function of the union is to make sure that neither the mineworkers nor government officials get bogged down in the administrative labyrinth that a welfare state could become.

There are also dangers to the union in the shift from "protest unionism" to "administrative unionism." If union officials become so concerned with the problems of the industry that they see things through management's eyes, they risk losing the confidence of their own membership. This might happen, for example, if union leaders resisted pressure for wage increases too long, if they accepted harsh disciplinary measures for wildcat strikers or absentees, if they pushed the revision of work-norms too energetically, or if they tried to break down local customs or to force foreign labor on all pits regardless of local sentiment. The union has already had its fingers burned on some of these issues.

How might a union walk the tightrope between arm's-length independence of management and a constructive approach to the success of nationalization? Only by equipping itself to educate its membership in the requirements of nationalization, preparing them to accept the changes that are necessary to efficient operation of the industry. This does not mean that the union must sacrifice its birthright, as in Russia, by providing a disciplinary arm and moral sanction for management. It does mean that the union must try to throw its influence on the side of change, while yet protecting its members' interests as changes are introduced. To fulfill this function will require, first of all, leaders with sufficient imagination to understand what changes are necessary and with sufficient courage and energy to undertake the new functions which seem necessary if the union is to play its part in encouraging their introduction.

Some of the general phrases one hears from responsible people in the union and the Coal Board as indicative of the directions in which the NUM might be acting with somewhat more imagination are these: more educational work among its membership; the institution of leadership training for the oncoming generation of officers; the publication of a journal or newspaper for better communications between leaders and the rank and file; the building up of a competent research staff to organize and analyze the mass of information about the industry's problems now available to the union; and the establishment of something like a management engineering staff to take the curse off scientific management and to spur on the Board's efforts to improve managerial performance.

Although a few of the Area unions (for example, Yorkshire, Durham, and Scotland) have done some educational work, it is fair to say that little thought has been given to the contribution which new staff specialists might make to the effectiveness with which the union functions. By and large, the NUM still operates as a "rule of thumb" organization, relying almost exclusively on the seasoned judgments of its line officials and on whatever information or misinformation they happen to carry about in their heads. The increased work-load which nationalization has put on these men has kept them preoccupied with short-term concerns and pressures; few of them have time for reflective thinking about the union's problems or potentialities. But, like many conservatively run unions, the NUM is not easily persuaded to spend money on what would admittedly be unproven projects, and it is safe to guess that the union will not expand rapidly into the new fields suggested. But if it does not do so, custom and tradition will continue to play too large a role in the making of decisions, and the reorganization of the industry will be made that much more difficult.

UNION RECOGNITION

The precise extent and nature of the NUM's jurisdiction in the industry were not settled matters at the time the Nationalization Act was passed; indeed, this act contained a provision which touched off a scramble for recognition among several unions which had lived together for varying periods in an uneasy peace. There is of course no general law in Britain which governs those knotty questions of union recognition and collective bargaining which are covered, in the United States, by the detailed body of statutory and administrative law which has grown up since the passage of the National Labor Relations Act in 1935. The Coal Industry Nationalization Act, however, contains the following provision:

(1) It shall be the duty of the Board to enter into consultation with organizations appearing to them to represent substantial proportions of the persons in the employment of the Board, or of any class of such persons, as to the Board's concluding with those organizations agreements providing for the establishment and maintenance of joint machinery for — (a) the settlement by negotiation of terms and conditions of employment, with provisions for reference to arbitration in default of such settlement in such cases as may be determined by or under the agreements.[24]

This "compulsory bargaining" provision in the Act, added at the insistence of the NUM as insurance against a Coal Board that might some day be less friendly than the Labor Government's Board, has not greatly affected the handling of questions which have traditionally been resolved through "voluntarism," which means, in practice, the spirit of compromise mediated by the pressures of power politics.

It would lead too far afield to review in detail the specific recognition disputes which have occupied so much of the time of union and Board officials since nationalization. It is possible, however, to identify four different types of situations that have arisen:

(1) Potential rivalries between the NUM and other unions, which have been successfully resolved by using the internal machinery of the union movement. The NUM's agreements with the Amalgamated Engineering Union, the Electrical Trades Union, the Building Trades Workers, the Transport and General Workers, and the National Union of General and Municipal Workers come under this heading. In these cases the unions themselves have saved the Coal Board the possible embarrassment of having to choose between them.

(2) Disputes between the NUM and other unions which fundamentally cater to a different class of employees, but where the two unions have had difficulty in defining the boundaries of their appropriate spheres of activity. The relatively unimportant but time-consuming dispute between the NUM and the British Association of Colliery Management has been of this kind.

(3) Head-on rivalries between the NUM and other unions which have not been solved by invoking the internal machinery of the trade union movement. Such was the dispute between the NUM and the Clerical and Administrative Workers' Union, which lasted from 1946 to 1951.

(4) Disputes between the NUM and break-away groups within the union. The dispute between the NUM and the craft of winding enginemen, which began in 1946 and may not yet be ended, has been much the most important of these. We shall devote most space to a discussion of this dispute, because more than any other it has had greater consequences on coal output, has more intimately involved the Coal Board, and has more to tell us about the internal problems of the NUM.

All the NUM's traditional disputes with "outside" unions (and this would exclude its disputes with the CAWU and the BACM, which con-

---

[24] Article 46 of the Act (9 and 10 Geo. 6, Ch. 59). Section (b) is omitted and refers to the establishment of machinery for joint consultation. The juxtaposition of these two provisions reflects and symbolizes the emphasis which British unions put on a twofold approach to the problem of securing more control over industrial affairs — collective bargaining and joint consultation. American unions rely almost exclusively on collective bargaining.

fine their jurisdiction to the coal industry) yielded to application of the Trades Union Congress' "Bridlington Formula." The relevant part of this asks TUC affiliates to "consider the possibility of joint working agreements with unions with whom they are in frequent contact . . ." [25] The NUM negotiated agreements with the Amalgamated Engineering Union, the Electrical Trades Union, and the Building Trades unions which settled long-standing jurisdictional claims in its favor.[26] These agreements provide that members of these three unions working at collieries shall be made members of the NUM, although they remain free to keep up membership in their former unions, thus protecting any equities in benefits the member might have and protecting him in the event he leaves the mining industry.

The settlement of the disputes between the NUM and the two general workers' unions (the Transport and General Workers Union and the National Union of General and Municipal Workers) was on slightly different lines. The NUM got each of these unions to agree to a standstill so far as further organization was concerned, while the NUM agreed that coal-industry employees who were members of these organizations should retain such membership. Such employees thus became, automatically, members of two unions; within the NUM, these two classes of members were kept separate for administrative purposes: the T&GWU section of the NUM is now known as Power Group No. 1, the NUGMW section as Power Group No. 2. In practice, these members look to the NUM to handle their representation with the Coal Board but look to their general union for "friendly benefits."

As a direct result of nationalization, many grades of managerial and technical personnel which had not traditionally been organized have now joined trade unions or associations. By far the most important of these new managerial unions is the British Association of Colliery Management (BACM), with which the Coal Board now has more than a dozen separate agreements covering such grades as pit managers, certified surveyors, uncertified surveyors, electrical and mechanical engineers, finance officials, rescue station officers, and others.[27] Although the NUM may have entertained, in the triumphant mood of 1946–1947, hopes of representing all Coal Board employees (including managerial personnel), the latter held different notions and opted for an independent organization of their own. Although the NUM has always regarded the

[25] The Bridlington proposals, which emanate from the TUC and take their name from the Yorkshire site of the TUC's 1939 Conference, are summarized in *British Trade Unionism,* by Political and Economic Planning (London, 1948), pp. 106–107.
[26] The text of the agreement between the NUM and the AEU is included as Appendix II. The agreements with the ETU and the Building Trades Workers are almost identical.
[27] Copies of the Board's agreements with BACM will be found in the NCB's *Memorandum of Agreements,* etc.

BACM as a sort of poacher, it has refrained from any attempt to organize the core managerial grades for whom the BACM was established. The NUM thus tacitly recognizes a "sphere of influence" appropriate for the BACM. But there has been great difficulty in agreeing where the jurisdiction of the one should leave off and that of the other begin; and since the BACM is not a member of the TUC, there can be no appeal to the "internal machinery of the trade-union movement." The only appeal, therefore, has been to the Coal Board, which has successfully taken the view that recognition would be defined nationally according to job classifications (that is, cokeoven managers in Yorkshire could not be considered NUM members while those in Scotland were regarded as members of the BACM); all jobs above and including undermanagers are now recognized as the BACM's — though this has not always been unambiguous in application. In future, the relationship between the NUM and the BACM is likely to get better rather than worse unless either union unexpectedly attempts to upset the jurisdictional lines that have been worked out during the first eight years of nationalization.

Jurisdictional disputes among the three unions interested in the Board's substantial number of clerical employees have been more serious than those concerned with managerial and technical employees. Here the unions concerned have been the NUM (through its clerical affiliate known as Group No. 3), the Clerical and Administrative Workers Union (the CAWU, a TUC affiliate), and the National Association of Clerical and Supervisory Staffs (an affiliate of the T&GWU and hence also of the TUC). These unions could not agree among themselves on which ought to have bargaining rights at pit, Area, Divisional, and National levels, where there were many overlapping claims.

In such cases where all the rival unions are members of the TUC, the Coal Board's policy has been to avoid making its own choice among the rivals and to ask the TUC to make the choice for it. In this case, the TUC was successful in resolving the dispute at Divisional and National levels late in 1946, but it could not persuade the CAWU to cede its pit and Area jurisdiction to the NUM by an agreement similar to that signed by the AEU, the ETU, and the Building Trades Workers. What then does the Board do when it learns that the TUC is not able to resolve rivalries among constituent unions? The Board muddles along without making clearcut recognition decisions, which means, in practice, that it waits until time clarifies the real strength of the contesting parties. In this case, the Board signed separate but identical agreements with both unions early in 1948 even though the jurisdictional issue remained unsettled. By 1950, the NUM top command had finally been able to persuade its own clerical section (Group No. 3) that it would not be able to oust the CAWU and had better settle for joint recognition for the contested pit and Area employees. The concept of joint recognition is wholly foreign to American experience, where union tradition and the law both require exclusive jurisdiction for all employees in a bargaining unit. Although the NUM and the CAWU are still rivals for members at pit and Area offices, this institutional rivalry is limited to organizational work and does not affect collective bargaining. Neither union has tried to manipulate the bargaining unit in its favor by holding out for exclusive recognition in those districts where it held a decisive membership lead. The Board has thus been sucessful, so far, in persuading the clerical unions (as with most others) to accept industry-wide bargaining units.

The Winders' Dispute with the NUM and the Board. The internal dispute between the NUM and the section of winding enginemen has been much the most serious recognition dispute since nationalization. The winders, it should be explained, are a special category of enginemen, a large class of skilled men whose work consists of operating the stationary motors which transport coal from the face to and on the "bank" wherever such transport is by means of a wire rope; winders operate the shaft-winding engines and perform their work in a special house on the surface adjacent to the shaft top.

The winders' bid for self-determination was beaten by the determination of the NUM not to permit one of its most important craft groups to go out on its own — a determination the NUM could back up with effective sanctions. A look at this dispute yields much insight into the handling of the recognition issue; it also gives us insight into one of the most important aspects of the NUM's internal organization.

The winders' restlessness has stemmed from a gradual fall in their relative status and pay during the past generation. About the time of the first World War, colliery enginemen in most districts had formed themselves into separate craft unions, most of which were nevertheless affiliated to the Miners' Federation of Great Britain. At that time winding enginemen constituted a large majority of all enginemen. Gradually, however, as transport systems became mechanized, other enginemen and associated craftsmen came to represent a much higher proportion of membership in the enginemen's and craftsmen's unions, so that by 1943 the winders had become only a small group in these broad craft associations. This development led to a fairly widespread feeling among winders that their interests were not being properly represented by the wider unions which had grown up around them. Furthermore, winders' wages fell, during the interwar years, in relation to those of deputies and piece-rate workers. The postwar difficulties have resulted from a determined attempt to regain their former status and pay.

The difficulty began to show itself in the fall of 1946, just as the Board was preparing to take over the industry. At that time the NUM had secured a promise from the chairman of the Coal Board that the latter would not recognize any breakaway group of winding enginemen so long as the NUM remained faithful to the 1944 wage agreement which froze district wages, including the winders'.[28] Such an understanding had something in it for both the Board and the NUM Executive Committee; but it held no attractions for winding enginemen who felt that their interests were being neglected.

The wages of most winders, of course, were governed by agreements held by the NUM, and changes in those wages could be made only if (1) the NUM could negotiate such changes with the Coal Board at national level or (2) the Board chose to recognize the new non-NUM unions as representing winding enginemen in those districts where their strength was great.

In January there was formed an organization known as the National Union of Colliery Winding Enginemen, a new group outside the NUM. The nucleus of the new body consisted of winders' unions in five districts where the winders had had their own craft unions, independent of the NUM.[29] When in March 1947 the National Union of Colliery Winding Enginemen asked the Coal Board for recognition and a general wage increase, the NUM protested strongly and the Board, despite a strike threat from the winders, turned down their request. At the time, it appeared that the Board would have a NUM strike if it recognized the new group or a strike by the latter if it did not. However, the Coal Board was able to refer the dispute to a special Court of Inquiry, set up under the Industrial Courts Act of 1919,[30] whose report early in 1948 contained three main observations:

(1) The main dispute was over recognition, not wages, and arose "because of the dissatisfaction of a section of winding enginemen with the existing machinery for negotiating their terms and conditions of employment." The Court felt that there existed "some justification" for these apprehensions and that "the arrangements which the National Union of Mineworkers have adopted from the past, whereby winding enginemen are normally organized in associations of enginemen and other craftsmen, do not always provide for adequate representation of their particular interests."

(2) Whenever the rates and terms of employment of winders are being negotiated with the Board by the NUM, either at district or national level, "the Union (should) include among their representatives persons regarded by the winding enginemen themselves as competent to represent their special interests."

[28] See pp. 232–233, 1949 NUM *Conference Report.*

[29] The status of windingmen's unions in various districts is described — not without ambiguity — in the *Report of a Court of Inquiry,* reprinted in *NCB Memorandum,* p. 257.

[30] Such a Court constitutes a public fact-finding body whose recommendations have no more than moral authority on the parties interested in the dispute. (An introductory description of the 1919 Act will be found in the Ministry of Labor's *Industrial Relations Handbook,* 1944, pp. 116–123, esp. pp. 122–123 on the "Courts of Inquiry.")

(3) The Court could not agree that the Board ought to recognize an organization outside the "existing conciliation machinery" to represent winders, "particularly in view of the effect which its institution might have on other sections of the industry." [31]

It would be a mistake to conclude that the NUM was not interested in doing something for the winders, partly to kill off interest in the breakaway movement and partly out of genuine sympathy for their wage grievance. The Coal Board, however, persuaded the NUM not to press the winders' claims until a broader agreement had been reached on the wages of all craftsmen. When the matter finally came to a head, the Board and the Union could not agree on a national minimum rate for winders and the NUM accordingly sent the winders' claim to arbitration. The NCB argued against fixing a national rate, though not a national minimum, which they felt should be the same as that previously agreed for craftsmen (who worked an 8½-hour day against the winders' 8).

The resulting Eighteenth Porter Award satisfied neither the NUM nor the breakaway groups. But an arbitration award is an arbitration award and the NUM was bound by it: the NUM was now powerless to secure any further improvement in winders' wages for some time to come — a situation the breakaway group realized only too clearly. Under these circumstances there was every incentive for the winders to push on for an independent organization outside the NUM, since such a group, if recognized by the Coal Board, would be free to negotiate without being embarrassed by the arbitration award which now bound the NUM. Within three months of the Porter Award, the breakaway group found new leadership from the Yorkshire Colliery Winders' Association, a NUM affiliate which, dissatisfied with the handling of its case during negotiations, had decided to leave the NUM and join up with the breakaway groups in other districts. The breakaways reorganized and were henceforth known as the Colliery Winders' Federation of Great Britain. The Winders' Federation pressed the Board for recognition in the late fall of 1948.

[31] *Report of A Court of Inquiry,* reprinted in *NCB Memorandum,* pp. 254–266, and dated Jan. 24, 1948. The Court's observations met an icy reception within the NUM. The NEC promptly instructed the secretary to prepare a memorandum for submission to the Minister of Labor, setting out its view of the facts regarding the extent to which the NUM's structure catered to the interests of the winding enginemen. The memorandum's main point was that the NUM is composed of forty constituent associations and it would be impossible to expect every craft in the industry invariably to receive direct representation during negotiations; granting this impossibility, the NUM felt that winding enginemen were not "swamped" by other enginemen in those constituent associations of which both were members but, on the contrary, the winders enjoyed an influence (as measured by membership on executive committees) out of proportion to their numbers. The document also corrects some of the facts set out by the Court of Inquiry with respect to the status of winders' union membership in the various districts. (The NUM Memorandum is reprinted at pp. 215–218 of the NUM's *1948 Conference Report.*)

All during 1949 the Colliery Winders' Federation pounded the Coal Board's door for recognition. But the Board, under pressure of energetic representations from the NUM, refused to listen and took the position that it had no intention of negotiating any changes in winders' wages except through the established conciliation procedure (that is, through the NUM). Any other course of action would "seriously undermine the integrity of collective agreements in the industry," since it would show that dissatisfied minorities could break away from the NUM and use the strike weapon to force independent recognition from the Board. The only course open to the Board, in view of the pressures it faced from both sides, was to lay the dispute before the Ministry of Labor, which eventually submitted it to the nation's highest arbitration body, the National Arbitration Tribunal.[32] The latter's award in October settled nothing: in effect, it sent the dispute back to the coal industry's own arbitration authority, the National Reference Tribunal. This body then ruled that since the dispute was not between the NCB and the NUM (the only parties to the conciliation machinery), the Tribunal lacked jurisdiction — and the dispute bounced back to the National Arbitration Tribunal for a hearing on its merits. The Tribunal declined to upset the Porter Tribunal's original award which now governed winders' wages.

With such a tangled legal situation it is little wonder that several unofficial strikes erupted among winders;[33] the wonder is that these were officially condemned by the leaders of the frustrated Winders' Federation. In practice, the latter were willing to sue for peace if the NUM would make it possible for their members to return to the NUM in some new district organizations other than those which had traditionally catered to winders; that is, they did not want to remain in administra-

---

[32] The role of the Ministry of Labor and National Service requires explanation. One of the key wartime controls was a Statutory Rule and Order No. 1305 (one of the "Conditions of Employment and National Arbitration Orders"), which made any work-stoppage illegal unless the individuals or trade union or employer involved first reported the existence of the dispute to the Minister of Labor. If the Minister referred the dispute to arbitration within a period of 21 days, no legal strike or lock-out could occur; only if no reference to arbitration was made could a stoppage be legally initiated. The Minister could either refer such disputes to arbitration bodies within the industry's normal conciliation machinery or to a newly created, top-level arbitration authority for the country, known as the National Arbitration Tribunal, an *ad hoc* five-man body to be appointed by the Minister from permanent panels of public and representative members. (Ministry of Labor *Industrial Relations Handbook*, 1944, pp. 129–131.) This Order 1305 had been retained into 1951 by agreement between British trade-union and employer federations at national level but was replaced by a modified order in the summer of that year.

[33] In Yorkshire 72 pits were affected (about two-thirds of all in the Northeastern Division); in Lancashire about 26. Some 90,000 men were kept off work by the strike and some 350,000 tons of output were lost — nearly a quarter of all tonnage lost through disputes in 1949. See NCB, *1949 Annual Report*, pp. 82–83.

tive units that were dominated by other craftsmen. If some of the winders who had remained loyal to the NUM wanted to remain in "mixed" organizations, the leaders of the Federation had no objection.

The NUM, however, did not feel that it could allow particular sections to exercise the right of self-determination; this could only lead to the "Balkanization" of the union. Hence, the NUM would take back the disloyal winders only if they would return as members of the appropriate district unions which were already recognized (which usually meant mixed unions).

The Federation's position began to look hopeless late in 1950 when one of its important district affiliates (the only one which held a contract with the Board, a contract the Board had inherited at Vesting Day) crumbled. Pushed off this beachhead, the Winders' Federation began to look around for a graceful return to the NUM: the signing of the general wage agreement of 1951 between the Board and the NUM, which canceled the Eighteenth Porter Award and gave winding enginemen a larger raise than any other craft group, created an atmosphere in which reaffiliation became politically possible. Though the dissident winders had failed to secure independent recognition, their attempt to do so had apparently brought home to the Board the necessity of doing something special for the winders as the price of peace between them and the NUM. As long as this key group did not feel it was getting sufficient recognition, something had to give — either the NUM's structure, the pattern of union recognition in the industry, or the industry's wage structure.

In most districts the winders came back into the NUM on the latter's terms. The unanswered question, of course, was whether or not the winders would feel at home now that they had been brought back into the house. The development of a winders' strike in Yorkshire late in 1952, a strike that defied the NUM leadership, showed that all was not well. Ironically, the strike was induced by what Coal Board officials admitted to be an "unfortunate blunder" in the East Midlands Division. Somehow the winders in Derbyshire and Notts had won a purely local increase from the Divisional Board (since nationalization all changes in winders' wages had been negotiated nationally); when Yorkshire winders learned that winders across the county line had disturbed their accustomed parity by getting an extra 3s. a shift, they naturally tried to get 3s. out of the Northeastern Divisional Board. But the Divisional Board felt bound by the national agreement and would give nothing. So the Yorkshire Colliery Winding Enginemen's Association not only decided to strike but decided to go out during the Christmas Bull Week when production reaches its annual peak and when miners have a greater need for money than at any other time of the year.

The strike was strongly denounced by the NUM national officials and the union Executive,[34] who blessed the Coal Board's successful efforts to use substitute winders recruited from loyal NUM members. The strike collapsed after five days, largely because of the pressure which built up among mineworkers themselves on account of the Christmas need for money. The Yorkshire winders failed to get their three bob; but they did not fail to show the NUM leaders, the Coal Board, and the public that the internal harmony of the NUM could only be maintained if the Board and the Union paid somewhat more attention to the political effects of their economic decisions.

The outcome showed the NUM leadership that, despite the reorganization of the union described earlier, all was not well with the internal organization of important sections of the union. Indeed, the union's officers had assured the leaders of the dissident winding enginemen that once they had returned to the fold they would be given an opportunity to examine, in coöperation with leaders of other craft groups, the organizational position of craft groups within the union's structure. When this examination was conducted in 1952, it became clear to all concerned that "it will take a considerable time to formulate sound proposals which will solve all the problems and at the same time meet with the approval of all the constituent unions concerned." [35] It remains to be seen whether or not, before this "considerable time" has passed, other craft groups within the NUM will emulate the winders' tactics in forcing sectional improvements from the Coal Board by threatening to embarrass it in its relations with the NUM. One thing seems clear: the Board will not recognize any breakaway group the NUM does not want it to recognize, even if this course involves the risk of occasional wildcat strikes.

[34] The NUM Executive did not deny the Yorkshire winders' right to the 3s.; they simply held that the claim should be pursued through the regular conciliation machinery and not through strike action. Bitter feeling developed between the leader of the Yorkshire winders, W. Pilkington, and Horner and Lawther. Early in December, after Lawther had denounced the exaggerated sense of self-importance which the Yorkshire winders were showing, Pilkington said: "The miners have not been concerned about us for four years. It happens to be someone else on this occasion who is striking. When it is the miners, it is all right; when it is not it is all wrong . . . We feel that we have done everything possible except go to the conciliation board, which is a long-drawn-out procedure Last time we went there it took two and one-half years." (NCB *Daily Press Summary,* Dec. 4, 1952.)

[35] *NUM Conference Report,* 1953, p. 227.

CHAPTER THREE

COLLECTIVE BARGAINING

There are more strikes in coalmining than in any other industry in the world. Nobody understands very clearly just why this is so, but it must certainly be rooted in the environmental characteristics of the industry and not in any abnormal endowment of original sin within the mining population. A series of thorough ecological studies would be necessary before these unique "environmental characteristics" could be identified. Among the characteristics would certainly be found the high proportion of labor costs to total costs, the effect of a nonstandardized working environment on incentive methods of wage payment, the difficulty of maintaining good communications between management and the widely scattered groups of underground workers, the physical isolation of most mining communities, and the relatively undeveloped state of accepted personnel practices and outlooks in coalmining.

The expectation that nationalization would lead to a great reduction in strikes was based on an incorrect diagnosis of their causes — even of their nature. Particularly is this true of those who naïvely assumed that the removal of the profit motive would remove the main cause of strikes: namely, disagreement over the proper division of the industry's income. Whatever the importance of unique environmental factors, the institutional arrangements for collective bargaining and the settlement of grievances have much to do in controlling the characteristics of strikes. So, too, do the policies of the NUM and of management; for the strike is, after all, the classical hallmark of a free labor movement and the ultimate pressure that determines the outcome of the bargaining process.

Since nationalization, there has been some reduction in the incidence of strikes; the collective bargaining process has been made more complete, more standardized, and more centralized; a new reliance has been put on voluntary arbitration of a most thorough kind; but, despite these significant shifts in emphasis, there has been no dramatic change either in the number of strikes or in the way they are handled. We shall discuss in this chapter the characteristics of the unofficial strikes which lie at the heart of the strike problem and the steps taken by management and union to do something about them. It will be necessary to begin, however, with a description of the constitutional machinery through which bargaining (including the settlement of grievances) is formally conducted.

Bargaining Machinery and Grievance Procedure

Before the early 1940's, the main negotiating authorities for the industry were the nearly twenty independent "conciliation boards" in each wage district. These boards were standing joint committees, usually headed by an independent chairman, and they did all the negotiating for the district outside those matters which were recognized as appropriate for bargaining at individual pits. The agreements they arrived at were often of no fixed duration and could usually be supplemented or interpreted by the Conciliation Board itself.

The informality of much of the bargaining in the districts is reflected in the following excerpt from the preface of a 1920 compilation of agreements published by the Northumberland Miners' Mutual Confident Association:

> The lack of a compilation of our County Agreements between the Coal Owners' Association and our own has long been a felt want, especially by our branch officers. But to make such a compilation has meant a large amount of research work, as the records of these agreements are scattered almost indiscriminately through our printed Minutes, Circulars, etc., and some of them only in scrap books . . .
> A number of agreements once existing have now become obsolete, and there are others that were never printed, but established by custom. We have here entered only the printed agreements, as the others, in several cases, are not acknowledged by the colliery owners. This only shows how necessary it is, in the case of either general or local agreements, to have them in written form.[1]

Early in the second World War the traditional method of district bargaining was almost completely superseded by national bargaining, which was dictated by the logic of national price and production controls. In particular, district wage rates were "frozen" by a 1944 agreement, still effective in 1953, which put "on ice" the main subjects — namely, wages and hours — over which the district boards normally bargained. Although, as we shall see, the district boards still exist, the important bargaining is now done at national level through machinery which did not exist before the war. Although the following description

---

[1] Northumberland Miners' Association, *Agreements*, 1920, p. 4. A useful historical review of the development of negotiating machinery in the industry (from the mid-nineteenth century down to nationalization) will be found in I. G. Sharp's *Industrial Conciliation and Arbitration in Great Britain* (London: Allen and Unwin, 1950), 466 pp., chap. ii, "The Coal Mining Industry," pp. 8–57.

is based on an agreement between the NCB and the NUM, it should be remembered that the same machinery (except for the "Pit Conciliation Scheme") has been in existence since the spring of 1943.

THE NATIONAL CONCILIATION BOARD

At the top of the national collective bargaining machinery is a body known by the confusing name of the "National Conciliation Board." It is confusing because the Board consists of two separate bodies which perform separate functions. These two bodies are (1) the Joint National Negotiating Committee (the JNNC) and (2) the National Reference Tribunal (known as the NRT or, more commonly, as the Porter Tribunal, after its chairman, Lord Porter). The JNNC is a negotiating body, pure and simple. The Reference Tribunal is, as its name implies, a body to which questions are referred when the JNNC is unable to settle them; it is an arbitration body, pure and simple. Thus, the Conciliation Board includes machinery for both collective bargaining and, failing settlement, voluntary arbitration.

Joint National Negotiating Committee.[2] The JNNC is made up of all members of the National Coal Board (in 1949 an additional member was appointed from each of the Divisions) and not more than fourteen members nominated by the NUM. It has two chairmen and two joint secretaries, one of each appointed by each side. The committee has a low quorum (four from each side), and the practice has grown up of allowing a small subcommittee to hammer out agreements until the final stages when the full JNNC is usually brought in. Voting within the committee is not by individuals but by blocs. The jurisdiction of the JNNC extends to "any question raised by the Board and the Union or either of them," provided that the question is a national and not a district one. Such national questions may be introduced to the JNNC at the top level, or they may be sent up to the committee from any of the districts, or they may be referred to the committee by the Minister of Fuel and Power. The committee may interpret agreements it has arrived at previously.

Since agreements do not usually run for specific time periods, and

[2] This description is based chiefly on the NCB-NUM Agreement of December 5, 1946, known as the "take-over agreement." Under this the NCB and the NUM agreed to take over the bargaining machinery which had been worked out in 1943 and incorporated in the Agreement of May 25, 1943, between the Mining Association of Great Britain and the MFGB. This 1943 agreement represented the adoption of the system of national or industry-wide bargaining machinery which had been recommended to the industry by the Greene Board, a special Government board of investigation set up in the fall of 1942. The Greene Board had been appointed when the Coalition Government found itself recommending the adoption of a national minimum weekly wage for the industry in 1942, and found that some form of national negotiating machinery would be necessary in order to work out the problems involved in this far-reaching reform.

since bargaining is on a piecemeal rather than an "all-in" or "package" basis, the JNNC is in effect a body which has a continuous existence— or, more properly, one which is subject to convening at will. There is not as much periodicity to bargaining in British coal, or indeed in most British industries as there is in the United States.

The National Reference Tribunal. This body is a permanent, part-time three-man arbitration authority for the coal industry only. It deals only with national questions — that is, its jurisdiction is the same as that of the JNNC and, indeed, it exists simply to resolve questions upon which the JNNC members are unable to agree. The three members of the tribunal are appointed by the Master of the Rolls, one of the three members of the highest court below the House of Lords.[3] They must be men who are neither engaged in the mining industry nor sitting in either House of Parliament (except for a member of the House of Lords who holds or has held a high judicial office). Tribunal members are appointed for five-year periods, subject to reappointment, and are paid such remuneration as the Board and the Union may agree upon (or as the Minister of Fuel and Power fixes if the principals fail to agree); this remuneration and all expenses of the tribunal are shared equally by the NCB and the NUM.

When any question has been before the JNNC for a period of five weeks (a period subject to alteration) without agreement, the question is automatically transferred to the National Reference Tribunal. Whenever the tribunal sits on any question, there must be appointed two "assessors" or advisers from each side (four in all) who must sit with the tribunal during the hearing of evidence and argument and who may sit with it, if invited, during the tribunal's private sittings. The assessors are intended to brief the tribunal on the course of the preceding negotiations and hence must have been present during such negotiations; but assessors have no vote in the tribunal and are not parties to the tribunal's decisions.

Voting within the three-man tribunal is by simple majority and, like most decisions emanating from official bodies within the industry, no dissents from the majority decisions are allowed to be expressed or even recorded. This custom of "collective responsibility" is quite unlike American practice, where both the recording and expression of dissenting viewpoints are the rule. The British practice is of course designed

[3] There was no change in the Tribunal's membership during the first eight years of nationalization. The individual members are: (1) Lord Porter, president, who is a distinguished jurist; (2) Sir (James) Frederick Rees, who is vice-president of the University College of South Wales and Monmouthshire; a native Welshman, he has had a university career as a teacher and writer of economic and social history; and (3) Professor T. M. Knox, Professor of Moral Philosophy at the University of St. Andrews in Scotland; born in England, Professor Knox began his career with Lever Brothers, left to become a lecturer in philosophy at Oxford, and served during World War II as a member of the Catering Wages Commission.

to enhance respect for the authority of legal or quasi-legal decisions and it undoubtedly has this effect; the system's obvious defect is that it does not ventilate minority viewpoints as thoroughly as the American custom does. The British system, which also characterizes the British Cabinet and the National Executive Committee of the NUM, tends to cut off the competition of ideas at the point where a final decision must be made, whereas the American system tends to continue debate beyond this point.

The industry's bargaining and arbitration machinery at national level is duplicated in each of nineteen separate districts which make up the British coalfields. Each has a two-stage system of negotiation by a District Conciliation Board corresponding to the Joint National Negotiating Committee and, failing agreement, arbitration by three-man panels of "referees" (the district arbitration authorities are not called "District Reference Tribunals"). The jurisdiction of these district conciliation bodies is limited to questions affecting the district involved; questions which involve several districts are considered national questions. The determination of whether a specific question is to be considered a "district" or a "national" question is left, in borderline cases, to the National Reference Tribunal. These several district conciliation agreements are not identical, but the national agreement requires that (1) every district shall negotiate a district conciliation agreement; (2) such agreements must provide for arbitration of all questions upon which the district conciliation boards are unable to agree; (3) the agreements of the district boards and the decisions of the district referees must be binding on all parties; and (4) all questions arising in the districts which ought to be transferred from the district to the national level are in fact so transferred (there have been only one or two such transfers during the eight years the Porter Tribunal has been in existence). In practice, the District Conciliation Boards have ceased being negotiating bodies and have become mainly grievance-settling institutions.

The industry's terminology in regard to the handling of disputes is likely to confuse Americans. It will be remembered that the term "conciliation" in Britain refers not to what we call conciliation in America but to collective bargaining. There is no agreed term to describe formal mediation in Britain — where it is less frequently employed than in America. Likewise, the phrase "grievance procedure" is not used generally in British industry and never in coalmining; instead the phrase "disputes machinery" is used. The point of greatest confusion, perhaps, is that the "disputes procedure" in the mining industry is known as the "Pit Conciliation Scheme." The terminology is justified not because much grievance-settling is a form of bargaining but because many local (that is, pit) questions in mining involve disputes over the negotiation of local issues as well as the interpretation of individual rights under local, district, or national agreements. Thus, a "dispute" is a somewhat wider concept than what is known as a "grievance" in American labor relations.

A national agreement of January 1, 1947 lays down the specific steps to be followed in handling disputes at pit level and for resolving such disputes if they cannot be settled at the pit; it also lays down a specific timetable according to which all disputes must proceed through the prescribed steps. There is thus a very high degree of uniformity throughout the country in the handling of pit disputes.

THE PIT CONCILIATION SCHEME

The pit procedure, unlike the national conciliation machinery, is a creation of the NCB and the NUM. It is specific enough to be free of ambiguity and contains time limits which prevent delays and set up specific expectations as to just when each step will be brought into play.[4] There are six distinct steps in the pit-level procedure:

Step 1: "Any question" in dispute between an individual or set of workmen and the management is discussed initially between the man or men concerned and "the immediate official of the pit."[5] If no settlement is reached within three days, the dispute is then discussed between the workmen and the pit manager or his representative.

Step 2: If no settlement is reached at Step 1, the workman must immediately report the dispute to the "appropriate trade-union official" (i.e., the Branch Secretary) if he wishes to pursue the matter further.[6]

[4] In the description which follows, we will refer constantly to "Disputes Committees" and "Umpires." These bodies correspond almost exactly to the "Conciliation Boards" and "Referees" of the negotiating machinery. It is generally true, and may well be invariably so, that any agreement in the industry which refers to disputes committtees or umpires is an agreement dealing with negotiations and grievances regulated by pit agreements or customs; any agreement which uses the terms "Conciliation Board" or "Referee" involves district matters which are negotiated on a district and not a pit basis. The national "Pit Conciliation Scheme" agreement makes it clear that the Disputes Committees and Umpires are to be appointed by the District Conciliation Boards.

[5] The "immediate official" would probably be the overman, undermanager, or even the manager — but not the deputy, as the latter has practically no supervisory duties.

[6] The union does not become involved in the handling of disputes until the second stage. It is a standard demand of many American unions that they shall be permitted to participate in the settlement of grievances at the *first* stage in order to forestall the possibility of individual settlements that may threaten union policies and to increase the union's prestige in the plant. At the NUM's 1949 Conference, Derbyshire delegates sponsored a resolution which would have amended the pit conciliation scheme so that branch officials could have a guarantee of being in on all disputes from the very beginning.

The NEC opposed the change. Speaking for the Executive, Arthur Horner said: "We do not accept that every dispute arising at a pit should be taken to the local trade-union officials before any attempt is made to deal with it . . . We believe that the trade-union official should not be called in when any mistake or error is made [e.g., in calculating pieceworkers' earnings]. He should be called in only when the workman himself has failed to find redress for his grievance. We think it important that our people should not have the opinion that they must be carried everywhere , , , There is now no justification for a fear of victimization which in the

Step 3: If the trade-union official believes that the question reported to him "is of minor importance" which may be settled by further discussions with the pit manager, he must hold such further discussion within three days from the date on which the dispute was notified to him. If, however, he feels the issue is more important, or if he has been unable to settle it within the three-day period, he must proceed in accordance with Step 4. In addition, any question raised either by the local branch itself or by the management must in the first instance begin at this next stage, Step 4.

Step 4: The branch secretary (or the manager, if he be raising any question) must submit a written request to his opposite for what is called a "pit meeting." In urgent cases, the written request may be dispensed with. The "pit meeting" is to be held "as soon as possible" and in no case later than five days after receipt of the request for such a meeting. At pit meetings, each side may call in higher officials from its respective side — but it must notify the other side before exercising this option. There may be more than one pit meeting to discuss a single dispute, but any adjourned pit meeting must reconvene within four days. The pit manager, in consultation with the local union officials, is responsible for seeing that minutes of such meetings are taken and that agreed copies are made available to both sides.

Step 5: Any dispute remaining unsettled 14 days after either side has first requested a pit meeting must be jointly referred by the local union and the manager to the District Conciliation Board, which then farms out the dispute to the appropriate Disputes Committee.[7] If a pit meeting desires to refer a dispute to the Disputes Committee before the 14 days are up, it may do so.

Step 6: If a dispute remains unsettled 14 days after its reference to a Disputes Committee, it is automatically referred by the District Conciliation Board to a one-man Umpire chosen from the panel of umpires previously appointed by the Conciliation Board. The reference to arbitration may occur before the 14-day period is up if the Disputes Committee so decides. The umpire's first duty is to assure himself that the dispute is in fact "a pit question." [8] If he determines that it is not a pit question, he must turn it over to the District Conciliation machinery for settlement. Otherwise he proceeds to render an arbitration award. As with the sitting of "Referees" under the District Conciliation machinery, umpires must be accompanied by "assessors" during the holding of their hearings — in this case one assessor nominated by each side. Again, the assessors do not vote on and are in no way parties to the umpire's decisions. These decisions, like those of the referees, are binding on the management, the union, and its members.

old days might have obtained. Anyone with an apparent grievance can express it now without fear of consequences." Without any further debate, the proposed change was defeated. (*NUM Conference Report,* 1949, pp. 33–34.)

[7] In County Durham, for example, there is one over-all Disputes Committee and three specialized Disputes Sub-Committees — one to deal with questions involving guaranteed wages and the "Porter Minimum"; a second, called the "Urgency Committee," which meets on very quick notice when any dispute threatens to stop the pit; and a third "appropriate committee" to deal with the fixing and altering of price lists. (Agreement of April 15, 1947, between the NCB and the Durham Area of the NUM governing the settlement of unsettled questions in accordance with the national Pit Conciliation Scheme.)

[8] The national "Pit Conciliation Scheme" agreement defines a "pit question" as "a question relating to terms and conditions of employment which are regulated by agreement between the Management of a colliery and the workmen at the colliery or their representatives."

The foregoing description reveals a local grievance and negotiating machinery not very different from that common in large sections of American industry. There are, however, some aspects of the machinery which differ from the typical operation of grievance procedures in American manufacturing industries.

The problem of selecting panels of referees and umpires, for example, does not seem to be nearly as controversial a problem as in many American industries: there is remarkably little concern for the "strategy and tactics" of selecting arbitrators, and nowhere in the industry's written agreement does one find clauses governing the privileges of striking off possible umpires. This fact perhaps reflects the inherent reasonableness, practical-mindedness, and tolerance of the British temperament, plus a less aggressive trade-union attitude. The British respect for law, probably higher than in America, is not born of awe for authority as much as of an appreciation of its necessity, however fallible the authority may turn out in practice. They can tolerate the slowness with which unjust legal or quasi-legal decisions receive subsequent remedy in a way that would be intolerable to the more aggressive, less respectful temper of American industrial relations.

Again, the Disputes Committees are joint committees made up of equal numbers (usually a quorum of two) from each side, and it might be assumed from American experience that such an arrangement would frequently, if not invariably, result in inability to produce decisions. This is not so. There are apparently very few deadlocks in the work of the Disputes Committees, which — like the National Reference Tribunal — render decisions as a committee, without the recording or expression of minority viewpoints. In America, the normal expectation (though not universal) would be that union representatives on any such joint committee would throw a dispute into arbitration if they could not secure agreement to their point of view from the other side. If this were not done, the presumption is that the rank and file (or the representatives' political rivals in the union) would seek their removal from the committee. It would be unrealistic to suppose that such considerations are wholly lacking from the British mining industry: the point is that they are not so much to the fore that the procedure is dominated by them, as so often occurs in America. Indeed, in Britain the temperament of the British people tends to curb the political dynamics of trade unionism, whereas in America popular temperament tends to reinforce that dynamic.

Another observation with respect to the work of the Disputes Committees is in order: the volume of work which has reached these Committees in many districts since nationalization is much greater than that which reached their counterparts prior to nationalization. This is attributable to the sheer number of changes which have come over the industry since nationalization, giving rise to more grievances than were produced under the more settled conditions of employment during the war and prewar years. It may also reflect, to a lesser degree, the unwillingness of some managers and agents to take responsibility for final decisions at pit level. On the other hand, in each year, 1949 to 1952, about 92 per cent of all cases referred to "pit meetings" were settled without reference to the District Disputes Committee or Umpire. But even in a big district with a bad strike record, such as Yorkshire, neither the number of stoppages nor the number of cases going to arbitration has increased significantly since Vesting Day, despite the increased number of formal disputes. Among the members of the Disputes Committee in the Northeastern Division, there was no feeling that illegal strikes were being systematically used to force favorable decisions from the Disputes Committee.

There is an almost total absence of the "management's prerogative" issue in the British industry. There is no "management's prerogative" clause in any of the coal industry's agreements and all arbitration is what Americans would term "open end" arbitration. This is partly attributable simply to the absence of the issue in the industry and partly to the fact that the "prerogatives" of management are legally defined by the Coal Mines Act in the interest of safety — that is, they are made "management's responsibilities." But even on such nonsafety questions as the manning of jobs, technological change, promotions, discipline, and the allocation of housing, there are no hard-and-fast written agreements governing the respective roles of management and union. This does not, however, mean that each side does not have a well agreed understanding of what the proper roles of both sides are; these understandings differ from coalfield to coalfield.

We have confined ourselves to a description of the conciliation machinery which covers the overwhelming majority of those employed by the Coal Board — the mineworkers. Separate but very similar conciliation agreements have been signed by the NCB and various other unions, agreements which cover most of the nonmineworker classes employed by the Board. There are, for example, separate conciliation agreements for (1) the coke oven and by-product workers, (2) workers employed at briquetting plants, (3) most of the managerial grades above the position of undermanager, and (4) the clerical grades.

The only substantial group of employees for whom no conciliation machinery had yet been established by 1952 was the overmen, deputies, and shot-firers; delay in establishing machinery for these grades was caused by a jurisdictional conflict between the NUM and the NACODS (National Association of Colliery Overmen, Deputies and Shot-firers) over the overmen and shot-firers — the NACODS has been recognized by the NCB as exclusive bargaining agent for the deputies, but the

NACODS turned down a conciliation scheme for deputies offered by the Board on the ground that it did not include overmen and shot-firers (for whom NACODS had not received exclusive recognition).

The subsidiary conciliation schemes were worked out and signed at various times between 1947 and 1951. This did not mean, however, that no substantive negotiations took place on behalf of a class of employee before permanent conciliation machinery has been agreed. For example, the Board has negotiated regular agreements covering overmen, deputies, and shot-firers even though no conciliation machinery had formally been established.

Strikes Since Nationalism

There has not been a single official strike in the British coal industry since Vesting Day; however, every year has seen 1500 to 1600 unofficial strikes or "go slow" campaigns that violate the disputes machinery. This hard fact suggests at once both the progress that has been made and the problem that remains.

Rule 41 of the NUM reads:

In the event of a dispute arising in any Area or applying to the workmen in any Branch or possible to lead to a stoppage of work the questions involved must be immediately reported by the appropriate official of the Area in question to the National Executive Committee which shall deal with the matter forthwith, and in no case shall a cessation of work take place by the workmen without the previous sanction of the National Executive Committee, or of a Committee (whether consisting of members of the National Executive Committee or of other persons) to whom the National Executive Committee may have delegated the power of giving such sanction, either generally or in a particular case and no funds of the Union shall be applied in strike pay or other trades dispute benefit for the benefit of workmen who shall have ceased work without the previous sanction of the National Executive Committee.

The absence of any "official" strikes implies only that no such sanction has been given at national level, which is the only level authorized to sanction any strike. It does not mean that several branches and even some constituent areas of the national union may not have promoted and blessed strikes in their districts — though they would have been in violation of national rules and would not have been allowed to use any national funds for strike-pay purposes. The complete absence of strike payments (and victimization pay) has greatly changed the unions' financial situation: they are in a much better condition than ever before.

The absence of official strikes in recent years reflects the NUM feeling that the strike has become an outmoded means of getting what it wants. Arbitration is easier, cheaper, and almost as sure. True, this informal change in union policy may not be permanent — the union need only cancel the basic conciliation agreement to throw off its obligation to arbitrate all issues. This step has been urged more than once by the more militant sections of the union when bargaining has progressed slowly or disappointing arbitration awards have been made. Some might suspect that the contrast between official union policy and the mineworkers' persistent use of this weapon bespeaks a lack of integrity and that union leaders may have been "playing both sides of the street." There is little or no evidence to suggest that this has been the case, but before further discussion on this point, let us review the strike record in more detail.

There are four quantitative ways by which to measure the incidence of strikes in a given period of time: (1) the number of strikes that occurred, (2) the number of people involved, (3) the number of working-days lost, and (4) the amount of output lost. The acute coal shortage in postwar Britain makes the latter measure the most useful one in suggesting the seriousness of the problem. What we want to know is how much difference these strikes have made in the supply situation and how the postwar strike record compares with previous periods. Table 1 states the number of tons lost for each year since 1938, the earliest year for which continuous figures are available.

TABLE 1

Tons of Coal Lost through Labor Disputes, 1947–1953

Year	Tons Lost Through Disputes (Nearest 000,000)	Year	Tons Lost Through Disputes (Nearest 000,000)
1938	943,000	1946	770,000
1939	675,000	1947	1,654,000
1940	501,000	1948	1,062,000
1941	342,000	1949	1,543,000
1942	833,000	1950	1,040,000
1943	1,091,000	1951	1,113,000
1944	3,002,000	1952	1,710,000
1945	957,000	1953	1,150,000

Source: Ministry of Fuel and Power, *Statistical Digest, 1946 and 1947*, Table 16, and NCB *Annual Reports,* 1947–1953. The figures refer to tons "raised and weighed"; they also include estimated amounts each year for losses arising from "go slow" tactics — an amount that rises to nearly a quarter of the total in some years.

These figures show that the tonnage lost each year since nationalization has run somewhat higher than in all but two of these prenationalization years (these two exceptional years being 1943 and 1944). However, there were so many unusual influences at work during these war and immediate postwar years that little useful interpretation seems possible. All we can say with certainty is that nationalization has not operated so as to reduce the loss of annual output as compared with this recent troubled period.

During the first seven years after nationalization, the industry has lost approximately one to one and a half million tons annually because of illegal strikes and "go slow" tactics. Against a total annual output of about two hundred million tons, strike-losses represent well under 1 per cent of production. But in a situation where every ton counts, a million or more tons is a lot of coal. Certainly very few people in the industry are satisfied that this figure cannot be substantially reduced, given time.

TABLE 2

Proportion of Total Number of Industrial Disputes and of Man-Days Lost
Represented by Disputes in the Coalmining Industry, 1928–1953

Year	Per Cent of Total Disputes	Per Cent of Total Man-Days Lost
1928	32	33
1929	36	7
1930	36	15
1931	35	41
1932	29	4
1933	36	42
1934	30	38
1935	39	70
1936	33	47
1937	41	44
1938	41	52
1939	43	42
1940	41	54
1941	37	31
1942	40	55
1943	47	49
1944	57	67
1945	57	23
1946	60	20
1947	61	37
1948	63	24
1949	61	42
1950	65	31
1951	61	21
1952	71	37
1953	75	18

Source: Computed from strike figures regularly reported in the January number of the *Ministry of Labour Gazette.*

A more suggestive conclusion can be derived from other figures. Table 2 gives the proportion of total industrial disputes in Great Britain represented by disputes in coalmining (1928–1953), plus the proportion of total man-days lost from all such disputes represented by the man-days lost in coalmining. Note first that the proportion of total disputes represented by mining disputes rose from roughly one-third (1928–1936) to two-fifths (1937–1943) and then to three-fifths (1944–1953). There thus appears to have been a fairly pronounced secular increase in the proportion of total reported disputes represented by mining disputes. But it will be observed that despite this tendency, which grew rather than diminished after 1943, the proportion of total man-days lost represented by those lost in mining has diminished since 1944. This suggests that, while the coal industry is having more disputes, the disputes themselves have tended to become of shorter duration and/or to have involved fewer men so that the proportion of man-days lost in coal to those lost throughout industry has fallen well below the figure that characterized the period from 1934 to 1944. The figures in Table 2 refer only to "reported" disputes and do not include those involving fewer than ten persons (unless their total lost time reached 100 man-days); thus the basis upon which the figures are collected probably understates the relative importance of coal disputes, since mining has a comparatively large number of short disputes by small groups of men.

It is doubtful whether this development has much to do with nationalization, but it does tell us something about the kind of strike which afflicts the industry. The British problem is not the American one of periodic long-drawn-out stoppages on a national scale and officially sponsored by the national union; it is rather the short, local, unofficial strike that has its roots not in trade-union policy (or even union structure) but rather in the psychology of small groups. True, sometimes these local issues can affect many pits in a district — as the winding enginemen's strike in 1949 affected much of Yorkshire and Lancashire, and the issue of higher rates for lower-paid men affected a great many Scottish pits in 1950 and Scottish and Welsh pits in 1952. But it is usually not even whole districts which get involved; it is more often individual pits — or only parts of individual pits. We shall have more to say about the nature of strikes in the industry when we come to examine in some detail the strike record of one big Division (Yorkshire). In the meantime, it is worth remembering that the industry has not experienced an industry-wide stoppage since 1926.

What kinds of issues most frequently underlie the industry's strikes? Table 3 gives us a general answer, expressing the seriousness of various causes in terms of tons lost rather than in terms of the number of disputes caused. Three things stand out most clearly: (1) the huge importance of disputes over "wages and price lists" every year; (2) the large role which random issues can play — for example, the issues surrounding the five-day week agreement in 1947 or the concessionary coal issue in 1949; and (3) the fairly stable relative importance of most causes. It is worth noting the complete absence of the issue of "nonunionism" — an issue that was a persistent source of friction and stoppages from time immemorial until World War II.

A word ought to be said concerning the issue that has ranked second only to "wages and price lists" as a source of trouble — namely, "methods of working and colliery organization." This does not necessarily reflect "syndicalist" tendencies in the industry, such as men trying to dictate who ought to be their manager, whether the High Main seam ought to be worked before the Five Foot, or whether the pit ought to be worked by longwall advancing rather than on the retreat. Nor does it represent a deep-seated refusal of men in the industry to accept technological change. While these issues may occasionally be present, the problem represented by this classification is not so much one of where

TABLE 3

Causes of Work Stoppages and Tons of Output Lost, 1947–1951

	Loss of Output (*Thousands of Tons, Raised and Weighed*)[a]				
Cause of Dispute	1947	1948	1949	1950	1951
1. Dissatisfaction with allowances and bonuses	42	59	52	51	74
2. Wages and price lists	493	585	813	702	575
3. Refusal to accept alternative work	37	23	18	36	36
4. Sympathy with men dismissed or suspended	22	24	17	15	39
5. Refusal to perform work left over from previous shift	11	8	8	7	26
6. Terms and conditions of Five-Day Week Agreement	838	15	—	—	—
7. Refusal to await repairs after mechanical breakdowns	12	11	8	10	10
8. Objection to, or disputes with, officials	8	13	12	5	14
9. Personnel and grading questions	26	13	11	15	25
10. Working conditions	25	44	63	55	50
11. Methods of working and colliery organization	80	182	128	104	230
12. Concessionary coal	—	—	383	—	—
13. Miscellaneous	60	85	30	40	40
Total	1,654	1,062	1,543	1,040	1,113

Source: NCB, *Annual Reports*. Publication of these figures was discontinued after 1951.
[a] Because the above table refers to tons "raised and weighed," the total figures are higher than shown for 1947–1953 in Table 2. Raised and weighed output is about 8 per cent higher than salable output. This table also includes an unspecified estimate of tonnage lost through temporary go-slow tactics, which the figures for "salable output lost" do not, thus magnifying the discrepancy.

the authority ought to lie in making these technical and managerial decisions as it is the development of better methods of "communications" so that the nature and wisdom of these changes is understood before they are made rather than afterwards through the educational function which strikes often serve.

We have emphasized the great importance of purely local or pit questions in the total strike picture. Two correctives are needed to this emphasis. In the first place, it remains true that nearly half the total tonnage lost in five of the first seven years was accounted for by a few major (unofficial) strikes in particular districts; in the second place, some districts show a much greater propensity to strike than others.

The problem of extensive district stoppages is not unlike that of a forest fire or an epidemic: the cause is often local but its effect tends to fan out because of sympathetic conditions in the surrounding area. If the trouble can be dealt with early enough, it can often be prevented from spreading; but once it breaks out, it often has to "burn itself out." The way to localize strikes is not to let them begin.

UNOFFICIAL STRIKES

Some people attribute the persistence of a large number of unofficial strikes to the growth of centralized bargaining since nationalization. The relationship between these two phenomena seems doubtful, at least as a general explanation. In the first place, the problem of "unofficial strikes has been a familiar one in coal for a great many years: what is new is their persistence in the face of a theoretically air-tight system of arbitration. Secondly, a very large proportion of strikes involves face-workers' dissatisfaction over wages — a question that is still handled on a pit-by-pit basis.[9] Thirdly, postwar British governments have tried to restrain wage increases without any official wage-control machinery: this has required the use of "delay" as a major tactic of wage restraint. There have unquestionably been strikes in protest against procrastination in the bargaining process, but these have been a necessary price of the wage-restraint policy. Fourthly, collective bargaining has not only become more centralized, but it has become vastly more complex than ever before. If the same kinds of agreements now made nationally were made on the old district basis, the problem of securing understanding of and compliance with these agreements would still arise. Fifthly, and finally, it is wrong to think that the prewar district agreements were always made more quickly than today's national agreements or that the rank and file (who, after all, do the striking) had more chance to participate in the making of the district agreements than they now do in the making of national agreements. Agreements before the war were negotiated by union officials on the district conciliation boards, not by rank-and-file miners. The centralization of bargaining has transferred some

[9] It is true that there can be some centralization of bargaining even though the final agreements are pit agreements. For example, the Board's Divisions try to maintain some uniformity in the level of facemen's earnings within their own Divisions (and sometimes within regions within Divisions); this may mean that managers will not settle until they have determined the limit of their offer-price from their Area office, and this process may lead to delays that invite strikes. But surely managers under private ownership must often have consulted higher authority before concluding agreements, at least when their bargaining power was weak. After 1926, "bargaining" could be conducted with a promptness limited only by the speed with which a manager could say "no."

functions from the district officials to national level, but not from the rank and file to national level.

These reasons suggest that we should be wary of general statements that link unofficial strikes to the centralization of bargaining. If bargaining on today's topics could somehow be "decentralized" (an unlikely prospect), the problem of unofficial strikes would probably not be much nearer solution.

TABLE 4

Tons Lost through Disputes, by NCB Divisions, 1946–1952

(Total Tonnage in Thousands and Tons-per-Man-Employed per Year)

Year	Scottish	Northern	North-eastern	North-western	E. Mid-lands	W. Mid-lands	South western	S. E.	Gt. Brit.
1946									
Total	271.2	33.6	293.0	50.9	71.6	2.3	88.0	13.2	769.8
per man	2.7	.2	1.9	.9	.8	a	.8	2.2	1.1
1947									
Total	375.2	113.3	869.2	140.0	29.8	6.2	109.4	0.6	1,643.5
per man	4.6	.7	5.6	2.3	.3	.1	.9	a	2.3
1948									
Total	198.8	19.4	471.1	42.7	56.1	18.2	78.7	14.9	899.9
per man	2.4	.1	3.4	.7	.6	.3	.7	2.4	1.2
1949									
Total	275.4	0.9–N 32.4–D[b]	407.0	420.6	24.2	20.2	83.8	1.7	1,266.2
per man	3.3	a–N .3–D[b]	2.9	7.0	.2	.3	.7	.3	1.8
1950									
Total	498.8	1.5–N 18.0–D[b]	180.5	57.2	12.6	16.3	63.1	4.7	852.7
per man	6.1	a–N .2–D[b]	1.3	1.0	.1	.3	.6	.8	1.2
1951									
Total	253.0	4.0–N 14.0–D[b]	278.0	44.0	49.0	7.0	151.0	3.0	853.0
per man	3.1	a–N 0.13–D[b]	2.0	0.8	0.5	0.12	1.4	0.5	1.2
1952									
Total	371.0	5.0–N 49.0–D	636.0	39.0	32.0	11.0	224.0	16.0	1,383.0
per man	4.4	a–N 0.5–D	4.5	0.6	0.3	0.2	2.0	0.3	1.9
Average per man per yr.	3.7	0.3	2.8	2.2	0.4	0.2	0.8	1.1	1.5

Source: Total tonnage figures from NCB *Annual Reports;* tons per man lost per year derived by dividing this total tonnage figure by the average numbers employed in each Division for the year in question.

a Indicates less than one-tenth of a ton per man.

b The original Northern Division was split in half; N refers to Northumberland-and-Cumberland, D to Durham.

Table 4 shows the number of tons lost through disputes for each of the Board's Divisions, 1946–1952 (1946 being the last year before nationalization). Because of the large variation in the numbers of men employed in each Division, the over-all tonnage losses are also expressed on a tons-per-man basis.

It will at once be noticed that there are "good" and "bad" Divisions. The industry rarely loses any significant amounts from the East or West Midlands or the Northern Divisions. On the other hand, the Scottish and Northeastern (Yorkshire) Divisions have accounted for a very large proportion of the industry's total loss in each year. The Southwestern Division, traditionally volatile, showed a surprisingly good record during nationalization's early years; but after 1950, losses have risen to serious levels. Every year these three Divisions (Scotland, Yorkshire, and South Wales) have accounted for 75 to 90 per cent of all the industry's tonnage lost through disputes.

Losses in the Northwestern Division (Lancashire) have been moderate. The Southeastern Division is so small that its relatively poor record, while not creditable, is not important to total output.

Putting the loss figures on a per-man-employed basis tells us less about which districts are contributing most to the country's supply difficulties but more about the industry's labor problems. The average output lost per man is given for the whole industry for each year and has varied from the low of 1.1 tons in 1946 to 2.3 tons in 1947. The average loss for the six-year period was 1.5 tons. Three of the nine Divisions have a five-year record much worse than the national average — Scotland, Yorkshire, and Lancashire. The two Midlands Divisions as well as the two Northern Divisions have markedly better records. And, what is perhaps most noteworthy of all, the South Wales Division — the storm-center of the industry a generation ago — has a loss that is less than half the national seven-year average. Nor are the seven-year averages misleading: the good Divisions have had a better-than-national-average record each year and the bad Divisions have had a worse record each year, with the one exception of South Wales. We are inevitably led to ask what accounts for these differences among the Divisions. Why are there so many strikes in Scotland and Yorkshire? Why are there so few in the Midlands, in Durham, and in Northumberland? Indeed, it is somewhat misleading to think merely in terms of divisional differences: for example, there are ten to twelve particular areas within the Scottish, Northeastern, and Southwestern Divisions which have persistently experienced much worse strike trouble than other areas in their own Divisions.[10]

There is no easy explanation of the considerable variations among the

[10] For a summary statistical review of this situation, see the Board's 1952 and 1953 Reports, pp. 52 and 15, respectively.

Divisions with respect to the incidence of unofficial strikes.[11] Indeed, the easy explanations sometimes given do not hold water. For example, it might be assumed that Scotland's poor record is to be explained by Communist domination of the union's Scottish Area. But the South Wales Area has also been Communist-dominated; yet South Wales has a relatively good record. A more likely cause of Scotland's poor record is that (1) wage levels in Scotland are substantially lower than in England and hence most of the pressure for increases (especially among the lower-paid groups) has come from Scotland, (2) most of the Scottish stoppages have been in the Lanarkshire area around Glasgow, where the type of miner is more "cosmopolitan" and volatile than in more typical mining communities, and (3) there is a religious conflict between Catholics and Protestants in the Lanarkshire coalfield which has been as responsible for local stoppages as has the more widely publicized Communist influence in the Scottish Executive of the NUM.[12]

None of the explanations suggested for Scotland, however, seem to apply to Yorkshire, where the strike problem is even more serious because of the larger output of the Division. In Yorkshire wages are very high; there are relatively few "cosmopolitan" elements; and there is no Catholic-Protestant strife. The two most common explanations one hears in Yorkshire itself are (1) that the Yorkshireman is inherently an "awkward" type of man who is just plain hard to deal with (this seems to apply especially to South Yorkshiremen in the southeastern half of the field; it is less true of the men who work in the northwestern half), and (2) the Yorkshire Area of the NUM is not only very big but it is also

---

[11] Some material on regional differentials in strike-proneness (in coal as well as in other industries) will be found in K. G. J. C. Knowles, *Strikes — A Study in Industrial Conflict* (Blackwell's, 1952), pp. 186–198. Knowles specifically compares South Wales and the Northeast, two regions with similar industrial structures but very different strike records.

[12] The President, Abe Moffat, and the General Secretary, Will Pearson, are both Communists; the Executive Committee of the Scottish Area of the NUM had a long-standing Communist majority until the spring of 1950. The unseating of this majority did not affect the positions of the President and General Secretary, as they (like all "permanent officials" of the NUM) enjoy life tenure.

There is a strong Irish Catholic element in Glasgow (the "Glasgow Irish" and the "Liverpool Irish" are well-known "types" in Britain). The Catholic Church, through an organization known as the Catholic Guild, is reported to have been more responsible than the Communists, for instance, in the numerous 1947 strikes for an increase for the lower-paid workers in 1947 in Lanarkshire. On the Edinburgh side, in the Lothians, there is an organization known as Protestant Action, which has been active in local branches on that side of the coalfield. The rivalry between the Guild and Protestant Action was described as "bitter," but I did not get close enough to the industry in Scotland to get any firsthand evidence. After being told that some of the Catholic Guild leaders at the local level are also Communist Party members, one hesitates to come down hard with an opinion as to the relative roles of the Catholic Church and the Communist Party in the highly volatile Lanarkshire coalfield!

very highly centralized, so that the union may not be able to give the prompt, informed service that is required to keep unofficial strikes from breaking out. It is also worth noting that many of the Yorkshire pits (especially the South Yorkshire pits, where stoppages are more frequent) are very large-scale pits, so that there may be poor internal "communications" within the pits themselves, allowing grievances to boil over into stoppages. But that this is not the sole explanation is evident from the fact that in Lanarkshire the typical pit is quite small. Certainly a mining psychology with an inherently low "kindling point," plus big pits, and a remote union, are important factors; but they may not be sufficient for a complete explanation.

In the East Midlands, labor relations are proverbially good. Wages have always been the highest in the country; a higher proportion of the men are said to own their own homes, motorcycles, and even automobiles (and hence to be more settled in their habits and under greater need for steady incomes). In the northern part of the Division, in Nottinghamshire, Spencer's independent union in the late 1920's and 1930's made very favorable settlements with the coal owners of the district. There has been a tradition of getting good settlements without strikes, attributable in large part to the combination of moderate trade-union leadership and a healthy financial condition on the employers' side. Since the mid-1930's (and indeed before Spencer returned to the fold), there have been four separate district unions in the Division, which may indicate better local service to the 98,000 men in the Division than the one Yorkshire central office in Barnsley is able to provide the 140,000 men in the Northeastern Division.

On the other hand, the Durham Area of the NUM, serving just over 100,000 men, is just as highly centralized as the Yorkshire Area: all the permanent officials, including the six Agents, live in the union's bungalows on Red Hill in Durham City. Yet the Durham Division is relatively strike-free.

The good record of the West Midlands Division is likewise partly explained by the peculiar psychology of the men in that Division — at least in parts of it. In "the Potteries" — the popular term designating the six towns that make up Stoke-on-Trent, in North Staffordshire — both union and Coal Board officials point proudly to the inherent good sense of "the North Staffordshireman." One manager explained that he had "given up" in Yorkshire simply because he found that over there he "spent three-quarters of his time meeting deputations and one-quarter running the pit," while in North Staffordshire he more than reversed the proportions. On the other hand, the Moral Rearmament movement is fairly strong in North Staffordshire (at least a third of the branch secretaries, but no officers of the North Staffordshire Miners' Association, were actively connected with it in 1950); and this has undoubtedly, at a few pits anyway, reinforced the inherently peaceful tendencies imputed to the North Staffordshire miner.

One well-known branch leader in the Area, Bill Yates, described the influence of MRA at his pit in the following terms:

> In the old days when we used to bargain over a price-list it was like "haggling in an Indian bazaar." I'd ask a "quid" (£) for something I knew was worth a "tanner" (6d.) and the manager'd offer me a "tanner" for a job he knew was worth a "quid." Of course this sometimes meant I'd get a fancy price for a job I knew the pit had to have done right away — so a few men got the "gravy" while some of the older men only got the crumbs. Then there'd be times when the shoe got onto the other foot — the manager had certain jobs he knew *could* wait and he'd make darn sure he got a low price to get back what he'd lost the last time I'd skinned him. I got to thinking that this kind of thing wasn't fair to the majority of the men, so one day I walked into Bill Archer's office and said, "Let's have a talk." We both got around to confessing our sins and we decided that from then on we'd both ask and offer just what we thought was a fair price. And you know we've been darn close on our initial prices nearly every time. Also, now we don't just get a copy of the price list: we get working drawings of the condition of the face as of the date of the agreement, so that if there're any changes in conditions [the usual excuse for renegotiating a price list] we can check it up on the drawings.

At this pit, both the pit manager and his immediate superior, the Sub-Area Agent, have been accepted into membership of the miners' lodge. In a 1949 lodge election, the Agent (a relatively high management officer) received more votes for membership on the union's executive committee than any other candidate!

Needless to say, there are many miners' officials in the West Midlands Division (including some in North Staffordshire) who are not sympathetic to the MRA movement, feeling that it can lead only to a watered-down trade-union movement. It seems clear that MRA is not primarily responsible for the low number of stoppages in North Staffordshire; they were already low before MRA appeared in the district shortly after the war.

The "South Wales miner" (and people in the industry constantly talk in terms of these "types") is the product of the peculiar environment of the isolated, cramped Welsh valleys — mining country unlike any other in Britain. This, so even many Welshmen tell you, produces a "clannish" mining community, which when combined with a national temperament that is less phlegmatic than the English, tends to make the Welsh miner strike more easily and to stick to his guns "right or wrong," as a former Welsh miners' official remarked. In addition, radical influences have traditionally been strong in the South Wales Miners' Federation, and even in 1950 the Area Executive Committee was Communist-dominated and one of the two permanent officials, the late

Alf Davies, Secretary, was a Communist. When one bears in mind that the Coal Board has closed more pits in South Wales than in any other coalfield except Scotland (a process which often produces stoppages), these factors make it appear that the Division's record on strikes is unexpectedly good.

It will be recalled that the Northern Division was split in two at the beginning of 1950, Durham constituting one new Division, and Northumberland and Cumberland a second; this raised the total from eight to nine. The Northumberland and Durham unions have always been close to each other and had long stood aloof from the old Miners' Federation of Great Britain because they wished to adhere to a system of sliding-scale wages (geared to the selling price of coal) and because they opposed the eight-hour day, which became law in 1908.[13]

The sliding-scale method of regulating wages had tended, historically, to make for a more coöperative relationship between the Durham and Northumberland Miners' Associations and the employers than was true of the more opportunistic tactics used in other coalfields. In addition, the mining communities of these old coalfields are more settled than, say, South Yorkshire or Kent, which have only developed within the past forty or fifty years.[14] A Yorkshire pit manager glibly explained that the reason they do not have as many strikes in Durham as in Yorkshire was because wages were lower up there, "and when you've got a few bob in hand you don't think so much about 'playing' now and then." But Sam Watson, General Secretary of the Durham Miners, said he put down his Division's good strike record since nationalization to the very good system of intraunion communications which he has been so largely responsible for developing in his Area. This means that branch officials may be better informed, and may get their information more quickly, than is true of some other large Divisions. But he did not offer this as a complete explanation, which he could not give.

Nobody, in fact, either on the trade-union side or the Coal Board side, has a confident explanation of the regional differences (and sometimes much more local differences) which the statistics show so clearly.

[13] Probably the personal differences between the leaders of the Federation and of the Northumberland and Durham unions had as much to do with the independent course followed by the latter from 1888 to 1908 (see R. Page Arnot, *The Miners*).

[14] People think in terms of decades, not years, when they tell you whether a pit has "settled down" yet. Thus, a twenty-year-old pit in Yorkshire was considered still prone to unofficial disputes because "the men there haven't really settled down yet." Undoubtedly there is a shred of truth in some cases: pits in South Yorkshire and Kent were opened only by collecting a labor force made up of "bits and pieces" from everywhere — Englishmen usually complain that it is the "castoffs" from the other English coalfields (plus the embittered Welshmen who sought work in England in the 1930's) who explain the (1) "inferior" type of men in some of the newer districts and (2) the unsettled community and pit relationships in these new districts. It is amusing how often Englishmen who complain about "cosmopolitan elements" end up by talking about Welshmen and Scots.

For this is one of the problems where the area of ignorance is still greater than the area of understanding.[15]

STRIKES IN YORKSHIRE IN 1949

By looking at certain characteristics of stoppages in the Northeastern Division, it is possible to gain a somewhat fuller understanding of the strike problem than by reference to national figures alone. Detailed figures on this, as on most questions, are not published by the NCB — and cannot be; for their bulk would be huge and their use easily subject to abuse. Detailed Divisional returns are of course submitted to the Board — in the case of stoppages, on a quarterly basis. The figures for 1949 show (1) the number of stoppages by causes and the number of tons lost, (2) an occupational analysis of those participating in the stoppages, and (3) the duration of the stoppages. We will review them in that order.

Causes of Stoppages. Table 5 shows the number of strikes caused by disputes over various questions, plus the number of tons lost because of each type of dispute. The number of tons lost during the year was exceptionally high, even for Yorkshire, since 70 per cent of the loss was attributable to the one-week stoppage of the winders. Eliminating this unusual cause, we are left with 269 disputes involving an average loss of 500 tons per dispute. Of the total of 330 separate disputes during 1949, almost one-third were caused by disputes over wages; this implies chiefly price lists, not general wage levels. Omitting the enginewinders' disputes (which were really one big dispute), the second most important

---

[15] The "area of ignorance" extends to the whole strike problem, not just to the explanation of regional differences. This ignorance can only be reduced by systematic, detailed study of labor relations at pits where strikes are frequent and at others where they are infrequent or unknown. So far neither the Board nor the Union has felt that such studies should be undertaken, though the possibility has been discussed by the National Consultative Council.

One of the most useful commentaries on the strike problem in coal has come from T. T. Patterson, Senior Lecturer in Industrial Relations at Glasgow University and a man who has spent considerable time studying the industry. In a BBC talk printed in *The Listener* (Jan. 29, 1953), Patterson argues that it is tempting to jump to *post hoc ergo propter hoc* conclusions. What is needed is systematic investigation of total situations, situations which lead not only to strikes but to such other "pathological" symptoms as absenteeism, high turnover, low productivity, accident-proneness, and the like. Until knowledge is built up that will allow correction of the underlying condition that produces such forms of protest as strikes, absenteeism, etc., there is more reason to be tolerant than indignant when strikes occur. Strikes at least have a useful cathartic effect and, as visible symptoms, they draw attention to the invisible conditions which produce them. See also the article "Unofficial Strike," by T. T. Patterson and F. J. Willett in *The Sociological Review*, vol. XLIII, sec. 4, 1951, pp. 57–94. The paper suggests an explanation of the origin, conduct, and abandonment of a one-day strike in a Scottish colliery in 1949. While stimulating reading, the article does not point to any specific steps that might be taken to reduce the incidence of the type of work stoppage that is characteristic of the industry.

cause was a refusal to perform alternative work when an individual's or group's own jobs were not available (53 disputes). Third in importance were disputes over working conditions (32 disputes).

TABLE 5

Number of Unofficial Stoppages, by Causes, Northeastern Division, with Tons Lost, 1949

Cause of Dispute	Number of Stoppages	Tons Lost
1. Dissatisfaction over wages	102	70,000
2. Refused alternative work	53	6,800
3. Objection to/or disputes with an official	5	600
4. Refusal to work contrary to Fatal Accid. Agreement	4	11,400
5. Working conditions	32	9,200
6. Refusal to perform work not completed by previous shift	6	1,000
7. Dispute over method of working	1	(40)
8. Dissatisfied with preparation of face	8	1,800
9. Refusal to await repairs after a mechanical breakdown	6	2,200
10. Objections to findings of Joint Disputes Sub-Committee	3	12,500
11. Sympathy with dismissed or suspended workmen	5	1,000
12. Go-slow policy adopted by winding enginemen in protest against wages and conditions	1	200
13. Refusal to do work outside contract	1	1,700
14. Methods of working and/or colliery organization	11	6,100
15. Refusal to start work after a breakdown when repairs had been completed	1	500
16. Enginewinders' dispute	61	306,500
17. Refusal to walk in after being stopped from illegal riding in a drift	1	200
18. Refusal to obey instructions given by deputy	1	(60)
19. Adoption of go-slow policy as a result of an Umpire's award	1	1,400
20. Miscellaneous	27	8,700
Total	330	441,900

Source: "Unofficial Stoppages," 1949, mimeo. résumé published by the Northeastern Division, NCB.

Occupational Analysis of Stoppages. Most unofficial strikes involve small groups of men, not whole pits — though often the whole pit goes out in sympathy if these occupational strikes last more than two or three days. Table 6 shows how many stoppages were accounted for by each occupational group. It will be noticed that the colliers (the disappearing group who work the remaining hand-got faces) and fillers between them accounted for almost 50 per cent of all the stoppages in the Division for the year; indeed, this percentage is abnormally low, because of the inflation of the number of separate stoppages which results from counting the widespread winders' strike as 56 separate stoppages. If the enginemen's strikes are eliminated from the total and all colliers' and fillers' strikes are lumped together, the latter accounted for 70 per cent of the disputes; and this is undoubtedly a more representative year-in-year-out figure for this division. This high a figure would not be representative, however, for Scotland, where stoppages by daywagemen have played a much larger role than in Yorkshire.

TABLE 6

Number of Unofficial Stoppages among Various Grades,
Northeastern Division, 1949

Occupation	Number of Stoppages
1. Colliers	79
2. Fillers	81
3. Colliers or facemen and haulage	14
4. Fillers, with others affected	23
5. Packers	17
6. Cutters	2
7. Packers, with colliers affected	1
8. Rippers	12
9. Wastemen	14
10. Duffers	2
11. Machine men	1
12. Pan turners	4
13. Faceworkers, mixed	17
14. Enginewinders and all others	56
15. Enginewinders alone	6
16. Repairers	2
17. Haulage, with others affected	3
18. Haulage alone	1
19. Datallers	1
20. Pony drivers	1
21. Screenhands	1
22. All grades	8
Total	336

Source: See Table 5. The small discrepancy between the two total figures is unexplained, but probably arises from different methods of classifying the same incidents.

Noteworthy is the fact that stoppages involving all grades, the haulage workers and other datallers, and surfacemen were all of minor importance. The conclusion is clear: the great majority of unofficial strikes begin with the coalface grades, particularly the "direct labor" groups (colliers and fillers), although stoppages among rippers, packers, wastemen, and pan-turners (conveyor shifters) are not uncommon. It is the highest-wage groups (those paid under incentive contracts) which produce most stoppages and the lowest-wage groups (those paid by time) which produce the fewest; it would appear that the prevailing method of incentive wage payment underlies the great majority of stoppages. This is not meant to imply either that the characteristic method of wage payment ought to be altered or that, if it were, the number of stoppages would greatly diminish. But clearly the wage area is one that ought to be, and is, receiving a great deal of thought.

Duration of Stoppages. Table 7 shows a frequency distribution of the number of days lost for each stoppage.

TABLE 7

Days Lost per Strike, Northeastern Division, 1949

Number of Days Lost[a]	Number of Stoppages
1	139
2	40
3	66
4	34
5	24
6	10
7	7
8	—
9	1
10 or more	—
Total	321
Returned same day	7
Grand total	328 [b]

Source: See Table 5.
[a] Date the dispute started until date work was resumed, exclusive.
[b] This total is slightly different from those found in Tables 5 and 6. These minor differences occurred in the original source.

Three-quarters of the disputes lasted three days or less; over 40 per cent lasted one day or less; practically none lasted more than a week.

The source of these figures also showed the numbers of men involved according to the duration of the strike: in almost all cases the figures were very small, for example, "11 colliers," "40 fillers," and so on. This indicates something about the nature of stoppages that is quite different from many factory strikes, even unofficial ones: in the first place, a pit is really a number of separate producing units, made up of various districts and various faces within separate districts. Thus, unless strikes occur "outbye" on the main haulage roads or at the pit bottom or among the winders, a strike by particular grades does not physically stop the pit; it stops only a local section of the pit. Secondly, the small groups who characteristically start the unofficial strikes do not, as a rule, try to bring the whole pit out with them when they go out; it is only if these men stay out more than a day or two that knowledge of their dispute becomes general and enough sympathy is generated to stop the whole pit. This means, naturally, that both the Coal Board and the district officers of the NUM have an interest in seeing that all such disputes are settled just as quickly as possible to avoid losing the whole pit's output. But this does not mean that the Board is continually giving in to pressure from unofficial strikers: the principle has been well established that no negotiations can take place until the men have returned to work. Exceptions occasionally occur, but it would not be correct to say that the "men have learned they can get away with anything by holding a pistol at the Coal Board's head."

Before turning to the important question of what the Coal Board and the NUM do about men who participate in unconstitutional stoppages, we should not leave the over-all picture incomplete by suggesting that stoppages are common at all pits, even in such a "bad" Division as Yorkshire. For the country as a whole, roughly half the pits have no strikes at all during the year.

TABLE 8

Number of Strikeless Pits, Northeastern Division, 1949

Area Number	Total Number of Pits	Number of Pits without Stoppages in 1949
1	12	2
2	12	1
3	12	nil
4	12	1
5	12	4
6	18	11
7	18	7
8	15	nil
Total	111	26

Source: See Table 5.

Table 8 shows the number of pits in each of the Yorkshire Division's eight Areas, and indicates the number of pits in each Area at which no stoppages occurred during 1949. In just under one-quarter of the pits in the Division, no strikes occurred during the year. By far the best records were in Areas 6 and 7 (West Yorkshire). There, many of the pits are old, small, and have been family-owned in the past; family ties have likewise been strong — the workmen and the local union leaders normally work in the pit, in contrast to the full-time officials characteristic of the big pits in South Yorkshire. Some people (union and Coal Board) feel that there are more grievances at pits with full-time officials than where there are none: they are more accessible to the membership and have more time to "bother the manager."

DISCIPLINARY ACTION AGAINST STRIKERS

It might be thought that prompt and perhaps strict disciplinary action against men participating in unofficial strikes would soon diminish their number and even stamp them out eventually. The first question is: just what can be done, what alternatives are open? So far as the employer is concerned, the penalties are five in number — four of them an industrial and one a legal penalty. The industrial penalties are these:

(1) Dismissal.

(2) Fine.

(3) The extraction of a pledge, often secured by a monetary bond, to comply with the conciliation machinery for a stipulated period.

(4) Closure of the colliery.

The legal penalty open to the Coal Board (as it has long been open to all employers in Britain, unlike America) is to prosecute offenders in the local courts for civil damages claimed for breach of a contract of service. British labor agreements generally (including coalmining) do not carry the "no strike" provisions so common in American agreements. Unofficial strikes are subject to control through the legal prosecution for breach of "contract of service" which the acceptance of a job assumes. Union-sponsored strikes during the life of the agreement are often prohibited by implication through the provision of voluntary arbitration. But the frequent absence of fixed-duration contracts enables unions to strike at short notice, simply by notifying the employer that they wish to terminate the agreement.

In addition to the penalties listed above, the Board formally recognizes (in its statistical reporting of unofficial strikes) two other steps which do not involve any personal penalties: (1) joint union-management investigations, and (2) resumptions of work without further investigation or penalties.

In contrast to the numerous contract clauses which have been written into American labor agreements during and, especially, since the war, there are no contractual penalties against unofficial strikers in any NUM-NCB agreements. This is explained chiefly by the parties' desire to rely on the traditional remedy for such offenses, namely, prosecution for breach of contract. The four industrial penalties — while not contractual — have been adopted by the Coal Board as matters of policy; none of them represents a new departure in the industry's approach to this long-standing problem. The present arrangement, which, as we shall see, leaves the specific type of action to local discretion, has the great merit of flexibility and experimentation — for both the Board and the NUM are going through a period of learning how best to deal with this tricky question. We should note that the national conciliation agreement between the Board and the NUM does contain a clause pledging the union to use "its best endeavours" to prevent unconstitutional stoppages and preventing it from giving support, "financial or otherwise" to persons who engage in such strikes. The same clause pledges the union to try to secure from its membership compliance with arbitration awards. At least some, and probably all, of the district conciliation agreements commit the district unions to a similar pledge with respect to agreements between the district unions and the Divisional Coal Boards.

Some notion of the degree of reliance put upon the various penalties

TABLE 9

Disciplinary Action Taken in Stoppages and Restrictions Owing to Trade Disputes, Northeastern Divison, January–June 1950

Action Taken	Number of Disputes Involved	Number of Men Directly Involved in the Disputes	Number against Whom Disciplinary Action Was Taken
1. Legal action through the court			
a. Results in favor of the board	1	8	8
b. Against the board [a]	0	0	0
2. Dismissals	2	264	11
3. Imposition of fine or mitigated damages by arrangement with Trade Union	9	110	110
4. Written undertaking to comply with Conciliation Machinery (sometimes coupled with monetary bond to take effect in event of breach within prescribed period)	9	155	144
5. Closure of colliery showing unsatisfactory results through labor problems	0	0	0
6. Joint investigation by management and trade-union officials involving none of the measures indicated in 1–5	47	1,683	—
7. Other measures, including			
a. Prosecution pending	11	461	461
b. Action not yet decided	2	31	
8. Disputes where no disciplinary action was thought necessary	99	6,531	645
Total	183	9,306	1,379

Source: NCB, Northeastern Division, Quarterly Reports (mimeo.).

[a] Interpreted by NCB to mean any settlement which "in any way ENHANCES the men's PRE-DISPUTE wages or conditions"; thus it excludes any compromise settlements, even though such settlements "may be less favorable than the men's full claim." (*Labor Relations Circular No. 44, 1950.*)

is given by Table 9, which shows the various actions taken during the first half of 1950 in the Yorkshire Division.[16] It will be seen that there were 183 stoppages and restrictions during this period and that in slightly over half of these no disciplinary action was thought necessary. In the 84 cases in which some form of action was taken, the action amounted in over half of them (47) simply to joint union-management investigations; these usually did not lead to any formal discipline. Where positive disciplinary action was taken, the most common forms were (1) legal action through the courts, (2) the imposition of a fine or mitigated damages by arrangement with the trade union, or (3) making offenders sign written undertakings to comply with the conciliation machinery, sometimes coupled with a monetary bond to assure performance. There were, as one would expect, no cases where the very drastic step of closing the pit was taken; the number of dismissals was also very low (11 men dismissed as a result of two separate stoppages).

The Board has — in a few rare instances — threatened to shut down pits where labor relations were so generally "soured" that the men's actual performance was considerably below what the Board considered reasonable. The Board's threat has always been coupled with the hope that the pit would be kept open if performance, usually measured in OMS (Output per Man Shift), improved. Such pressure was used, for example, at Betteshanger in Kent, the Monckton pits in central Yorkshire, and at certain pits in South Wales. In all but perhaps one or two cases, there has been subsequent improvement and the pits have been kept open, though it would take much more detailed investigation to find out whether or not the subsequent improvement was attributable to the Board's threat of closure.

It would be wrong to conclude that the measures adopted by the Northeastern Divisional Board are the same, either in detail or in emphasis, in all other Divisions. Local judgment and tradition play their parts in this as in almost all the affairs of the industry. Indeed, even with the Northeastern Division itself, there is no hard and fast rule as to which penalties shall be imposed in specified types of cases. Responsibility for deciding what action to take is left up to the Area General Managers; consequently, the approaches have varied somewhat from Area to Area. But generally speaking, the AGM's and the Area Labor Officers (whom the AGM's consult on disciplinary action in those Areas where the two individuals are on good terms) are reluctant to rely on legal prosecution as a method of controlling illegal strikes, as they feel that such proceedings "don't settle anything" and inevitably produce a

[16] The National Board began collecting detailed information on this subject from each of the Divisions in March 1950. (L. R. Conciliation Circular No. 42, March 27, 1950.) The discussion of "action taken" is confined to Yorkshire because a special study of the problem was made in that Division; it is fairly safe to assume that experience in most other Divisions has been similar.

residue of bitterness which lesser forms of discipline do not carry. Even though Board officials are reluctant to invoke the legal sanctions available to them (and the Labor Government did nothing to alter this long-standing remedy of employers), it is often useful to invoke the threat of such action in persuading trade-union officials to sit down and talk over particular situations. The threat is effective because the union knows that it cannot defend unofficial strikers in court or in public; yet it is so against legal remedies as a matter of principle that it is willing to discuss alternative measures.

The National Board agreed to suspend all legal prosecutions of men for breach of contract in the spring of 1947 when the Five-Day Week Agreement was introduced. This was done on a trial basis as a demonstration of good faith. Experience during the weeks following the introduction of this reform was so disappointing that in the last week in June 1947 (seven weeks after the Five-Day Week Agreement went into effect), the Board wrote the NUM that it would resume legal prosecutions in flagrant cases unless the union could suggest any effective alternative method of dealing with unconstitutional stoppages. The devices of "bonding" men and fining them were subsequently readopted, but these have not served as alternatives to legal prosecutions; they are simply supplementary forms of discipline for use in cases where legal prosecution may be thought inadvisable.

Shortly after the Board authorized the return to prosecutions in mid-1947, the Production Director of the Northeastern Board confidently predicted that such steps would soon stamp out illegal stoppages. This "get tough" policy turned out to hold much lower curative powers than had been hoped and subtler forms of discipline had to be worked out. The problem in the British coal industry is much less amenable to legal control than in the American, since the British problem is one of many small unofficial stoppages and not large-scale union-sponsored stoppages.

Since an unofficial strike is simply a strike not officially approved by the union, and since the NUM has been officially against all strikes since nationalization, it follows that the NUM invariably "disapproves" of all illegal strikes, at least officially. There seems little or no hypocrisy in the NUM attitude, at least at national level and certainly in most of the districts. Even if local instances could be cited where branch or district officials unofficially encouraged or approved illegal strike action while publicly disavowing the strikers, it is certain that such situations (of which a certain minimum number would seem inevitable in any economic system) were not developed into a new technique of subterfuge by the NUM, designed to reconcile their loyalty to the Labor Government and to the success of nationalization with their traditional reliance on the strike weapon as the midwife of industrial progress.

The district in which this presumption might seem applicable is

Scotland, where there have been many unofficial strikes of a semipoliti-
cal nature during 1947–1950 and where the union was Communist-dominated until mid-1950; but there is not, even within the industry, any
widespread feeling that Scottish NUM officials (at least at district level,
as opposed to local officials) have been talking out of one side of their
mouths in public and out of another side in private. The "wink-and-a-
nod" technique which John L. Lewis was presumed to have followed in
the 1947 coal stoppage in America does not seem to apply even in
Scotland.

But if the NUM may be represented as sincerely opposed to stoppages, unofficial as well as official, it has been baffled by the continuing
readiness of many of its members to engage in "quickie" strikes which
flout the disputes machinery. The crucial questions concern (1) what
the union is prepared to let the Board do by way of disciplining offending members, and (2) what steps, if any, the union itself is prepared to
take in imposing forms of internal discipline. The answer to the second
question is easier than the first, since it is wholly negative. There are,
of course, very few rewards or penalties which a union like the NUM
holds within its power to grant or to withhold from its members. All it
could do, in effect, would be to declare a member "unfinancial" for a
specified period, thus depriving him of benefits during this period; or it
might reduce by a specified amount the union retirement benefit (in
districts where such are in effect); or it might fine members just as the
employer might; but beyond such steps there is little or nothing a union
can do by way of penalties. Consequently, the union's attitude toward
managerial discipline is the only significant aspect of the discipline problem in connection with strikes.[17]

This attitude may be characterized as one which acknowledges the
necessity for disciplining "the irresponsible minority," which prefers to
leave this disagreeable duty up to management (more for political reasons than out of any well thought-out philosophy of "management's
needs"), and which discourages resort by the Coal Board to legal sanctions without really approving of (or indeed without inventing) alternative measures. However, this should not be interpreted as a union "veto"
of any steps toward effective action by the Board to suppress illegal
stoppages; it is rather the use of persuasion to dissuade the Board from
invoking penalties which the union feels cannot be effective and which
it feels solves one dispute only by laying seeds for others. The prevailing
philosophy is that "in time" (a generation or so) old memories and old
habits will fade out and men will develop the habit of relying on the
legally established grievance procedure as a matter of course. If a criti-

---

[17] No one in Britain would suggest that deprivation of state social security benefits should be used as a method of industrial discipline, as has been done in Russia
(see Isaac Deutscher, *Soviet Trade Unions*, pp. 91–92).

cism can be made, it is not that the union has successfully intimidated the Coal Board into a position of impotence in the face of illegal stoppages but that it places too much emphasis on the salutary effect of the mere passage of time without having sufficient appreciation of the importance of specific educative steps (including both penalties and more positive forms of education and research) to bring about the changes in attitudes which many union leaders privately acknowledge to be necessary.

CHAPTER FOUR

JOINT CONSULTATION AT PIT LEVEL

Among the many proposals for improving the industry's labor relations, joint consultation, or the use of labor-management committees as they would be called in the United States, has held a leading place. Unlike labor and management in the United States, the British have retained throughout much of their industry the interest in consultation which, originally started during the first World War by the Whitley Committee, had been revived by the second World War. One reason for this is that the postwar British Government (partly, no doubt, because it was a Labor Government) encouraged the spread of consultative machinery throughout industry generally and saw to it that all the nationalized industries were required by statute to set up special consultative machinery. There has been, however, a distinct shift in emphasis with respect to the purpose of consultation: the wartime Joint Production Committees (as they were then called) existed mainly to increase output and to prevent the waste of scarce materials; since the war, the main emphasis has shifted to the part that consultation might play in improving relationships between management and employees.[1]

The desire to make something of joint consultation in the nationalized coal industry is not, however, a development peculiar to the socialist sector of British industry. Nevertheless, a special importance has attached to the development of consultation in coal; partly this is because trade-union and Labor Party people had felt that joint consultation somehow had a unique role to play in a nationalized industry (it was one of the principal devices for extending a "nationalized" industry to a "socialized" one), and partly because coal was widely thought to stand in greater need of this prescription than almost any other industry.

While the first seven years' experience with joint consultation did not produce unrelieved failure, it should be admitted that many persons

[1] See "Joint Consultation in Industry," Supplement No. 3, Dec. 1949, *Industrial Relations Handbook*, issued by the Ministry of Labour and National Service, 100 pp., HMSO, 1950, 2s. This contains a useful summary of the arrangements in almost all industries which practice joint consultation.

who expected much of it have been disappointed. This disappointment stems from a double deception into which the more ardent advocates of joint consultation fell: (1) they often underestimated the extent to which consultation is limited by the particular context of the coal industry's labor relations, which had not usually been good; and (2) they tended to expect from consultation, wherever it was tried, more than the device itself could yield. The root difficulty is that joint consultation tries to make an institution suited to legislative purposes serve ends which can be effectively pursued only through better administration. This estimate of joint consultation is not wholly pessimistic, but admittedly it is modest. There is undoubtedly much room for experiment and improvement in the "techniques" of consultation (entirely apart from the more fundamental attitudes which inspirit the machinery).

Consultation has four interrelated objectives: to give the rank and file a fuller sense of "participation" in the affairs of their enterprise; to open up new "channels of communication" between management and men; to improve the respect which union leaders and managers have for each other; and to bring a larger measure of "industrial democracy" into the conduct of the industry. Joint consultation is not the only road to these four vague and distant goals; but it is the main road down which British labor relations have turned. To understand this development, we must bear in mind its origin. On the one hand, the unions' demand for consultation is descended in part from the militant, emotional demand for "workers' control" which was part of the radicalism of the first quarter of the century; on the other hand, the willingness of modern managements to try consultation is descended from that tamer, much more rational interest in better employer-employee relations which is a main concern of professional labor relations as it has developed during the second quarter of the century.

The earlier radicalism was not, of course, a single-minded challenge to the industrial order: anarchy, syndicalism, guild socialism, and Marxian socialism all competed for the right to lead the battle. But a common denominator uniting all these solutions was the assumption that industrial and economic salvation depended more on the negative operation of sweeping away the allegedly faulty institutions of economic life than on any positive reconstruction of industry. Professor John Mack, of the University of Glasgow, aptly emphasizes the naïveté of the early propagandists for workers' control. After pointing out that pressure for workers' control had decreased in the 1930's because the unions were reluctant to compromise their independence and because the Labor Party came to believe in the supremacy of Parliamentary control, he comments:

> These profound objections to workers' control are well-known. The third is much less insisted upon, and yet it seems to me the most important of them all. It is the recognition of the primary importance of management and ad-

ministration. The more long-headed Labour leaders grasped it a long time ago: but the movement as a whole is only beginning to wake up out of a beautiful dream in which socialised industries manage themselves in a general haze of mutual good feeling. The old-fashioned Syndicalist genuinely believed that managers did nothing but appropriate surplus value for his bloodsucking employer. The Guild Socialist looked the problem squarely in the face and then swerved aside, leaving behind a purely verbal solution. He assumed that an all-comprehensive Guild would include all the managers, who would now be called manager-workers, and who would submit readily to following the instructions of a committee of their own workers, by whom they might even be appointed and if need be dismissed. Even assuming that this psychological miracle could happen, a further insoluble problem would remain. In a socialised industry, the manager of each undertaking would be responsible to a higher director for that particular district or division, who would be in touch with the central planning agency or agencies. How then could he be responsible to his own workers at the same time? [2]

The subsequent development of "professional labor relations" has embraced both an academic interest in the frustrations of industrial life (and what can be done about them) and a new recognition by industrial management of the crucial importance of the personal relationships that get formed in any industrial undertaking. The democratic demand for greater rank-and-file participation in many industrial decisions has come to be viewed by many managements not only as a condition of survival but even as "good business." By now there is enough evidence to say that in some cases, anyway, this assumption has proved to be true; in many more cases this assumption is still untested, though it has become a goal toward which both management and unions have agreed to work. Joint consultation is thus a cautious attempt to realize a greater degree of industrial democracy by making use of some of the institutions of political democracy.

To be sure, this development is not a conscious adaptation, but the similarities between the structuring of democratic government and the machinery of joint consultation is too plain to miss. It is much less clear whether industrial decisions are likely to be as well served as are political decisions by this ordering of relationships. Nor is it clear that the ideal of industrial democracy, which includes the separate notions of where authority is formally seated and how power is actually exercised, is best advanced by thinking in terms borrowed from civil government.

It is helpful to analyze the level of control over industrial decisions by grouping them under three main headings. Under factors normally beyond the control of the firm one would list:

Ruling market prices of capital, raw materials, and often of the product.
Governmental laws affecting: safety regulations governing factory or mine operation; product standards (e.g., in food and drug manufacture); minimum

[2] From an article entitled "The Meaning of 'Workers' Control,'" 3 pp., mimeo., undated, but about 1948.

wages and maximum hours, child and female labor; recognition of trade unions and conduct of collective bargaining; zoning of industrial location in the community.

Existence of a demand for the firm's products.

General level of business activity in the economy.

The number and decisions of other firms in the industry.

Under control normally retained by management alone one would list:

Choice of management personnel, at all levels.

Level of managerial salaries; regulation of expense accounts.

Form of company organization (partnership, corporation, etc.); internal organization into divisions, departments, subsidiaries, etc.; line of authority within the organization (who is responsible to whom and for what).

Sources of capital; methods of raising capital; amount of capital required; level of interest or dividends to be paid.

Products to be manufactured; their design; their engineering specifications; their inspection and packaging.

Location of plant; design of plant; plant layout; choice of manufacturing process and equipment.

Use of an internal or external sales organization; choice of markets to be exploited; method of remunerating salesmen.

Whether to ship by rail or truck; whether to maintain one's own trucks or use outside carriers.

Pricing of the product(s).

Advertising policy; internal or use of an agency; media to be used; whether or not market research is needed.

Amount of research on new products and processes; timing of introduction of new products.

What sources of supply to use; appropriate sizes of inventory items; judging the market on "when to buy"; types of capital equipment to purchase and their specifications.

Scale of the enterprise and changes in scale.

Control over expenditures through budgetary controls.

What risks to insure against and what levels of insurance to carry; what carriers to insure with.

What system of accounting to employ; methods of costing.

What banks to use to safeguard company funds; the number of banks to use; how to invest cash surpluses.

And under control normally shared between management and employees one would list:

Rates of pay for different jobs.

Changes in wage levels.

Scale of such benefits as vacations with pay, paid holidays, pensions, insurance benefits.

Hours of work; shift starting times; premiums for afternoon and night work.

The amount of work to be performed by individual employees.

The number of apprentices; the length of apprentice training.

Length of lunch periods, smoking breaks, "personal time," etc.

Procedures for the settlement of disputes or grievances — i.e., a "judicial machinery" for assuring plant justice.

Whether or not employees must be members of a trade union; whether or not the employer is free to hire labor on the open market.

Payment for time lost because of machinery breakdowns, interruptions in materials supply, etc.

The Regulation of layoffs in the event of redundancy.

Transportation arrangements to and from work.

The right of promotion to higher jobs below the management level.

The provision of personal equipment such as tools, safety glasses and boots, and other items of protective clothing.

The provision of food or light refreshment through canteens.

The right of the trade union to utilize a plant bulletin board.

These lists illustrate the many different kinds of questions which arise for decision (or which must be "taken into account") in the conduct of any industrial enterprise. Even such a crude classification of the questions that arise in running a firm, and of the recognized loci of authority for deciding them, can offer a useful perspective from which to view consultation and collective bargaining — for both these approaches are bound up with the ideas of "workers' control" and "industrial democracy."

The labor agreement which emerges from collective bargaining clearly results in a high degree of control over those aspects of economic life which touch employees' lives most immediately. A union is an employer-regulating device, within limits, and the limits have shifted significantly in recent decades. Yet the classical demand for "workers' control" has always meant much more than the achievement of satisfactory labor agreements: it has meant the achievement of control over precisely those questions which labor agreements can never hope to reach and a much fuller participation in the process of executing decisions. To be sure, these aims are rarely expressed rationally, and it would be misleading to endow what is essentially an emotional drive with too much specificity. Furthermore, the militant mood which produced the drive for "workers' control" before and after World War I is certainly not strong in present-day Britain, though the mood may be dormant rather than dead. This is not to deny that there remain certain individuals in whom the old ideals are very much alive, but very few of these are influential in either the Labor Party or the NUM.

Joint consultation, then, represents a form of workers' control that stops far short of the historical movement that bears that name, though it continues to represent a desire for something more than collective bargaining can hope to supply. The boundaries that separate bargaining from consultation are, in the nature of things, arbitrary: should there be bargaining over whether a cup of tea in the pit canteen ought to cost 1d. or 1½d., or should this be determined (or only "explained?") through joint consultation? As we shall see, the difficulty of establishing substantive boundaries is made infinitely more difficult when the individuals who bargain are the same individuals as those who consult.

Consultation presents many of the problems which beset nearly all democratic institutions, whether they be parliaments, local councils, or the representative levels of voluntary bodies such as trade unions, church congregations, or university faculties. Such representative bodies almost always must struggle with

The apathy and indifference of constituents in the absence of some "crisis" (usually a crisis over money).

The difficulty of defining constituencies — which groups shall qualify for representation (or how much) and which shall not.

The difficulty of communicating decisions (and the *reasons* which lead to them) to "the public" — e.g., through newspapers, through periodic visits to constituents by elected officials, by the attendance of constituents at meetings, by letter, by posted announcement, and so on.

The reluctance to risk public criticism on the part of some elected officials.

The reconciliation of the expert knowledge and personal responsibility of appointed line executives with the lay viewpoints of elected representatives.

The abuse of democratic forums for the advancement of personal ambitions.

The critical importance of effective personal leadership in the conduct of a committee's work.

These and many other problems common to the work of representative groups have intruded themselves into the working of the Colliery Consultative Committees, though the particular problems and their intensity vary greatly from pit to pit.

The Machinery of Joint Consultation

The Nationalization Act of 1946 distinguishes between collective bargaining and consultation, and imposes a legal obligation on the Board to engage in both.[3] The Board's obligation is to consult on matters of "safety, health or welfare" and on "the organisation and conduct of the operations . . . and other matters of mutual interest." More specifically, the Board is intended to secure "the benefit of the practical knowledge and experience" of its employees with respect to "the organisation and conduct of the operations in which they are employed." This latter qualification suggests that consultation at pit, Area, Divisional, or National level should be limited to topics appropriate for treatment at the level concerned; pit committees are not intended to discuss how the National Board ought to finance capital expenditures. Although consultation has been established at all four Coal Board levels, our concern will be almost exclusively confined to pit level.

Although there had been a history of joint consultation in coalmining before nationalization, much of this history had been unhappy.

[3] Coal Industry Nationalization Act, 1946 (9 & 10 Geo. 6. Ch. 59), sec. 46.

Committees established during the war were called "Pit Production Committees," though in their early days they dealt almost exclusively with absenteeism. Later efforts to shift their emphasis to production problems were only partially successful. The chief problems in securing effective work by the PPC's were: the uneven skill and enthusiasm of the colliery managers as chairmen of the committees; the reluctance of technically trained managerial personnel to accept advice on production problems from uncertificated rank-and-file representatives; the limitation of participation to active trade-union members and the physical difficulty of publicizing the work of the committee even at the minority of pits where their work was worth publicizing, so that the committees operated in great isolation from the body of workmen on whose interest and coöperation their work largely depended.[4]

Since there was no reservoir of accepted tradition and practice to guide the pit committees' organization or work, the Coal Board, in consultation with the interested trade unions, worked out a written constitution for the guidance of local colliery and union officials. This constitution gave a common form and authority to each Colliery Consultative Committee and minimized the confusion and misunderstandings that would have developed if each pit (or Division) had been left free to design its own machinery. Joint consultation is thus not something that has grown up as a "grass roots" development; it has been imposed from the top according to a uniform national pattern.

THE COLLIERY CONSULTATIVE COMMITTEES

With the exception of very small pits, the Colliery Consultative Committees (CCC's) are composed of thirteen individuals. Membership is determined as follows: three are ex officio members (the pit manager, the district "miner's agent," and the secretary of the pit's local branch), three are appointed by the pit manager (two underground and one surface official), seven are elected (one deputy, and six others from the five main grades of colliery workmen — two faceworkers and one each from among the underground haulage workers, contractors not employed at the face, surface workers, and craftsmen). All nonmanagerial employees are entitled to vote without any service qualification, and each person is given six votes — not just one for the candidate(s) representing the voter's classification. The elected representatives are thus not to be regarded strictly as representatives of their own grade, but of a much wider constituency instead; the occupational basis of representation (which is not retained in the voting, except in the case of deputies voting only for deputies) merely serves to assure that the CCC will include men with knowledge of the main grades of work at the pit.

Although nominating procedures are scarcely the weakest aspect of

[4] See W. H. B. Court, *Coal*, pp. 322–323.

the consultative machinery, they have raised the chief constitutional issue. When the committees were originally established in 1947, nominations were called for by posting at the pithead a general invitation for the submission of nominees. The latter and their proposers both had to have been employed at the pit for a year and had to be a member of a recognized trade union. Beyond the ex officio membership enjoyed by the lodge secretary and the miner's agent, the local branch played no formal part in the nomination process.

The lack of union control over the nominating process quite quickly became a source of irritation to the NUM and late in 1948 the Union demanded that its branches should be designated as the exclusive agency for making nominations (the NUM did not assert a right to appoint the workmen's representatives). The reason for asserting this right was the necessity felt by NUM branches for integrating branch decisions with the workmen's attitudes as expressed in consultative meetings. If rank-and-file representatives on the CCC said or voted for things that embarrassed the local branch (even though these representatives had to be branch members), the branch would obviously not like it. If CCC members refused to speak up on an issue for lack of knowledge as to branch sentiment, the committee's work would become ineffective. If personal rivals of the branch's leadership turned the committee meetings into forums for ventilating political differences within the branch, this could weaken the branch's bargaining position with the manager and could provide political rivals with a fortnightly opportunity to embarrass the incumbent officials.

The National Consultative Council (the industry's top-level consultative body) agreed to the NUM's demand after consultation with Divisional Consultative Councils. The Board's side was not enthusiastic about the change, but consented to it in the face of demonstrated difficulties at a number of pits. The 1949 change required the NUM to nominate "at least two candidates" for each vacancy, a requirement that prevented the Union from "appointing" the membership by refusing to nominate more than one candidate.[5] The branch is further required to

---

[5] At some pits it has been found difficult for the union to find more than one nominee and the question of what to do in such a case (which is not provided for in the Model Constitution) has been referred up to the National Consultative Council by at least one of the Divisional Councils (the Scottish, in May 1949). Abe Moffat, President of the Scottish Area of the NUM, said that "Scotland were not trying to break the Constitution, but it was necessary to be realistic." Replying to his fellow-Communist in a way not uncommon in Britain but unheard of in the United States, Arthur Horner said of the Scottish proposal (to recognize unopposed nominees as "elected") "struck at the basic conception of consultation. The intention had been to give the whole body of men in the colliery responsibility for the elections, not just a small body of the Committee" (i.e., the Lodge Committee). He said that the NUM should "persist in their efforts to get two nominations." (Minutes of the Seventeenth Meeting of the NCC, London, May 10, 1949, reprinted with the Minutes of the Meeting of the National Executive Committee of the NUM on

post at the pit an open invitation for anyone with more than a year's service to make individual nominations — a further safeguard against the temptation to nominate a single candidate on the excuse that the branch could find no more. Thus, while it is correct to say that the NUM sought and secured a significant degree of control over the process by which individuals get onto the CCC's, it would not be correct to say that the NUM was given full control over committee membership.[6]

Did the integration of the NUM and the pit committees lead to a great change in the membership of the pit committees? The original elections to pit committees had been held "as soon as practicable" after Vesting Day in 1947 and, as the term of office for committee members is set at three years,[7] the next elections were not due until December 1950. This would have meant a wait of some two years before the NUM could have exercised its new control over committee personnel; consequently in the Northeastern Division (it is safe to assume similar procedures were followed in some other Divisions) the NUM was allowed to decide whether or not it wished to hold fresh elections. Significantly, only eight of the 114 branches in the Division exercised this option, the remaining 106 branches expressing themselves as sufficiently satisfied with the membership of the pit committees to allow their membership to remain unchanged until the normal election period in December 1950.

The *Model Constitution* sets out the elementary questions of committee organization and procedure, which are necessary but insufficient conditions of effective committee work. Most of these questions concern such details as the frequency of meetings, payment for attendance, and the preparation of agenda and minutes,[8] but the question of the chairmanship is so important that a few comments are necessary.

In all cases, the pit manager is the chairman of the Colliery Consultative Committee. This is an almost inevitable arrangement in view of the fundamental conception of the committees as purely advisory or

---

June 23, 1949, p. 29.) There may be local instances where the NUM branch has negated the spirit of the double-nomination requirement by pleading inability to find more than one candidate, but this tactic has not been widespread.

[6] It will be recalled that in some coalfields certain grades of men are primarily members of unions other than the NUM, though the "dual membership" agreements make them members of the NUM as well. So far as can be ascertainted, these other unions are given no role in nominating CCC candidates.

[7] One-third of the six elected members retire at each annual election, so that in practice there are only two elective positions to be filled in each year's election. The first couple to retire were those receiving the lowest number of votes in the 1949 election (the special election authorized after the reform of the nominating procedure early that year); the next lowest pair were to retire in 1951. Members returned unopposed are considerd to have received the highest number of votes and would therefore be entitled to remain on the committee the longest — a privilege which may in a few cases have tempted NUM branches to nominate only one candidate for some or all of the positions. (Model Constitution for Colliery Consultative Committees, *Guide to Consultation*, p. 36, 1949 ed.)

[8] Model Constitution, *Guide to Consultation*, 1949 ed.

staff bodies, whose advice or criticism the pit manager is free to accept or reject. Not only is the manager's absolute line authority required as a matter of sound industrial organization; it is also required by the Coal Mines Act. It would be legally possible to hold consultative committee meetings with someone else in the chair, but — if the manager were present at such meetings — it might often undermine the authority of his position.

The difficulty in protecting the authority of the manager *qua* manager by making him chairman of the pit committee lies in the fact that many managers are much better at running pits than at acting as chairmen of committees. And poor or indifferent chairmanship almost invariably makes for an ineffective committee. This problem is more critical in the conduct of pit committees than is true of committee work in other groups because of the roles in which the manager and many of his committee members are cast outside the committee. For this reason the quality of the general relationship between the pit manager and the local branch officials has an almost conclusive effect on the success or failure of joint consultation at any particular pit.

Indeed, the central problem of consultation is whether the bad labor relationships so common in the industry will persist, thus fatally contaminating joint consultation; or whether the process of consultation itself can be made an agent of gradual change in the quality of those over-all relationships at the pit.

There has been no widespread, militant demand from the NUM branches to alter the constitutional arrangement which makes the manager chairman of pit committees. But a survey of branch opinion on the functioning of joint consultation after the first year's experience showed that there was "the general complaint in nearly all reports that the Manager ought not to be the Chairman of the Consultative Committee in every instance, but that there should be interchangeability of Chairmen." [9] This seems to be a proposal to raise the workmen's status by giving them a right to share in the chairmanship rather than an indictment of how most managers conduct their meetings.

Rotation of chairmen is customary in many of the entirely separate joint committees which govern welfare matters in the industry. There are probably several higher-level NUM officers who, understanding the "line" position in which pit managers stand, would not alter the present arrangement — though they might not be willing to oppose a popular movement for a rotating chairmanship.

For example, one does not come across instances where local branches have withdrawn from the consultative committee at the pit until they have been given the right to share in the chairmanship. Without the authority of a Model Constitution, many such instances would

[9] *NUM Conference Report*, 1948, p. 149.

have been likely. Indeed, the Constitution has undoubtedly been of almost crucial importance in affording leadership to the organization of the consultative machinery and in establishing certain "rules of the game." Without such a document, much anarchy would have been likely and managers would have become even more demoralized than they have been.[10]

The CCC's are constitutionally barred from discussing any questions "which are normally dealt with through Trade Union negotiating machinery." Under the terms of reference laid down in the *Model Constitution*, committees would commonly devote their attention to problems such as these: interviewing, warning or fining absentees (often done by a subcommittee); altering conveyor styling to prevent spillage; stocking one type of safety shoe instead of another; distributing copies of the NCB house-organ, *Coal*; investigating why the cost of workmen's gloves is so high, how to speed up the supply of empty tubs to a particular district on Monday mornings, what kind of publicity to put out to encourage a "bond drive," why a particular accident occurred; the allocation of colliery housing to those members of the pit on the waiting list; securing better lighting arrangements underground or on the surface; the amount of steel and timber losses; whether or not to work a Saturday shift in view of the high rate of absenteeism among faceworkers, and so on. Any such list must of necessity be only illustrative;[11] but it can perhaps convey a feeling for the type of question with which the committees mainly concern themselves.

Needless to say, different pit committees have worked out different conceptions of their effective authority on an informal basis, depending on local relationships and traditions. At some pits, for example, housing is allocated by the CCC — the manager would not think of violating their wishes, even though he might have formal authority to do so. At other pits, the manager would not think of letting his pit committee participate in house allocation. Again, at some pits, the committee has assumed sole responsibility for disciplining absentees, while at others it has preferred to leave this unwelcome task to the manager.

[10] On the other hand, withdrawal from consultative relationships has been invoked as a sort of symbolic strike over issues that lay wholly outside consultation — e.g., in Lancashire in 1949 over the concessionary coal issue in that coalfield.

[11] Coal Board and trade union offices are rapidly filling up with files of mimeographed minutes sent in by the colliery committees. Each Division has a staff person who oversees the administrative side of consultation in the Division and who is responsible for analyzing the work of the committees in the Division so that it can be appraised by the Divisional Consultative Council or by the Board's line officials. In most Divisions, extracts are taken from the Minutes of each pit's committee and are mimeographed for circulation to all the Area General Managers in the Division once every month or two. These Minutes, then, provide a very useful and new channel of communications within the industry — though the extent to which use is made of mimeographed reports would require a study in itself. But for the student of the industry, these reports are a huge help.

No mention has been made of consultation on financial matters. In theory, the Coal Board has laid down a strict rule that colliery committees are not to be given and may not demand financial information (that is, profit and loss statements; discussions of the cost of specific items of equipment or supplies are, of course, routine).

At most pits, interested workmen active on the CCC or in branch affairs can usually get enough information to know whether or not the pit is "paying," especially if it is not. Each Area draws up a profit and loss statement on a pit-by-pit basis and it would be most uncommon for those statements to be sent or shown to local NUM officials in the district, though they might be told in a general way that a certain pit was "not paying." On the other hand, the National Coal Board prepares a monthly financial statement by Areas and Divisions (not by pits) which is circulated throughout the industry on a restricted basis: they do not, I believe, go regularly to all Area offices of the NUM, but copies are sent to the NUM headquarters in London, where they are seen by members of the NUM Executive Committee.

NUM officials are by no means unanimous in wanting ready access to intimate financial information on the part of local officials and members: they share the Coal Board's fear that such a policy might create more misunderstanding and frustration than it would clear up. A prominent miners' leader in the north believes that "it's wrong for unions to expect all the information there is to get, since a union ought not to wish for any more information than it's willing to take responsibility for." He does not want local unions to go "fishing" for information with which they can then turn around and "flog the Board." Also, if the union asks for full financial information, he fears that the Coal Board may begin to expect the union to "do its work for them." (For example, to educate members on costs.) He thought it most unwise for the men at good pits to know just how much profit was being made on their tonnage, "because they'd naturally want some of that profit for themselves." On the other hand, when a pit is in serious trouble and is threatened with closure, this man believes that if the Board expects the union to support justified closures, the union must — at that point — be given all relevant financial information.

Undoubtedly, some of the workmen's representatives on some committees chafe under this restriction and there have been staff people high up in the Board's organization who have deplored the policy of treating such information as confidential. Nevertheless, there is a difference between the full disclosure of financial information under a system of collective bargaining where the bargains cover employees working only for the unit for which the information is disclosed and where there is industry-wide bargaining and unit financial accounting. In British mining both situations exist: pieceworkers bargain on a pit basis but daywagemen bargain on a district or national basis. If the profitable pits are to carry the unprofitable, employees at the former must not coöpt the profits of their pits before they can be transferred into wages at the latter. "Full disclosure" might run the risk of tempting pieceworkers to run off with the profits or might tempt daywagemen to wish for local instead of district or national bargaining. The policy of treating financial information as confidential is designed in large part not to raise these unsettling temptations.

Consultation in Action

It is impossible to catch the flavor of consultative meetings from abstract accounts of their functioning. In an attempt to convey a more meaningful impression of the work of the CCC's we will examine briefly the issues involved in two very different meetings which happen to lie at opposite ends of the wide spectrum along which the various shades of pit relationships fall.

DUKESBURY FIVE-FOOT PIT

The pit was one where relationship between the branch and the management had been described by the president of the district miners' association as "very good." This was not a universal judgment: the president is active in the Moral Re-Armament Movement (the Buchmanites) and the "militant" trade unionists in the district — outside MRA — assert that the "good" relations at Dukesbury are in fact too peaceful simply because the local branch is "weak" and not self-assertive enough. Thus there is no common agreement as to what kind of pit relationships constitute "good" relationships: if these be conceived in terms of maintaining a precarious tension between local union and management personalities, so that neither fully accepts the other and each operates with some uncertainty with respect to the other's intentions and reactions, then it would seem rather difficult to expect a completely good atmosphere for consultation, where each side does fully accept the other and treats it as an equal partner engaged in a common search for the best way to handle specific problems. Thus the difficulty of defining what is meant by "good labor relationships" spills over and affects how one can define what is meant by "good consultation."

The atmosphere of the following meeting was calm and noncontroversial; the manager was always fully in control of the meeting and no personal strains were apparent. It was held — as so many such meetings are — in a very small room, with union representatives at one end of a rectangular table, the management representatives at the other. The secretary was a woman. All the men but one were in their street clothes, the odd man being still in his "pit dirt." The manager did not call the committee members by their first names, nor they him.[12]

[12] Two very competent branch secretaries from Yorkshire told me on another occasion that the mark of a good manager is "a man who can come down to the men's level without losing his dignity." "It wouldn't do," they agreed, "to have him called by his first name, even though it might sound democratic. You've got to show respect for the position, even if you don't respect the man."

The meeting began by considering matters arising out of the minutes of the last meeting, two weeks earlier. The following points were discussed:

Gloves: The men were not being issued the type of gloves they preferred, according to the branch secretary. Could the matter not be referred to the Area Consultative Council so that the Area Purchasing Officer might be asked to order the kind which the men do like? The manager and the rest of the committee agreed that this would be a good thing to do. The course chosen seemed too circuitous to get a completely satisfactory and quick decision. Was this a case where the machinery of consultation — an exaggerated constitutionalism — was dominating sound administration on a minor, routine matter?

Nomination of men for an NCB Summer School at Oxford: A letter had come in from Hobart House (London headquarters of the Coal Board) inviting the Colliery Consultative Committee to nominate a pit representative to attend the NCB's annual summer school to be held at Oxford. But the letter required the selection to be made by May 29th — three days before the next regular meeting of the CCC, at which the individual would be selected from nominations received during the next two weeks. This was the second year Hobart House had submitted the request for nominees on too-short notice. In order to comply with the deadline and to respect the procedure by which the man was to be nominated, a special meeting of the CCC would have to be called. It was agreed to do this.

The meeting then turned to a review of the routine statistical reports on the colliery working for the past fortnight:

Absenteeism: The records of some of the worst individual offenders (6–8 men) were reviewed; only one man had a really bad record. The manager suggested that the undermanager see this man and warn him — a procedure agreed to by the Committee. The manager noted that voluntary absenteeism for the pit was down to 4–5 per cent, which he considered "very low" — indeed, "we can't expect it to get any lower." Hence, he considered it unnecessary to do anything about the remaining 5–7 men whose offenses were less serious: "It's no use dealing with men just for the sake of dealing with them."

The manager then proceeded to read out the percentage of men who had lost their attendance bonuses during the previous two weeks — the figures were quite high, indicating that a large proportion of men had lost a few shifts.

Output: The pit target had been exceeded during both the previous weeks, but no one raised the question of whether or not the target ought to be raised.

A failure of local transport: A group of men had failed to turn up for work during one shift of the preceding week, claiming that the local bus had not shown up at their bus stop. The manager wanted the Committee's opinion as to whether or not he should pay these men their attendance bonus. The Committee decided to do so, but the lodge secretary then got the Committee to agree not to do so again in the future, since the bus company had promised that it would send out a special bus if the men would only come to the company's offices, which were near the bus stop in question. The men should take more responsibility for getting to work than merely standing at their stop "hoping the bus might not turn up."

Dirty coal: The manager expressed satisfaction at the "very good" position at the pit. Union spokesmen pointed out that sometimes the men on the faces even "ask the officials if this coal is too dirty to throw on the belt!" Therefore, the Committee decided not to implement a suggestion of the Area Consultative Council to set up a "dirty coal committee" to visit the faces in an effort to persuade the men to fill cleaner coal. Such a step would "only get the men's backs up."

OMS: The figures showed it was 37 cwts. (1.8 tons per manshift, over-all). This was the highest ever recorded for the pit and compared with 26 cwts. ten years previously. There was general satisfaction with the steady climb up from the 1940 figure.

The Committee then turned to any "new business":

Short-circuiting of line officials: A union representative mentioned that some workmen were asking members of the Committee to raise questions about supplies at the Committee meetings: he thought these suggestions ought properly to go first to the regular officials in the pit, as it meant a loss of time and a violation of regular procedures to try to channel such complaints to the CCC. He thought the Committee ought to be kept from becoming just a "complaints committee" for routine matters. (This was obviously a good comment and it met with general agreement from other members; it was agreed to urge men to use proper channels for such complaints.)

People who entertain high hopes for consultation may well wonder just how good a meeting this was; certainly there was nothing dramatic or exciting in the substance of its deliberations: no discussion of financial results, of future developments underground or on the surface, of possible new machines to install or new methods of lighting, or of why a particular face had done so well during the preceding fortnight, or of the markets to which the pit's output had been sent. The meeting was good largely because it was not a clear failure — because the men around the table respected each other, and because the approach to problems was not tactical and partisan but open-minded and problem-centered. These elementary standards — standards that refer only to what occurs in the consultation room — largely determine present judgments about "how well consultation is working." A well-conducted meeting is surely a necessary condition of successful consultation; but, equally surely, it is not a sufficient condition: indeed, the crucial questions concern what has happened before a meeting and what will happen afterwards.

Now let us turn to another meeting with a different flavor — at a different pit, but within the same coalfield.

ROWLEY COLLIERY

This meeting was stormy, with constant wrangling and resort to the tactics of embarrassment, threat, and point-scoring in debate. The pit is located on the outskirts of a Midlands manufacturing city and has a cosmopolitan labor force — that is, men brought in from many other districts. The manager was fairly new, having come to the pit a few months previously from a position as undermanager at a pit about five miles distant; the branch secretary was a colorful Welshman in his middle fifties with considerably more years at the pit than the manager; a vigorous member of the men's side was a Scot with a radical, left-wing temperament of the emotional, not the rational, type. The undermanager was a man older than any of the foregoing, patient and unruffled by the bombast from the branch. These four dominated the meeting.

The manager opened by asking others how they felt about permitting the sale of the NCB magazine *Coal* on the pit top, there being a national campaign to increase the magazine's sale within the industry. The branch secretary was "for" the proposal, saying confidently that "he could sell 100 at Lodge meetings alone if he had the support of the Lodge Committee." He considered the canteen a poor place for sales, as too many men do not use the canteen. The manager feared that if permission were given to sell *Coal* in the colliery yard, then requests would be submitted to sell many other articles. (An inconclusive discussion left the impression that no decision had been reached and that the manager would probably not permit the magazine to be sold in the pit yard.)

The undermanager then asked the men's representatives if there had been any more complaints about men using other men's tools. (An offense, as each man "finds" his own hand tools.) No one had any further complaints.

Discussion on new length of cutting jibs (recently increased by the manager from 3 ft. 6 in. to 4 ft. 6 in.): When asked why the manager had made this change, he stated that he had done so in order to secure better roof control. The branch spokesman said the men don't like it at all and that it would have been better to get the men to take another yard on their stint (i.e., go from 6 to 7 yards) instead of asking them to shift a deeper pile — which means a longer throw to the conveyor. This change, claimed the branch secretary, would cause a loss of face manpower — since the men simply would not work with such a long throw. Then there followed a rambling discussion on the general manpower problem, the union secretary claiming that it was "anomalies" such as these that are driving men out of the industry. At this point the undermanager intervened to remind the group that he was interviewing each man who left; he sent out for a notebook in which he had recorded the reasons men gave for leaving. But the branch secretary charged that these were not the real reasons, that the management only asked men the questions it wanted to ask — while the lodge secretary heard of different problems. If the trade union could ask the questions, the explanation of why men were leaving might be very different. The conversation was aggressive, with each side talking in terms of the men's side and management side as though this was the normal state of affairs — as indeed was only too obvious. There was clearly great confusion as to precisely why in fact men were leaving the pit.

The manager then got back to the jib by commenting that the men were "just prejudiced" against the 4 ft. 6 in. jib. The Scot then said he would try to "stop the pit" if the manager persisted in trying to use the longer jib — a threat which the manager asked the Committee's secretary "to please record in the Minutes." The branch secretary then complained that with every change of management at the pit they try to get the 4 ft. 6 in. jib introduced, whereas the men know from experience the long jib would not work (in the present case, it had been on trial only three days). To the manager's claim that he's taken good technical advice on the question from the Agent, the Scot shouted that "you're listening to some technical advisors right here; we're telling you from the point of view of men with practical pick and shovel experience!" The branch secretary admitted that there was some place for theory — but only if "coupled with experience." He then added the comment that "the men don't talk to you the way they talk to the Lodge Secretary . . . If the manager will agree to take the Lodge Secretary into the pit when he goes down, the management won't have any trouble."

It was obvious that the manager was not in good control of the meeting and was, in fact, simply "going through the motions" of consultation in order to meet the requirements of his job. Despite vigorous language — though not blasphemous, as the committee secretary was a woman — the manager controlled his temper. A typical remark during the foregoing discussion on the length of the cutting jib was the following from the undermanager: "Mr. Chairman, I hope these gentlemen will be good enough to let the management run the pit." To which the branch secretary replied, "We know they have the right, but we have our organization!"

At the end of the formal agenda, the branch secretary raised three additional items:

The removal of shelters on the pit top: Why had this been done? The secretary wanted to know. The manager claimed that men had been loafing in them and "he would not have it." The secretary claimed that the manager could — if this were true — penalize the men under the Five-Day Week Agreement. The secretary claimed there had been no loafing, that the men had only gone in the shelters to get out of the bad weather. Furthermore, he thought the management ought to provide protective clothing "in accord with the national agreement." There then followed a heated exchange as to whether or not there was a national agreement covering this subject. No one suggested that there be any reference to the printed volumes of NCB-NUM national agreements which have been supplied to each pit. Later the manager indicated that he was aware of a national agreement, but said he thought it applied only to "selected" surface workers.

Grievance hours: Next, the branch secretary asked the manager if some definite time could not be established during which men with grievances could have interviews with either the manager or the undermanager. Considerable quibbling took place on just how many men had been frustrated in their attempts to see the manager recently. ("Three men were waiting when I came into your office last Friday . . ." "There weren't three . . ." and so on.) Discussion followed as to precisely what hours for interviews ought to be painted on the manager's door; the undermanager saying he was too busy to be bound by any specific hours every week, since there might be "an emergency." The branch secretary then stormed that "we won't coöperate until you are prepared to coöperate!" To which the undermanager replied, "If the men want a scrap, I'll give it to them." Branch secretary rejoinder: "Is that a challenge? Put it in the Minutes, Mr. Chairman, put it in the Minutes!" No decision was made on the question of grievance-hours for interviews.

Housing: The branch secretary said that some huts had recently been declared redundant in the city: could not the NCB buy them and put them up on colliery property to house pit workers? The manager agreed to look into the matter.

The meeting, having lasted about an hour and a half, then closed.

Several deficiencies of this meeting are obvious: the lack of decision on matters that admitted of decision (for example, how to distribute *Coal*); the difficulty of establishing whether or not a particular technical question should have been left entirely in management's hands (the cutting jib); the reliance on strike threats by the union officials and on "management's prerogatives" by the manager; the intrusion of grievance-procedure questions into the meetings; and resort to the tactic of embarrassment by threatening to "put it in the Minutes." The meeting had ended as it began, with mistrust and disrespect on both sides, each of which knew that the gloomy drama would be replayed again two weeks hence. Given these personalities and their underlying relationships, joint consultation could scarcely produce agreement on any issue. At pits such as Rowley, there seems little contribution for consultation to make until there is a complete replacement of the leading union and management personalities.

These two examples suggest the range of atmosphere within which joint consultation takes place. What we now need to ask ourselves is this: does the achievement even of good consultation realize the aims which consultation is supposed to serve — a greater degree of participation by the rank-and-file, better communications upwards and downwards in the pit, and a gradual improvement in the pit's labor relations? To answer these questions, we will turn to a general appraisal of the problems raised by joint consultation.

The Main Problems of Consultation

The first six years' experience with joint consultation in the coal industry has shown that this device has worked neither as much mischief as most managers originally feared nor as much magic as its advocates had expected. This estimate would probably apply to the device itself and not merely to its operation in coal.[13] But confining our present attention solely to the experience in coal, I think we may fairly doubt whether joint consultation has successfully reached the heart of the problems of worker-participation, better communications, or the general improvement of union-management relations. We may go even further and question whether this particular approach is inherently capable of making a fundamental contribution to these problems, even given the time for improved performance which all agree is necessary.

THE HOSTILITY OF "LINE" OFFICIALS

By and large, production officials in the industry do not like joint consultation; they go along with it simply because they have to, but rarely are their hearts in it. There are exceptions, of course; but they

[13] See the report by Helen Baker, "Joint Consultation in England — An American's Comments," *Journal of the Institute of Personnel Management,* March–April 1951, pp. 1–9. Miss Baker's point of view is indicated by the following quotation: "Four months' observation . . . did not make me an enthusiastic advocate of such plans. It did, however, give me a fuller appreciation of the conditions in Great Britain which created the need for and gave greater emphasis to joint consultation in the postwar years and influenced its particular form" (p. 1). She goes on to enumerate "four outstanding weaknesses . . . (i) the frequent failure of joint committees to be successfully integrated with established inter-management and supervisor-worker relations; (ii) a similar but perhaps less frequent failure in the integration of joint consultation and union-management relations; (iii) the continuation of radically different concepts among managers and workers' representatives of the desirable and ultimate scope of joint consultation; and (iv) the failure to interest the mass of workers and to create a sense of participation in a common enterprise throughout the department, plant, or company" (p. 7).

are not common. Not all production officials are militantly opposed to consultation (though there has been strong, active opposition in some coalfields — for example, Scotland in 1949); but it remains true that the single most important source of enthusiasm and leadership on which the success of consultation depends frequently fails to furnish that enthusiasm and leadership.

There are several explanations for the attitude of grudging toleration with which so many pit managers (and higher Area and Divisional production officers) conduct their consultative meetings: (a) Pit managers work extremely hard, with long hours, great responsibility and very modest salaries. Any activity which impinges on their time must seem fruitful in order to seem worth while. In too many cases, the two hours of consultation every fortnight seem a pure waste of time. (b) Many managers (especially those of working-class origin) have won their positions by virtue of long study in their leisure time; the attainment of their position constitutes a personal victory and testifies to a degree of knowledge which the Coal Mines Act recognizes as belonging only to them. It is easy to feel that the lay criticism which consultation often means, explicitly or implicitly, threatens their hard-won status. (c) These facts make some pit managers jealous of sharing their superior knowledge with laymen. A manager who has denied himself beer, football, and cinemas during his younger years in order to become a manager may not always accept gratefully the "lay" suggestions or criticism of workmen's representatives who have not sacrificed to get ahead.

THE "COMMUNICATIONS" PROBLEM

Furthermore, the method of disseminating information about the work of pit committees, even where the work is fruitful, has not yet been worked out satisfactorily. It is undoubtedly true that consultation has improved communications somewhat, upwards and downwards, as compared with previous conditions. But it may be doubted whether the improvement is substantial enough to justify the huge effort that has gone into consultation. This problem is, of course, central to modern industry; it is not peculiar to joint consultation nor to coal mining. Nevertheless, communications are particularly difficult in mining and particularly important to the success of joint consultation. This is a problem which requires much detailed study and experiment; it is sufficient here merely to note that no dramatic or ingenious developments seem to have occurred during the first few years. This difficulty seems also to be present in the work of the postwar "enterprise councils" which are the Swedish counterparts of Britain's consultative machinery. Many of the comments made here on the experience in British coal seem equally applicable to the Swedish experience.[14]

[14] See Charles A. Myers, *Industrial Relations in Sweden,* chap. v on "Labor-Management Committees (Enterprise Councils)," Cambridge, 1951, 112 pp.

Sam Watson, in his *Annual Report for 1949*, had this to say about the communications problem: "The principal bottleneck in the rhythm of consultation is that of channeling the decisions and spirit of the Consultative Committee to the workmen and officials in the pit." He felt that the CCC "should be looked upon as a parent body to more effective working parties of Consultative Sub-Committees" and went on to outline the type of subcommittee he had in mind. He noted that "there is a big field of experimentation in Managerial function and responsibility in the Mining Industry . . . Where we have good Management (and we mean 'good' in the sense of understanding the vital problem of human relationships as well as mastery of production technique) we have the basis for industrial democracy, and what we need to add in order to assist and play our part in helping such Managements is to accept our responsibility as a Trade Union, and work together with the Management, on the basis of 'We' and not 'They.'" Thus the NUM, he felt, has a responsibility for inculcating more coöperative attitudes among its own workmen's representatives, in giving them a minimum of technical education to inform their participation in committee discussion, in seeing that the best qualified union members are elected to the CCC's and not simply the most popular men. But practically no Area Unions of the NUM have done anything along these lines.

THE OVERLAP BETWEEN CONCILIATION AND CONSULTATION

The leading union and management officials at each pit must play two roles. They meet sometimes as bargainers under the conciliation scheme and sometimes as discussants under the consultation scheme. It is often impossible for individuals to make the radical adjustment in attitudes which underlie these roles. The difficulties have been well put by E. V. Day, a Divisional officer with several years' experience of consultation in Yorkshire:

Quite often the meeting of the Consultative Committee follows a "stormy" Pit Meeting under the conciliation scheme. The memory of that meeting is fresh in the minds of both the manager and other members of the committee. It might well be that there was keen bargaining — hard things said on both sides — in short, an unfriendly and hostile atmosphere. In such circumstances, it is somewhat difficult for the same people to sit round a table holding each others' hands and discussing the various problems in a friendly, helpful, and constructive atmosphere.

He goes on to note that if there should occur a significant improvement in the relationships between the parties at pit meetings under the conciliation scheme, "it will have a very important effect upon the work of the Consultative Committees." [15]

[15] "The Strength and Weakness of the Consultative Machinery," an unpublished paper delivered to the conference of NUM branch secretaries at Grantley Hall, Yorkshire, in August 1950.

THE "STAFF" POSITION OF THE COMMITTEES

It is not uncommon for at least some of the workers' representatives on the CCC's to regard their committee as holding "line" rather than "staff" authority. This means that they regard consultation as effective only if the manager invariably takes the committee's advice. However, this misconception is becoming less common as a result of the educational work done by the Coal Board and, to a much smaller degree, by the NUM.

This educational work consists mainly of week-end conferences put on by the Divisional Coal Boards, jointly attended by pit managers and union officials in the Division. The frequency and effectiveness of these "week-end schools," as they are usually called, undoubtedly vary considerably from one Division to another. These "schools" normally consist of lectures by experts on consultation from the London headquarters staff or from Divisional headquarters, by senior production officials of the Division, and by trade-union leaders in the district; but main reliance would be placed on small discussion groups held between lectures. A primary aim of such conferences is to provide an opportunity for union and management officials to meet outside the working environment (they are usually held at one of the local country houses which have passed into public use in recent years) in the hope that representatives can develop a personal respect for each other which they will take back to their pit committees.[16]

"HOPELESS" PITS

At the small but not insignificant minority of pits where consultation has obviously gone badly (where it is "cut-and-thrust all the way"), nobody outside the pit expects to see any improvement until there is a complete change of personalities on both sides. This is not something that either the NCB or the NUM can move in on easily; in most cases death or retirement will be the occasion of change, rather than the transfer of managers or the failure to reëlect union representatives. The industry seems obliged to tolerate a portion of outright failures while it seeks to improve consultation at those pits where it is reasonable to look for improvement. Ironically, it would doubtless be politically impossible to abandon consultation even at these "hopeless" pits: consultation was launched as a mass "shot-gun" marriage under conditions that make "divorce" all but impossible.

JEALOUSY BETWEEN THE CCC AND THE BRANCH

On quite a few questions there is an unavoidable jealousy between the trade-union branch and the CCC. Wherever the latter discusses ques-

[16] A brief, general account of work along these lines begun in Durham in 1948 will be found in Sam Watson's *Annual Report for the Year 1948,* pp. 80–82.

tions which involve union policies on industrial affairs, the branch may almost invariably be expected to secure control over the discussion, either by insisting that the question be formally referred to the branch meeting or by securing indirect control through the mandating of the men's representatives at branch meetings. Such control obviously limits the freedom of discussion in the consultative committee. In practice, whenever the branch secretary is persuaded that a question is a union question, it becomes a union question. Though the practice of mandating union spokesmen is not uncommon, it does not seem to constitute a serious abuse of the consultative process: in an industry where collective agreements are often loosely drawn (or are of the administrative rather than the legislative type), this practice is the way the union defines what is to be negotiated and what is to be discussed. Doubtless at some pits "mandating" may be resorted to so frequently that the CCC is little more than a rubber stamp for the branch, but this is not common. An awareness of this practice, however, often enables one to understand what is going on in a consultative meeting when otherwise things make little sense.

DISCIPLINE AND CONSULTATION

A great many consultative committees have fallen from grace in the eyes of their constituents because of their participation in disciplinary cases, particularly absenteeism cases. It might be wise to relieve the committees of all responsibility for disciplinary action. At present, the relation of pit committees to discipline is ambiguous. This means that the pit manager's responsibility for discipline is also ambiguous: both the CCC and the manager are able in many cases to "pass the buck" to each other. It is certainly desirable that there be some "workers' control" over the definition of offenses and the scale of penalties, and a right of review of specific cases. Such control might be much more effectively administered if it were done in accordance with a set of "pit rules" (comparable to the Plant Rules customary in American industry) formulated by management and approved by the local union branch. This would greatly improve administration without in any way giving arbitrary authority to management. It would lift a curse from which a great many pit committees have long suffered. The reason the pit committees have found themselves in this difficult position lies in the intimate relationship between absenteeism and production, in which the committees are chiefly interested. There is no reason why the committees need to be made responsible for controlling absenteeism.

SHORT-CIRCUITING OF NORMAL "LINE" CHANNELS

Not infrequently, the pit committees find themselves discussing routine questions (shortage of particular items on a face, failure of a previous shift to perform certain operations, and so on) which ought to have been raised initially between the affected men and their immediate supervisor. Not only does this waste the committee's time but, more important, it weakens the use of the normal channels through which such matters should go. By the time such matters reach the committee the trouble is often a matter of history, the committee has no function to perform except to note the matter, and the deputy or overman is annoyed that the man did not come to him with the problem instead of taking it to someone on the committee.

The specific problems identified in the preceding section are real ones, but are much less impressive to an observer on the spot than two more general aspects of consultation. These wider aspects are: the inability of joint consultation (1) to substitute for a basic reorientation of management methods in the field of human relations, and (2) to engage men's enthusiasm without an appeal to "interest."

Nearly all the Anglo-American Productivity Teams that visited the United States (including the coalmining team) have noted two major differences between the conduct of British and American industry. These differences are (1) the more conscious cultivation of job-planning, in all its aspects, through the use of management specialists and through greater emphasis upon effective first-line supervision, and (2) the achievement of relatively harmonious human relationships through the use of informal consultation "on the job." [17] It may be said that many British observers have begun to feel that in these two crucial areas they may have something to learn from American practice, however imperfect and uneven this practice may be. The key question is whether the development of better managerial practices may not contribute more to the improvement of human relationships than the development of rep-

[17] Note the following observation on the American coal industry by the Anglo-American Coal Productivity Team: "Satisfactory labour relations in the American mines are due mainly to . . . [inter alia] the intimate informality of 'consultation' on the job, as part of the normal process of management; we consider that the more widespread adoption of a similar approach would add greatly to the value of the existing consultative machinery in Britain." (*Productivity Team Report on Coal,* p. 23.) It is ironic that Britain, characterized by the efficiency of its informal and organic institutions in the field of political democracy, should place so much emphasis upon the highly formalized method of industrial democracy which joint consultation implies. Americans, on the other hand, have a political weakness for pursuing democratic ends in their political life simply by "passing a law" — while in our industrial life we have recognized more widely than Britain the importance of informal, organic relationships. One reason why Britain has invoked the formal approach to "industrial democracy" represented by joint consultation may lie in her traditional social structure: the more rigid social stratification of British life may have led to more "social distance" between the employer and the employed than is true in America. Hence, joint consultation may represent an attempt to span the wider crevices of British society in a leap, in the hope that a changed society will eventually narrow the crevices and permit closer informal relationships.

resentative institutions modeled on political democracy. One major difficulty with joint consultation is that it demands so much of the time and energy of pit managers and higher officials that the task of improving line management tends to be neglected or lost sight of.

This is not to suggest that joint consultation should be abandoned in favor of an exclusive drive to improve managerial practices. Such a step would be politically impossible; it would look like a confession of failure and an attempt to reëstablish prenationalization habits of management. The improvement of management practices must proceed within a continuing framework of joint consultation that is itself being improved by local experimentation.

Regarding the workers' interest in consultation one wonders whether experimentation with the techniques of consultation can arouse the continuing interest of the rank-and-file, unless they are given a financial "interest" in the committee's work. Joint consultation as currently practiced does not seem to have touched large proportions of the men at any significant number of pits (if any).[18] Rank-and-file participation in trade-union affairs is normally confined to a quite small group of active individuals and not unless large issues (normally involving earnings) are at stake do significantly large numbers of the constituents play an active part in the life of the union; given this experience with trade-union "participation," there is a presumption that similar considerations will apply in operating the machinery of joint consultation. Finally, experiments which seem to have evoked the highest degree of worker-participation in industry have tied this participation to earnings.[19]

The great difficulty in tying consultation to a new system of wage incentives (incentives which would operate over and above the piece-workers' incentives now characteristic of large sections of the industry) would be to reconcile the trade union's desire for "equal pay for equal work" among all pits with the new differences in earnings to which the above suggestion would give rise. The union would have to be able to regard uniform wages as "standard rates" or as the "union rate for the job," while any extra earnings that might arise from the operation of

[18] The extent of the impact of joint consultation on the rank-and-file has been studied in one particular Area of the East Midlands Division by Gerhard Ditz, an American who visited England from about October 1949 until August 1950. A person in the Ministry of Fuel and Power said that one of Ditz's findings was that nearly a fifth of the workmen he questioned were unaware even of the existence of joint consultation.

[19] Note the high degree of participation and the great improvement in internal communications which have been achieved under the so-called Scanlon Plan. This plan combines the technique of "working parties" (which has the advantage of bringing consultation to specific working-places during working hours, through the use of departmental committees) plus the incentive of higher earnings if production is increased. (See "Enterprise for Everyman," by Russell W. Davenport, *Fortune,* January 1950; also George P. Shultz, "Worker Participation on Production Problems," *Personnel,* Nov. 1951.)

joint consultation would have to be regarded as bonuses, subject to periodic fluctuations.

Thus the use of pit production incentives as part of the consultation process would confront the NUM at the outset with a problem that some people foresee the United Steelworkers of America might eventually have to face if the "Scanlon Plan" should be widely adopted in the American steel industry. This problem is the threat which widespread deviations from a "standard rate" pose for the latter, since few people may actually earn it and different patterns of actual earnings tend to become established in people's minds as the true "standard." The adoption of the "Scanlon Plan" as the prevailing method of payment in any industry (instead of as a marginal, exceptional method) involves problems which have not yet arisen in the operation of the plan.

Such an adjustment in union thinking may well be too much to expect, especially within the immediate future, when the union's primary concern is the rationalization and standardization of the wage structure. In addition, a form of wage-bonus scheme was attempted during World War II without success, and it is perhaps hoping too much to expect that any new proposal for bonuses on an individual pit basis would not be discredited at the outset by unpleasant memories, no matter how different the two schemes. Nevertheless, this is one important area where experimentation might some day be undertaken, preferably on a local basis. The chief difference between the feasibility of such schemes in coal and in, for example, manufacturing operations is the fact that underground conditions are so likely to change in ways that cannot be foreseen or controlled that an incentive system based on them is likely to become unrealistic.[20]

So far there has not appeared any serious disposition to criticize the consultative machinery as such, even though many persons acknowledge the limited results it has yielded. There are three main views as to what is required if consultation is to yield the results originally hoped of it:

First, many union and management people ascribe the early unproductiveness of the committees to "teething trouble." They feel that "in four or five years, consultation should be making a real contribution." There is thus considerable faith in the mere passage of time, a faith that appears based at least as much on hope as on wisdom. The element of wisdom is the recognition, to quote John Mack of Glasgow University, that "the effective operation of a formal structure of committees depends on the existence of attitudes and practices favourable to consultation throughout the undertaking . . . It may be and usually is necessary to set up committees before changing attitudes and informal organization, but the former will only begin to work when the latter has

[20] For a summary discussion of experience with the wartime bonus scheme, which was on a district, not a pit, basis, see W. H. B. Court, *Coal*, pp. 225–228.

come about." [21] To bring about this change in attitudes — a change often as necessary on the trade union side as on that of management — top union officials must realize the need for a large-scale educational effort "on the part of that minority of trade unionists who are capable of such an effort" (Mack). Even such an ardent supporter of workers' control as Professor G. D. H. Cole has written that "the present opportunities for the development of joint consultation are not properly used, because so few trade unionists know what they really want to make of them." [22] While a few of the Area Unions have made a small beginning on this educational task, there has been no imaginative approach to the problem on a scale large enough to promise much for the future.

Secondly, the NUM will undoubtedly be very reluctant to coöperate in any attempt to deëmphasize the work of the CCC's; many management people, however, may well shift their concern for better communications techniques away from formal consultation and toward better informal consultation through the regular channels of managerial authority. The NUM would undoubtedly welcome such an improvement in managerial practices, but only if it came as a supplement to and not in substitution for the established consultative machinery.

Thirdly, there is some interest among the more thoughtful NUM leaders in decentralizing the present consultative machinery, through the use of subcommittees, to bring it into closer touch with the life of the rank-and-file down in the pit. Some managers may welcome experiments along these lines, but in general production people are not likely to do so; many would feel that "we've trouble enough with the committees we have."

[21] John A. Mack, "More Education for Industrial Democracy," *The Highway,* Dec. 1950, pp. 52–54.
[22] G. D. H. Cole, *Consultation or Joint Management: A Contribution to the Discussion of Industrial Democracy,* Fabian Tract No. 277, Dec. 1949, p. 27.

CHAPTER FIVE

GENERAL WAGE MOVEMENTS

The subject of miners' wages is both fascinating and formidable. This chapter will review general wage developments since nationalization; the next will deal with the various kinds of incentive contracts used for work at the coalface. Our main interest here will be the wage objectives and problems with which the Coal Board and the NUM have struggled since nationalization, the way in which general settlements have been reached and the pressures on the parties in reaching them, and the effect of these wage changes on both geographical differentials and the relative position of miners among Britain's wage earners. It is not possible to understand recent developments without some knowledge of the way in which wartime conditions changed the traditional methods of handling wage questions.

The Traditional "Basis Rates" and "Percentage Additions"

For a generation or so prior to 1921, the wages of both pieceworkers and daywagemen had consisted of two elements, a basis rate and a percentage addition thereto. These two components were retained until the second World War, although the important 1921 Agreement changed the principles by which the percentage addition was calculated. In the last quarter of the nineteenth century, the percentage addition was determined exclusively by the selling price of coal in nearly every district. This yardstick was impersonal and automatic, though there must have been many an argument over which coals to price and the frequency with which price changes ought to be reflected in wage changes. Growing dissatisfaction with this "sliding scale" led to its general abandonment around the turn of the century; thenceforth the percentage addition was determined by collective bargaining through the conciliation boards which existed in each district. These boards were free to take into account not only the price of coal but any other factors they considered relevant. Thus flexibility was substituted for the former objectivity, a change initiated by the Miners' Federation but one not strongly opposed by many owners. However, the miners were not able, until the 1920 Agreement, to put the percentage addition on a national basis — like any union, they wished to eliminate interdistrict competition arising out of different percentage additions in different districts. The Federation did not attempt, however, the formidable task of equalizing basis rates among the different districts.

In the 1921 Agreement, automatic profit-sharing (expressed as a percentage addition to basis rates) within thirteen separate wage districts was substituted for "bargained" percentage additions determined in twenty-two wage districts. The 1921 Agreement also provided a minimum daily wage below which the lowest-paid daywageman could not fall in any of the thirteen wage districts (the minimum varied according to the district). The 1924 Agreement extended this daily minimum to pieceworkers as well as to certain daywagemen not at the bottom end of the wage-ladder.

When the coal trade collapsed in 1925, most districts quickly fell "on the minimum" — a minimum which the owners had to pay but which in most districts was not warranted under the district ascertainment. After the 1926 strike and until the outbreak of the second World War, the "district minimum" and the basis for determining any percentage addition above this minimum were regulated, as the owners wanted, by regional bargaining conducted through district conciliation boards (or by arbitration in accordance with district agreements).

Thus there were two fundamental principles which governed general wage determinations during the interwar years. One principle governed the amount of wages and was composed of two parts (a fixed daily minimum wage and a fluctuating "economic" element that varied with the state of trade), allowing owners and employees to share these vicissitudes in agreed proportions. The second principle, which grows out of the first, governed the method by which changes should occur, that is, in automatic response to objective standards and not by periodic power bargaining. This was accomplished through the quarterly ascertainments, which operated as follows: Each district was normally to pay a premium over the district minimum wage. This premium was expressed as a "percentage" over the minimum. The percentage was fixed by deducting certain agreed costs from the total sales revenue of firms in the district and dividing the remainder into the proportions of 83 per cent for wages, 17 per cent for profits. The percentage which the labor share represented of the district's total minimum wage bill was the percentage which each individual was to receive during the next wage period (usually one or three months) on top of his minimum wage. The determination of industry's proceeds was done by neutral accountants and the process of doing this and arriving at a final percentage was known as "the district ascertainment."

One effect of this system of wage regulation was that everyone's wages fluctuated by similar proportions, so that little encouragement was given to "sectional bargaining." Changes in pieceworkers' rates were regulated by the same rules that governed changes in the earnings of datallers; this must have given more stability to the wage structure than we shall find during and after the second war. Nevertheless, despite the logic and ingenuity of the ascertainment system, the miners became increasingly suspicious and resentful of it.

Wartime Developments

World War II not only upgraded the relative level of mineworkers' earnings (just as it did in the United States), but it also introduced several major changes in the principles by which miners' wages were regulated. These principles, embodied in four successive wartime agreements, have figured prominently in wage bargaining since nationalization. Unlike wartime developments in the United States, "fringe items" did not rise to an important place during the war, although the miners did get a week's vacation with pay from a 1943 arbitration award.

THE 1940 COST-OF-LIVING AGREEMENT

The earliest wartime problem arose from the union's demand for increases to offset the 16 per cent rise in the cost of living which occurred between September 1939 and March 1940. *Ad hoc* flat-rate increases had been agreed to on a national basis late in the autumn of 1939, but no guide to future adjustments was laid down until the signing on March 20, 1940, of an industry-wide agreement called the "War Additions to Wages Agreement."

This agreement provided for the continuation of the traditional district wage arrangements (the ascertainments system was not suspended until 1944, under the last of the four wartime agreements we will review); it also stipulated that any "increases . . . necessary to take account of the special conditions arising out of the war, and particularly the increased cost-of-living, shall be dealt with *on a national basis by means of uniform flat-rate additions*" (italics mine).[1]

---

[1] The text of all national agreements, arbitration awards, and interpretations can be found in the National Coal Board publication, *Memorandum of Agreements, Arbitration Awards and Decisions, Recommendations and Interpretations Relating to National Questions Concerning Wages and Conditions of Employment in the Coalmining Industry of Great Britain* (referred to hereafter as NCB, Memorandum of Agreements). A new volume is published annually and contains a cumulative index covering all agreements, etc., since 1940. The agreement quoted above will be found at pp. 8–9 of *Memorandum of Agreements (I)*.

During the war, then, it was agreed that wages ought to rise at least as much as living costs; the use of flat-rate "War Additions," however, was bound to disturb the customary differentials embedded in the industry's wage structure. The continued application of this 1940 Agreement was a major issue between the Board and the Union until the end of 1950.

THE GREENE AWARD OF JUNE 1942

Wages changed automatically under the 1940 Agreement until a new set of pressures built up in the winter of 1941–42. Miners' relative earnings had not improved during the early years of the war, especially in comparison with wage rates in the large number of war plants that sprang up in mining areas. By the spring of 1942 the MFGB had put forward a demand for a general increase of 4s. per shift and for a national minimum weekly wage of £4 5s. Mindful of growing dissatisfaction in the coalfields, the government set up an unusually distinguished fact-finding board, known as the Greene Tribunal, to assess the advisability of granting the union's claims. The Tribunal's recommendations were accepted by both sides of the industry and by the Government; the principal ones were these:

1. A flat-rate increase of 2s. 6d. per shift for adults (defined as all men over 21 and all underground workers over 18), plus appropriate increases for juveniles.
2. A national minimum weekly wage of £4 3s. for underground adult workers and £3 18s. for surface workers. This differential of 5s. a week between surface and underground minimums represented a premium of 6½ per cent for underground work.
3. The freezing of then-current district ascertainment percentages as minimums but not maximums.
4. An output bonus scheme on top of the foregoing increases and not as a condition of their receipt.

This was the first time in the history of the industry that national weekly minimum rates had been established; hitherto, minimum wages had been on a daily basis and had varied from district to district. This feature of the industry's wage structure has become a permanent fixture though the Greene Tribunal explicitly stated that it could not say how appropriate such an arrangement would be after the war. The output bonus scheme was not to prove a success and was abandoned in 1944.

The "Greene Recommendations" produced a six-month lull on the industry's wage front. The increases it gave raised mining to a much higher relative rank in the country's industries than it had ever enjoyed since the drastic reductions of the early 1920's: mining rose from 59th to 23rd in a list of some 100 industries. Unfortunately, it was to fall back to 40th during the eighteen months following Lord Greene's Award.

THE "FOURTH PORTER AWARD" OF JANUARY 1944

By early 1943 the miners' union had placed before the industry's negotiating body demands for an increase in the national weekly minimums and for a national agreement on overtime rates and practices. The union also wanted an increase in piecework prices, but did not seek a general increase in daywage rates. Since the demands could not be agreed to by the Joint National Negotiating Committee, they went to the National Reference Tribunal (the "Porter Tribunal") for arbitration. The Tribunal's Award, its fourth, contained the following provisions:

1. It raised the national weekly minimums to £5 for underground workers and to £4 10s. for surface workers. The differential between surface and underground work was thereby increased from 5s. to 10s., the percentage being raised from the previous 6½ to 11.

2. The previous low juvenile rates were raised, particularly in the age-range of 18–21, where the old rates had had to be kept low to prevent them from exceeding or too closely approaching the low national minimum rates for adults.

3. Any increase in piece-rates was turned down, the Tribunal declaring that any change of this importance (i.e., changes in piece-rates with no general change in day-rates) should be brought about only as part of a general overhaul of the industry's wage structure.

4. Minimum rates of overtime pay were laid down on a national basis for the first time. These were set at time-and-one-third for overtime during the week and double-time for week-end work and for work on specified holidays. Districts where the overtime rates were already more favorable would, of course, retain their existing practices.

The announcement of the award was met by fairly serious strikes in several districts, particularly in South Wales and Yorkshire. These were attributable to two main problems produced by the award: (1) the considerable disappointment felt in some districts that the Tribunal had not awarded a weekly underground minimum of £6 as the men had been led to expect by the Union's original demand, and (2) the anomalies produced by the telescoping of rates which the higher weekly minimums implied, especially among datal workers. This telescoping had wiped out many long-standing if not very large skill differentials among the lower-paid grades, and the groups who suffered sought to undo the "anomalies" thus created by restoring the differentials — which could only be done through a general rise in daywage rates.

The owners' representatives agreed to relieve these anomalies through appropriate district negotiations provided that the cost of relieving them would be met by the government out of the Coal Charges Account. This special wartime device enabled the industry to be treated as a single financial concern, so that the profits of the good pits were made available for subsidizing the bad pits — very much as has occurred

since nationalization. However, several district agreements were negotiated before the Government had agreed to underwrite this blank check
— and the sharp, steep rises in both price list rates and district daywage
rates (on the order of 10 to 15 per cent) during the two months following the award spread a "Christmas spirit" throughout many coalfields
which was, to say the least, an entirely new approach to the industry's
bargaining habits. The Government did in fact agree that most of the
costs of the Porter Award should be met out of the Coal Charges Account — though it would not extend this privilege to the cost of increased price-lists for pieceworkers (increases which the Porter Tribunal
had explicitly turned down). This immediately threw cold water on the
price-list negotiations then in progress in several districts and managements resumed their custom of hard bargaining, some even withdrawing
offers previously made on the assumption that they would be met out
of the Coal Charges Account. It was this denial of aroused expectations
which was primarily responsible for the strikes which followed the
award.

THE REVISION OF WAGES AGREEMENT, APRIL 1944

When the Government had agreed to meet most of the Porter Award
costs out of the Coal Charges Account, it suggested to both parties
that they sit down, under the leadership of the Minister of Fuel and
Power, to examine the problems involved in an overhaul of the industry's
whole wage structure. The reasons why a thoroughgoing overhaul of the
industry's wage structure was felt necessary at this time were that the
previous wartime settlements had considerably complicated the way in
which a man's wages were built up, had reduced the customary differentials between daywagemen and pieceworkers through the imposition
of flat-rate advances, and had given pieceworkers so many flat-rate increases that a large proportion (roughly a third) of their earnings were
no longer dependent upon effort.

The April 1944 "Revision of Wages" agreement restored pieceworkers'
earnings to their customary relative position and "froze" wages for a
four-year period. The intent was to reëstablish a sensible wage structure
and to reassure mineworkers that their wages would not fall during the
deflation that was widely feared after hostilities ended. However, it was
essential that pieceworkers be given an "escape" clause that would allow
them to renegotiate pit price lists if their working conditions changed.
In practice, pieceworkers were able to force a liberal interpretation of
this provision and by 1946–47 they had established (in nearly all districts) a position of preference over daywagemen that was much greater
than had traditionally prevailed before the war or that had been set in
1944. This situation gave the daywagemen a grievance and the union
a tactic for wage-bargaining after nationalization, bargaining which took

as its keynote an improvement in the position of the "lower-paid men."
Just who these were was left conveniently vague.

The combined effect of the Porter Award in January and of the new
wage agreement in April was to put miners in fourteenth position among
the wage earners of the country (measured by adults' average weekly
earnings). The weekly minimum for underground workers (£5) was
the highest guaranteed minimum in any British industry and that for
surfacemen was higher than the weekly minimum of all but three or
four other groups.

The Revision of Wages agreement was the last one before the Coal
Board and the NUM signed their basic "take-over" agreement of December 5, 1946 — the agreement by which the Board and the union took
over all the existing agreements between the former owners and the
union, including the important 1944 contract. We have emerged, therefore, into the postwar period, and all future discussion of general wage
settlements will take place within the context of nationalization.

Wage Changes since Nationalization

The war had ended with miners' wages governed by the 1944 agreement, which was not due to expire until June 1948 (the earliest date
the six-month "notice of termination" was permitted was December
1947). The main intent, it will be remembered, was not to prevent wages
from rising over this long period but to prevent them from *falling* — it
was widely feared that deflationary postwar forces would threaten the
gains made during the war. There was thus an extended three-year
period, from April 1944 until the signing of the important Five-Day
Week Agreement in April 1947, when no major national or district negotiations of any kind took place. At the time the first wage bargaining
between the Board and the Union took place late in 1947, the Labor
Government, alarmed by rising wages and costs in the face of Britain's
serious trade position, was in the process of developing a national wages
policy that would keep wages down. This policy, which was ushered
in as Sir Stafford Cripps's "wage freeze" of February 1948, lasted until
the end of 1950, and since NUM-NCB bargaining during this period
had to be conducted within the confines of the "wage freeze," some
account of the government's policy is necessary.

The "freeze" was imposed by the publication of a Government White
Paper entitled "Statement of Personal Incomes, Costs and Prices"; this
laid down four general principles:

1. There was to be no government interference with normal collective bargaining.

2. The wage provisions of collective agreements had to be strictly
observed — there could be no pirating of labor.

3. There were to be no increases in individual money incomes until more goods became available for the home market.

4. Exceptions to the "wage freeze" would be made where necessary to man essential industries.

At no time during or after World War II did Great Britain set up any formal machinery, or any explicit policies, for the administration of a wage stabilization program. A wartime order established a system of compulsory arbitration, and this was retained until the summer of 1951, but the arbitrators (who were frequently, as in coal, the regular arbitrators for an industry) were given no specific rules to apply; they were simply bound morally to observe the government's wish that increases be kept to a minimum. Such a procedure ran the inevitable risk that unions which were turned down in their wage demands would lose faith in the arbitration process. Nevertheless, the Trades Union Congress supported the Labor Government's policy for more than two and a half years, though it had no power to compel compliance from affiliated unions.

The effectiveness of the "freeze" did not depend, fortunately, on union forbearance in presenting demands. As *The Economist* wrote in 1950, "The main practical expression of restraint was the unions' acceptance of the long delays, and small results, which the processes of negotiation and arbitration have secured since 1948 . . . This policy has worked well. Its weakness is that of all government by postponement. It requires an increasing effort to produce a constant result." Some measure of the policy's success is indicated by the fact that the government's wage-index rose five points during the nine months preceding and only another five points during the thirty months following adoption of the policy (indeed, in the year following devaluation in September 1949, the index rose by only one point). By the late fall of 1950, however, it was no longer possible to mount the effort that would have been necessary to "hold the line."

Indeed, in January 1950, a tight TUC vote had barely upheld continued endorsement of the "wage freeze"; the favorable outcome was considerably weakened by the fact that the railwaymen, the engineers, and the miners all voted *against* the "freeze."

It is against the foregoing background that the 1947–1951 wage issues of the coal industry must be understood. Since the Coal Board was a Government corporation, it could scarcely act contrary to the Government's official policy. Since the NUM leaders were all anxious not to embarrass the Government, they, too, had pressures on them not to challenge that policy directly. But the miners' leaders — men who live uncommonly close to their membership — were also under strong pressures that pushed in the direction of higher wages. The following pages describe the four major wage settlements from Vesting Day until the end of 1951. There were no wage increases won in 1952 (the union's attempt to secure an increase ended in an arbitration award that denied any rise), but in March 1953, and again in January 1954, wage-increase agreements were concluded which followed closely the pattern set by the four negotiations we shall review.

WAGE OBJECTIVES OF THE NUM AND NCB

When nationalization arrived, neither the NUM nor the Coal Board were satisfied with the wage structure bequeathed to them by history; neither, however, had a clear conception of the direction in which it wanted to move. Nevertheless, by 1946 the union had its eye on two principal objectives, neither of them precisely defined. One of these, contained in the "Miners' Charter," was the following: "5. The average wage standards shall not be permitted to fall below those of any other British industry." Very little weight, I think, deserves to be put on the word "average": the NUM simply had a general urge to "put miners at the top of the list."

The NUM's second broad objective was the development of "a new wages structure." This proposal was loosely debated at the union's 1946 Conference, where the Executive Committee was instructed to prepare a memorandum on what this new structure ought to contain. The kind of considerations the Executive was asked to "take into account" were these:

(a) The substitution of the day-wage system of payment for that of the piece rates system for all classes of piecework.

(b) The necessity of amending the minimum wage arrangements and the provisions governing the payment of the cost of living allowances with a view to the reduction of the adult age from 21 years to 18 years.

(c) That men on the night shift should be paid at the rate of time and a third.

(d) An increase to lower paid day wage workers.

(e) An increase in the Porter weekly guaranteed wage, and that when the memorandum is prepared, arrangements be made for it to be considered by a Special Conference of the Union.[2]

As we shall see, it required three years to prepare even an introductory memorandum on a "new wages structure." Finally, in March 1955, the Board and the Union announced agreement on a new wage structure. Its main characteristics are summarized at the end of this chapter.

As for the Coal Board, it agreed in a general way that "the wages structure of the industry would sooner or later require an overhaul." The big question was whether this "overhaul" could be accomplished through a single piece of surgery or whether it would have to be done

[2] From the text of the 1946 wage resolution, reprinted in the NUM's 1946 *Conference Report*, p. 53.

on a piecemeal basis extending over several years. Initially, the Board expected that a rational new wage structure (or at least agreement on what it ought to be) could be worked out with the union fairly early in their relationship, as a single-shot operation. This would prevent immediate, short-run wage problems from being handled in a way that might prove embarrassing in trying to work out a sensible wage structure later on. In practice, however, the pressure of short-run *ad hoc* wage problems swamped the attention of both the Board and the union, and instead of wage decisions being made to fit into a new wage structure during the first nine years, the new wage structure has drawn heavily on the *de facto* changes which have already been adopted in response to short-run pressures. On balance, the succession of unplanned *ad hoc* settlements probably helped rather than prejudiced the development of an improved wage structure.

INDUSTRY-WIDE WAGE AGREEMENTS: 1947–1952

The first major bargaining issue to arise between the new Coal Board and the NUM was the five-day week. Though many unions agitated for this reform right after the war, only the miners were able to persuade the Government that they could be given it. Negotiations for the five-day week began soon after the Coal Board was appointed in 1946 and extended throughout the winter of 1946–47 (when bad weather gave Britain its worst coal shortage since 1926). The loss of one working day every week, and the granting of six days' pay for five days' work if men worked the full five days, was a controversial and risky step, and the Coal Board did not take it without first securing the approval and encouragement of the Government. The Agreement, which caused considerable difficulties in application and interpretation, was signed in April 1947 and became effective on May 5. Thereafter, the miners' leaders were free to concentrate on wages.

In a speech to delegates to the Union's 1947 Conference, Arthur Horner commented indirectly on a problem which was to dominate the wage picture for the next five years; this was the problem of the "lower-paid workers." Mr. Horner explained the difficulty in these words:

> There is not any doubt that as a result of what happened in 1944 — and as a result of the 16 per cent addition to piecerate earnings — pieceworkers are earning wages far in excess of the day-wage men, and far beyond that which previously they had the ability to earn. I am sure one of our main difficulties with our day-wage men on the surface and underground is not so much the absolute wage as the relation of that wage to pieceworkers' wages in these days as compared with what obtained some time ago.[3]

Union leaders were thus confronted by pressures from their day-wage members, pressures generated in large part by the datallers' feeling that

[3] *NUM Conference Report,* 1947, pp. 69–70.

the pieceworkers (their relatives and neighbors) had "got ahead of them."

The 1947 Conference instructed the National Executive Committee to try to obtain "a substantial increase in the minimum rates and to press for such rates for adults to apply to all workmen of 18 years of age and over." The term "minimum rates" was certainly ambiguous: was this intended to refer to the national weekly minimum, the various district minimums which were sometimes higher than the national minimum, or the minimum rates for each grade? If the latter were not intended, any raising of the national weekly minimum would involve the union once again in a telescoping of the wage structure at its lower end. These ambiguities were explicitly noted in the Executive Committee's Report to the 1948 Conference. By October 1947, the Executive had formulated the following proposals for presentation to the NCB:

An increase in the national minimums to £6 and £5 10s. for underground and surface workers, respectively.
An increase of 3s. 4d. per shift on all district day-wage rates.
An increase in the scale of juvenile rates.
An upward adjustment of piece rates in view of the higher minimum.

The first three points could all logically be brought within a wide construction of the term "minimum rates" — but this could scarcely be said of the last. Once the union petitioned to reopen piece rates, the proposals constituted an application for a general wage increase. The Coal Board doubtless reminded the union that inasmuch as its original application for an increase in minimum rates was prompted by pieceworkers' having got too far ahead of them, it would not put things right to increase piece rates further in the same operation!

It is significant that the resolution was moved by the Nottinghamshire Area, the highest-wage Area in the union. The 30,000 men in the Nottinghamshire Area received nothing from the Greene Award which had originally fixed a national weekly minimum in 1942, nor from the Fourth Porter Award of January 1944, which had raised it. These facts suggest that the intent of the 1947 Resolution was to raise the minimum rates for all jobs — not simply the national minimum which affected the lowest-paid jobs.

This presure to do something for the lower-paid workers resulted in the Agreement of December 18, the main provisions of which were these:

1. The national weekly minimum for five days' work was increased by 15s. for underground adult workers, from £5 to £5 15s.; the surface minimum for adults was increased by 10s., to £5. This differential increase gave men more incentive to accept underground jobs.
2. All underground daywagemen earning less than £6 15s. were granted increases of 2s. 6d. a shift; surface daywagemen earning less than £6 5s. were given 1s. 8d. These increases were not to bring men above certain

stipulated ceiling rates (19s. 10d. and 18s. 2d.), so that some got less than the full 2s. 6d. or 1s. 8d. No daywagemen over the ceilings got anything. This inevitably telescoped the existing differentials among lower-paid workers.

3. The Board and the union agreed to conduct a joint investigation as to which classes of workmen should be termed craftsmen, what names ought to attach to their different jobs, how many grades of craftsmen there ought to be, what wages ought to apply to such grades, and how individuals ought to be graded at each pit. (See below for a discussion of the "Craftsmen's Agreement" of 1948 which grew out of this preliminary action in late 1947.)

The Board refused any increases for pieceworkers, as it had for the higher-paid daywagemen. Worried by frightening losses during its first year, it was unwilling to grant any general wage increase but it was willing to give increases to its lower-paid employees in view of the fact that miners' wages had been frozen since the Fourth Porter Award of 1944, while the wages of other groups in the economy had crept higher. With the agreement signed, both parties hoped that they could now devote more attention to the development of a new wage structure. The logical starting point was to follow up the 1947 Agreement by examining the wages of skilled craftsmen.

The Agreement of December 1947 called for a joint inquiry "with a view to agreeing on the various classes of persons employed at collieries who shall be termed craftsmen and agreeing on the nomenclature of such classes and grading them in appropriate grades." The Board and the union were then to negotiate wages for the grades and the classes of jobs within grades; finally, individual workmen were to be classed and graded on a pit-by-pit basis in each Division. Traditionally, there had been little uniformity among the various coalfields as to which jobs should be regarded as "craftsmen's" jobs and there was very little uniformity among the rates for such jobs from one district to another.

The parties experienced no great difficulty in agreeing on the classes of men who were to be considered "skilled craftsmen." In some of these trades, there were to be two grades (Grades I and II, the latter carrying a lower rate); in others, there was to be only a single grade — Grade II.

When it came to negotiating prices for these jobs, the union had to admit that the Board's offer, of 21s. 0d. for Grade I jobs was in line with the earnings of skilled craftsmen in other industries. The union's problem was not that the Board's offer was low but rather that the adoption of a single national minimum rate would give a fairly substantial increase to the low-wage districts and little or nothing to the high-wage districts (such as Nottinghamshire, Derbyshire, and Leicestershire). The only way the union could "do something" for its high-wage districts was to force the Board to offer a higher national rate, something the Board refused to do. The union then had to decide whether or not to take the case to arbitration. After seeking the advice of its craftsmen's representatives, the union decided (1) to throw the winders' rate into arbitration and (2) to accept the Board's proposals for all other craftsmen. Why did the union adopt different tactics in the two cases? "Because," as Arthur Horner explained to the 1948 Conference just a month after the Craftsmen's Agreement was signed, "there are no other winding-engine-men who can be compared in any other industry. We did not go about the craftsmen because we knew we had reached the stage which could not be surpassed, having regard to what was paid as minimum rates in other industries, for engineers' craftsmen, and the rest of it" (p. 78, *NUM Conference Report*, 1948). The Board had thought that winders ought to be treated the same as Grade I craftsmen; the union plumped for an additional 10d. per shift — and got it from the Porter Tribunal's Eighteenth Award.

In its report to the union's 1948 Conference, the Executive Committee hailed the "Craftsmen's Agreement" as:

the first step towards a national wages structure for the industry; it provides, for the first time, for a national rate for the job, irrespective of the coalfield in which the job is performed, and places the craftsmen employed in the coal-mining industry in a favourable position when compared with craftsmen employed in industry generally.[4]

The rates were not yet national rates — they were national minimums; but by the beginning of 1952, after three successive adjustments of the wage scale, nearly all craftsmen would be on either the national Grade I or Grade II rate because the national minimum would catch up with the minimum of the highest district for these grades. The signing of the 1948 Agreement led to considerable dissatisfaction among the craftsmen of the high-wage coalfields of the East Midlands, since their men got nothing out of it. Some of these men complained that they were inadequately represented among the NUM negotiators (most of whom were ex-miners), but the craftsmen's own sectional representatives denied the charge. At the 1949 Conference, a craftsmen's spokesman tried to require that all future NUM agreements be submitted to the branches for their approval before the agreements could take effect, but Arthur Horner persuaded delegates that such a rule would tie the hands of union negotiators.

The administration of the grading provisions was handled differently in different Divisions. In Durham, for example, automatic standards were developed for assigning men either to Grade I or Grade II; in the West Midlands and Yorkshire, men were classified by less impersonal methods — usually the pit manager and his chief engineer would nominate classifications for specific men and the local branch would try to get as many men up into Grade I as it could. The Durham and East Midlands Divisions, at least, worked out automatic progression

---

[4] *NUM Conference Report*, 1948, p. 231.

standards for moving from Grade II to Grade I; many other Divisions had not by the end of 1950.

The Union's 1948 Conference met in a July chilled by Chancellor Cripps's February "wage freeze"; delegates paid no more attention to wages than to ask the Executive to get on with the drafting of the memorandum on a new wage structure which it had been instructed to prepare two years previously. There were no new wage agreements for miners during that year, but there did arise one issue which was sub-sequently to give great difficulty — the issue of what to do about the industry's cost-of-living agreement in view of the Government's intro-duction of a newly constructed index in the middle of 1947.

Although the NUM approached the Board late in 1947 with a view to revising the 1940 cost-of-living agreement to bring it in line with the new index, neither party was ready to get down to brass tacks on this issue until the end of 1948.[5] The Board's position was that while it was willing to continue the cost-of-living bonus of 2s. 8d. allowed by the level of the index at June 1947 (when the old index was replaced by the new one), it did not wish in future to have any automatic tie between the index and mining wages. The union wanted to retain some kind of automatic link-up with a cost-of-living index — and to treat all cost-of-living increases as independent of any other increases which it might negotiate. This was not, of course, explicit union policy; but subsequent negotiations revealed that this was what the union wanted.

Why was it that the 1940 cost-of-living agreement was allowed to lapse, in effect, when the new index was established in June 1947, when many other industries successfully transferred their escalator clauses to the new index? The reason the matter was not faced promptly may be put down to the extreme pressure on both the Board and the Union during the spring, summer, and autumn of 1947, when so many more important issues occupied the attention of leaders on both sides. Even when the Union did get around to specific negotiations with the Board on the issue in January of 1949, the former asked for more time to study the effect of reapplying the dormant 1940 agreement. Under this agree-ment, it was possible that an increase in the cost-of-living bonus might raise the wages of those at or above the minimum without giving any increase at all to those whose shift rates fell below the national mini-mum. This possibility arose from the method of calculating the amount of "make-up" wage necessary to bring the lower-paid workers up to the national weekly minimum rates. The amount of this make-up was a residual figure, determined by adding the cost-of-living bonus to a man's basic shift rate: the amount by which this sum fell below the shift equivalent of the weekly minimum represented the amount that had to be "made up." If the cost-of-living bonus increased, the amount of

[5] *NCB Report for 1948*, p. 49.

make-up would be decreased by the same amount — and the man would get no increase in pay so long as the increase in the cost-of-living bonus was less than the former amount of "make-up." Since such a result would increase differentials without raising the lower paid, the Union could not come out for a literal application of the cost-of-living agreement at a time when its main concern was to raise the lower-paid and to narrow differentials.

The Union could not abandon the 1940 agreement completely because it knew that the Coal Board did not want to agree to any kind of escalator clause for the industry. Hence the NUM had to look around for a way of "interpreting" the 1940 agreement so as to make it serve a policy quite different from that for which it was originally designed. The Board's attitude — to quote the Union — was that "nothing was due to the workmen under the terms of that Agreement because the basic principle of the Agreement no longer existed." The increases in 1947 far surpassed anything that could have been obtained under the 1940 Agreement, and the Board contended that the 1947 Increases in Wages Agreement had taken into account both the increased cost of living as well as the necessity to raise the wages of mineworkers and improve their status and conditions of life.[6]

By the time of the Union's 1949 Conference in July, its leaders were once again primarily concerned with the gap between pieceworkers' earnings and those of the "lower-paid workers." Obviously the 1947 increases in wages agreement either had failed to close the gap sufficiently (as the parties had thought it would) or the pieceworkers had once again crept ahead by steady pressure on pit price lists. At any rate the main pressure on Union leaders was somehow to secure an increase in the wages of the lower paid — and it would have to attempt this at a time when the policy of the Labor Government was one of strict wage restraint.

No blunt demand for a substantial increase for large numbers of the mining population could be defended before either the Government or public opinion. Hence the determination of the Executive Committee to reëstablish the continuing applicability of the 1940 cost-of-living agreement (which would throw a cloak of legitimacy over its pressure for substantial increases). With considerable resourcefulness, the Committee persuaded its own membership to limit any increases due under the agreement to lower-paid men on the ground that pieceworkers' earnings had gone further ahead of daywagemen's than ever before.

This was the background to the wage resolution passed at the 1949 Conference, a resolution which formally set in motion negotiations which were to last for a year and a half. The resolution read as follows:

[6] *NUM Conference Report,* 1949, p. 254.

This Conference of the National Union of Mineworkers instructs the National Executive Committee to endeavour to secure a substantial increase in minimum rates for all lower-paid workers.[7] [Moved by Scotland; seconded by Cumberland.]

It was fifteen months before an arbitration award finally gave the lower-paid men a small increase (the 23rd Porter Award of October 1950). This meant that the Executive Committtee had not realized the aims of the 1949 wage resolution by the time of the 1950 Conference, where it could do no more than report the history of a year's abortive negotiations with the Coal Board. Nor had the union dared press the issue to arbitration, for fear that the National Reference Tribunal (the "Porter Tribunal") would hand down an award which applied the 1940 cost-of-living agreement literally — which would have produced just the kind of unbalanced wage increase which the union was seeking to avoid.

It is important to realize that the TUC was the main agency through which the government could persuade individual unions to exercise the policy of wage restraint which was the keystone of the government's policy. If the TUC should cease to support such a policy, then the government "had had it," economically, and wages would begin chasing prices in the familiar, futile flight. The question of how long and how effectively the TUC could continue to counsel wage restraint to its constituent unions came to a head in January 1950, when member unions had to vote on a resolution approving the continuation of this policy. The vote went narrowly in favor of continuing restraint — but the miners' union voted against the policy, despite the wishes of its Executive Committee. This action testified to the amount of pressure which had built up in the coalfields for some action on the question of an increase for the lower-paid men.

After a winter of spasmodic attempts to persuade the Board to meet its wishes by a generous interpretation of the old cost-of-living agreement, the NUM received a letter from the Board reiterating its desire to see the agreement formally abandoned; the letter closed with this sentence: If this course were adopted a claim for an increase in wages to the lower-paid workmen could be considered on its merits and with due regard to all relevant circumstances, including the cost-of-living.[8]

The trouble with this suggestion was that it would require the union to give up the cost-of-living agreement without any guarantee that the Board would in fact make some concessions to the lower-paid men; all the Board would guarantee to do was to "consider" their claim "on

[7] *NUM Conference Report,* 1949, p. 184.
[8] Reprinted in the NUM pamphlet, *Wages,* being itself a reprint of Arthur Horner's speech on the history of the union's attempt to secure an increase for the lower-paid workers, delivered at the union's 1950 Conference at Llandudno, North Wales.

its merits." But the union was reluctant to let go of the bird in hand without being absolutely certain there was another one in the bush. The NUM's attitude was one of elementary common sense if interpreted from the point of view of normal bargaining relationships. But some influential miners' leaders felt that the union's relations with the Board were so much better than those of a "normal bargaining relationship" that the union should have "trusted" the Board, let go of the 1940 Agreement, and set to work negotiating directly for the lower-paid in the confidence that the Board would be reasonable and not take advantage of the union's temporarily reduced bargaining strength.

The refusal of the Board to give any hint as to what in fact they would be prepared to do for the lower-paid men before the union surrendered the old agreement could not help making the union reluctant to take a bold socialist plunge into the uncharted sea of faith, especially when it reflected, in the words of its secretary, that "a nationalised industry is a Government agency, and . . . it cannot easily pursue policies which are contrary to the policies of the Government."

The union had not become very specific in its negotiations with the Board until the middle of May 1950. At that time it put forward a specific claim for 2s. 6d. per shift, or 12s. 6d. per five-day week for all adult daywagemen — the union's bargaining definition of "lower-paid men." This would amount to a 10 per cent increase for a man on the underground weekly minimum of 115s. and about 11.5 per cent for a surface worker on the minimum of 100s. There was no clear information available as to how many persons were actually on the minimum, so that the cost of the more likely increase in the national weekly minimum alone was not publicly known. The Board, as expected, turned down this initial claim, and the union subsequently reduced its demand to 2s. per shift. The Board then turned down flatly a claim for any increase — and the refusal was delivered to the union's Executive Committee on the day before the annual Conference opened at Llandudno during the first week in July.

The Board's guiding consideration at this point was to wipe out the remaining £12 millions of its huge 1947 debt and to begin building up a surplus against a "rainy day." The Board had just reported, with the publication of its *1949 Annual Report* late in June, that it had made a handsome profit during 1949 and it was known that 1950 had continued a rate of profit that would have cleared the debt by the year-end if there were no further wage increases.

At that Conference, Ebby Edwards, NCB member in charge of labor relations and for thirteen years General Secretary of the miners' union, said that the union's proposal would give 2s. a day to all underground workers earning up to £7 14s. for five shifts and up to £7 3s. for surface workers. "It means," he pointed out, "that 83.5 per cent of all workers would benefit. It is not an application for the lower-paid men; it is an application for a general wage increase and it would cost around £8,000,000 a year." In a speech which was hard-hitting (though delivered and received in good temper), Mr. Edwards went on to point out that since nationalization the Board had granted concessions and improvements to the men in the industry costing nearly £125,000,000 a year. "If goodwill can be bought, the Coal Board have paid for it — sometimes out of future production that we did not get." He then tried to shame the delegates into a sense of perspective by reminding them of the persistence of unjustifiably high absenteeism and unofficial stoppages — an unveiled hint that responsibility for the industry's future must cut both ways.

The Conference renewed its instructions to press on for the increase, but when negotiations were resumed with the Board late in July, the Board once again turned down the substance of the union's claim, but did put forward some very modest proposals for lifting up the "lowest-paid" workers.

This modest offer from the Board late in the summer of 1950 was, as the union put it, "the first break we had had from the National Coal Board since we were called — the Coal Board and ourselves — to a joint meeting in October, 1948, when Mr. Hugh Gaitskell, then Minister of Fuel and Power, informed both of us that there could be no further concession to the miners unless it came about as a consequence of increased output. We have to bear in mind that, let the National Coal Board claim independence as much as it likes, it is a national organisation acting on behalf of the Government of the day, and it is inconceivable that an organisation such as the National Coal Board can pursue a policy which is diametrically opposed to the policy of the Labour Movement, which determines that of the Government of the day." [9]

Nevertheless the Board's proposal was not acceptable to the union, which then decided to put its case before the Porter Tribunal.

The National Reference Tribunal's award in October did not bring the long dispute over lower-paid workers to a successful conclusion. Indeed, the award proved so unsatisfactory that a much larger increase was negotiated by the Board and the Union only three months later, in January 1951. What did the Porter Award contain and why was it so unsatisfactory?

The Tribunal simply told the Board and the Union that they could have a global sum of "approximately £3½ millions" for distribution to lower-paid men. It suggested some extremely broad and ambiguous principles to govern the parties in determining just how this sum should be allocated, but it left the details wholly up to the parties themselves. This was, to say the least, "a new principle in wage awards," as the

[9] Arthur Horner to the York Conference of Nov. 2, 1950, *Report*, p. 7.

conservative *Daily Telegraph* commented; the award was roundly criticized in the press. The Tribunal had quite obviously declined to tackle
the most tricky questions which had been up for decision — the questions of "how much" to "how many." But the award at least created a
defined area within which the Board and the Union could bargain, and
the split-up of the global sum was made within a fortnight.

The settlement established higher national weekly minimum rates:
both the underground and surface rates were raised 5s., bringing the
adult underground minimum to £6, the surface minimum to £5 5s. In
order to maintain traditional differentials "to some extent," men underground earning under 24s. per shift (22s. 4d. on the surface) were given
6d. or such lesser amount as would not throw them over the ceilings
just named. Men above those ceilings got nothing — and there were
very few men in the Midlands and South Yorkshire coalfields who were
not already above those ceilings. The existing cost-of-living bonus of
2s. 8d. per shift was merged into the rates on which overtime would be
calculated, so that overtime pay would rise somewhat for all workers.
Juvenile rates were also raised slightly. To meet these new increases
(which were estimated to benefit about 225,000 men), the NCB added
nearly £500,000 to the £3½ million awarded by the Tribunal. At a
special delegate conference at York on November 2, the settlement was
ratified though none of the delegates was pleased with it. The absence of
fixed-duration contracts in the industry made the burden of acceptance
much lighter than it would otherwise have been, for there was nothing
to prevent the union from starting fresh negotiations at any time it
wished.

The union's problem, of course, was not simply that its leaders were
disappointed at getting less than they felt they deserved and had expected: as one of them put it, the award also "presented us with a
problem that could create divisions in our own ranks." Nevertheless, the
union could scarcely afford to turn down the award, an action that
would have discredited the whole conciliation machinery and which
would have run the risk of holding out for a larger amount without any
assurance that they could get it or any subsequent settlement would be
retroactive. So the union did what it had to do and grudgingly accepted
the award. But as Horner noted, "We have said the consequences of an
inadequate Award like this will be disastrous to recruitment to this
Industry . . . and it may be that experience and time will have to
teach the authorities the lesson that that is the case."

It did not take much more experience or much more time for events
to validate this convenient prediction. A major coal crisis quickly appeared on the horizon and extraordinary measures had to be taken to
minimize or, if possible, to avert it. The memory of the 3,000,000 people
put out of work by the 1946–47 crisis was not very old and a second

crisis in the life of the Labor Government would be politically explosive and economically disastrous.

The threatening crisis was European, not just British, for the basic reasons behind the threatening shortages in Britain were the same on the Continent. The cause was an unexpectedly large increase in home consumption and a disappointingly low output, two developments which gathered momentum from the end of July until the end of the year.

In the fourteen weeks following the end of July, internal coal consumption rose by an annual rate of 6 million tons, while output dropped by an annual rate of 2½ million. By mid-autumn, Britain was consuming more coal than at any time during the previous thirty years, including the late war years. Booming export industries were more responsible for this upsurge than rearmament resulting from the Korean War, the economic effects of which had not yet begun to be felt.

The lag in coal production was due chiefly to the inexorable drift of manpower from the industry, a drift which had begun in March 1949. This had been realized, of course, but no one had become seriously alarmed until mid-November, when everyone became alarmed at once. In the face of gleeful jeers from critics of nationalization, the Minister of Fuel and Power, Noel-Baker, authorized the Coal Board to import coal as insurance against a hard winter — for weather could make the difference between crisis and no crisis.

The Coal Board called in the full Executive Committee of the NUM late in November to outline the seriousness of the threatened crisis and to discuss ways and means of averting it. A joint union-management committee was set up to see what might be done to decrease the high wastage from the industry and to increase recruitment — for manpower was the crux of the short-run problem of increasing output. By mid-November, manpower was down to 686,400 men, the lowest figure in British mining since about 1900. As we shall see in Chapter X on manpower, the Board's suggested solution of the manpower problem was to import foreign labor. It was also proposed that all recruiting for the Armed Forces in mining areas be stopped, that no more reservists be recalled, that those already in the services be demobilized if they would return to the pits. The NUM's views on the sudden manpower crisis were that wages and conditions in the industry must be made sufficiently attractive to retain and attract native Britons to it and that the importation of foreign labor might constitute a threat to the establishment of such conditions. The union was plainly in about as strong a bargaining position as possible: events were arguing the union's case more strongly than its most eloquent spokesmen.

Between the acceptance of the Porter Tribunal's 23rd Award early in November and the end of the year, the mounting threat of a winter coal crisis (plus grumbling in the districts over the lean award) con-

vinced the NUM Executive Committee that they were justified in returning to the Board with new demands. The union drew up a program
containing sixteen proposals and entered into "informal talks about the
situation and about this programme" with three members of the Coal
Board.

During the early stages of these "informal talks" the Prime Minister, on January 3, 1951, requested the full Executive Committee of the
NUM to meet with him and other members of the Government at No.
10 Downing Street. This meeting was held to impress on the union's
Executive Committee the political importance to the Labor Government
of getting every bit of coal the miners could produce. More specifically,
Mr. Attlee requested the union to coöperate with the Board in producing 3,000,000 more tons of coal by the end of April than had been produced in 1950. He also assured the miners' leaders that their renewed
negotiations with the Board would be continued "in a sympathetic atmosphere" — which seemed to say that the Government would back up
the Board in finding the money to give the miners something more than
the meager award of three months earlier. By way of follow-up, the
Prime Minister sent a letter to every miner in the country stressing the
urgent need for increased coal production.

The NUM readily accepted the Prime Minister's "invitation to bargain" and on January 10 presented the Board with a formal claim for
a pension scheme, a second week's vacation with pay, a general wage
increase for all daywagemen, and other improvements. The general wage
increase and pensions formed the core of the proposals. The Board
promised to give the miners an initial reply the following day; in fact, a
complete agreement was reached the next day, the shortest period for
negotiating an agreement in the history of the industry. The new agreement was to cost the Board another £10 million a year, compared with
the £4 million involved in the November 1950 Agreement. It will be
appreciated that not only was the threatened coal crisis upon the
country, but the government's policy of wage restraint had by now been
officially abandoned.

By the terms of the January settlement, the national minimum rates
were raised, the differential in favor of underground work was increased
slightly, craftsmen and winders got specially large increases, some
anomalies relating to miners' coal were ended, and a contributory pension plan was agreed to "in principle." [10] Pieceworkers, however, got no

[10] Pension details were worked out during 1951 and the scheme took effect January 1, 1952. Membership in the plan is voluntary and as of January 1, 1952, less
than half the 670,000 eligible mineworkers had elected to join (the percentage varied
greatly from one coalfield to another — e.g., in Northumberland 85 per cent had
joined, while in Yorkshire less than 40 per cent were in). This disappointing result
led *The Economist* to refer to the "brief, sad history" of the miners' pension scheme,
a history which "shows once again how out of touch the workers' leaders often are

wage increase and the union's demand for a second week's vacation with pay was turned down because it threatened too great a loss in production. In return for its generous terms, the union pledged itself to the following seven measures, designed to increase coal output in the immediate weeks ahead:

1. The union will do its best to secure the willing acceptance of foreign workers "in every pit where there is at present a shortage of men."

2. The union will "continue to intensify" their efforts to ensure the fullest possible attendance on Saturdays and the working of the greatest number of Saturday shifts.

3. The union will, through its branches, initiate an all-out effort to reduce voluntary absenteeism and to suppress unofficial stoppages.

4. The union will coöperate fully in the effort to ensure the clearing of faces regularly so that the cycle of operations can be completed every twenty-four hours; this means that the union will encourage overtime working wherever necessary to complete the cycle.

5. Pits now working the extra half-hour per day instead of a Saturday shift (mainly Northumberland and Durham pits) will be requested to go over to the working of a Saturday shift, or, alternatively, to work one Saturday in every two as well as the half-hour each day. (The practical effect of this was small, as all Northumbrian pits and 126 of Durham's 144 pits were already working Saturday shifts.)

6. The union will call upon its branches to support fully the reassessment of tasks in accordance with the Five-Day Week Agreement of May 1947.

7. The union will confer with the Northern Divisional Boards to see if the system of "cavilling" customary in the north can be suspended for a six-month period.[11]

Finally, a joint NUM-NCB committee was set up to watch over the progress made on each of the foregoing measures, the committee to have authority to "invite" the union's headquarters to take action at specific branches where performance at any pit left room for improvement.

Several of the pledges made by the union as its part of the January Agreement were simply renewals of pledges which it had made three and one-half years earlier when the five-day week was negotiated — indicating that the execution of the earlier pledges had been disappointing. Sanctions or conditions which attach to the union itself in a general way,

with the workers themselves"; by this comment the writer implied that the mineworkers' behavior had shown this reform to hold a much lower priority in their thinking than their leaders had for years represented it. Such a comment neglects the deliberateness with which British mineworkers approach almost any change affecting their lives, making them take an exasperatingly long time to make up their minds about any new proposal. In 1952 Coal Board and NUM leaders made a strong effort to persuade more men to join the scheme (a higher participation was indeed necessary on actuarial grounds) and by the end of 1952 about 80 per cent of the eligibles had joined.

[11] The quarterly or semiannual cavil (pronounced "kay-ville" in Northumberland) is a "draw" by sets or teams of men for working places during the following period. It means an inevitable upheaval in pit life for a week or two while men get used to their new places and deputies and overmen get accustomd to the new men under them.

and which require positive administrative action by the union, have not proved effective remedies for specific problems confronting the industry: there needs to be a somewhat more realistic appreciation of the inherent limitations of the union as an executive agency which might help management perform its managerial functions and a greater willingness to design conditions which fall on individual employees.

During the weeks following the signing of this January Agreement, there was a marked rise in output and manpower, with the immediately favorable output results attributable almost wholly to the fact that more pits agreed to work on Saturdays and attendance improved at pits already working this voluntary shift.

When the January 1951 settlement was signed, the Board secured from the union a moral commitment not to put forward any new demands until June 1952. The union accepted this commitment with the important reservation that it would not be barred from reopening the wages question if there was "a substantial change in the value of money." [12] Between the signing of the January Agreement and the opening of the union's conference in June, the cost of living rose more than 7 per cent; this gave delegates ample cause to test the Executive Committee's willingness to press the Board for a fresh increase. Scotland introduced a resolution calling for a new weekly underground minimum of £7 10s., "with corresponding adjustments in the wages and rates of all other grades employed in the mining industry." A second resolution asked the Executive Committee to seek an increase of "at least 3s. per shift" in the wages of all pieceworkers. This was the first occasion since nationalization when a Conference resolution had requested action on behalf of the pieceworkers; the constant attention to the wages of the "lower-paid men" (that is, the daywagemen) in national negotiations had generated some feeling that the interests of pieceworkers were being neglected. This misleading impression was fostered by the wholly different methods of negotiating wage changes for the two groups. As Arthur Horner pointed out to delegates, and as we shall see in the next chapter, "The pieceworkers have another avenue for dealing with their problems" and since 1947 "the pieceworkers have, by one means or another, gained twice as much as the daywage workers."

The Executive Committee did not oppose either resolution, but it requested delegates not to bind the Committee to the suggested figure of £7 10s. — it wished more freedom "because we have to build a case" (Mr. Horner). Nevertheless, the Scottish resolution on behalf of the daywagemen was carried unanimously (though no clear decision was reached as to whether it bound the Executive Committee to the suggested figure); the separate resolution for a 3s. increase in pieceworkers' rates was only just carried, by 353,000 to 329,000.

[12] *NUM Conference Report*, 1951, p. 158.

The Executive Committee waited until November before submitting to the Board claims for 23s. per week for underground workers, 20s. for surfacemen, and 3s. per shift for pieceworkers. With manpower once again falling off alarmingly after the encouraging rise six months earlier, the Board and the union were able to reach agreement within a month, and without arbitration. The national weekly minimums were raised 11 per cent for underground workers and 10 per cent for surfacemen (the largest increases since nationalization); juveniles were granted even larger proportionate increases; craftsmen and winders were given equal raises; and pieceworkers, in the first national increase given them since nationalization, received nearly all of the 3s. asked by the union. The pieceworkers' increase was coupled with a special provision designed to limit future changes in their rates to those that could be justified on the grounds of changed methods or conditions of work.

An important part of the settlement was the establishment of a new joint NUM-NCB committee "to examine the whole procedure for the regulation of the wages and conditions of pieceworkers." Piecerates were to be "frozen" until this joint committee reached agreement on a revised procedure or for twelve months, whichever period was shorter.

The task of this committee was: "To consider the existing wage structure of the industry with particular reference to pieceworkers, and to submit a report to the Joint National Negotiating Committee on the procedure which would be most suitable for the future regulation of the wages and conditions of such workers, so as to achieve a more rational wage structure with greater uniformity in wages and emoluments for similar work and effort." [13]

As it turned out, the joint committee did not come up with any agreement before the stabilization period ended in November 1952. One reason the committee had difficulty making progress is that negotiations for a new general wage increase were initiated by the NUM just six months after the December 1951 Agreement had been signed. These fresh negotiations, unlike any that had preceded them, had their origin in national politics: they constituted the union's reaction to the first Tory Budget since the war. This budget included a proposal to reduce the food subsidies from £410 million to £210 million; the government estimated that the resulting increases in retail food prices would add about 1s. 6d. per week to the food costs of each person in a family.

The 1952 negotiations saw a repetition of the events of 1950–51: the union pressed its demands to arbitration; the Porter Tribunal rendered a decision that created more problems than it solved; and the Coal Board had to recognize realities by agreeing to things which had been denied by the arbitrators. Briefly, the union asked the Board in May for a "substantial" increase for all grades — an increase without any of the pre-

[13] *NCB Press Release,* Dec. 19, 1951.

vious "ceilings" which had become so unpopular with the union's higher-paid members and which had telescoped the wage structure. The Board made no offer, and in June the union put forward a specific demand for a 30-shilling (20 per cent) increase in the national weekly minimum and 5s. per day for everybody, without ceilings. The negotiation of wage increases with "ceilings" (i.e., limiting the increases to the lower-paid men) aided the eventual development of a national wage structure, but too many successive such increases created political strains within the union that periodically had to be relieved by negotiating increases in which the higher-paid datallers could participate. The 1952 negotiations was one of these times.

The Board rejected the claim outright and the union sent the case to arbitration early in the fall. When the Porter Tribunal turned the union down flat, even the Coal Board was surprised (there had been a "pattern" of 6–8 shilling increases per week in the economy during 1952). The Tribunal's reasons for denying any increase at all were two-fold: (1) economically, the Coal Board was operating at a loss and the cost of living had not risen significantly since the December 1951 Agreement; and (2) administratively, the Coal Board and the union were not too far away from substantial progress on their new wage structure, and the Tribunal wanted to create more pressure on the parties to get on with this long-delayed task (it was expected that the introduction of the new wage structure would itself be accompanied by wage increases).

The unacceptability of the October Award was made evident by sporadic strikes (especially, and typically, in Scotland) and by widespread proposals in the coalfields that the miners stop working overtime, particularly the voluntary Saturday shift. In December, the Board's position was that it would add 6s. to the national minimum if the union would agree to stabilize most piecerates for another year and would pledge itself then to accept the "Voluntary Saturday Work" agreement for another year (this agreement had been negotiated annually in April; the Board hoped for an advance commitment from the union, since the Board was already negotiating its export contracts for 1953 and the 12 million tons produced by Saturday working was just about equal to a year's exports).

The unacceptability of the Tribunal's October Award led to renewed negotiations, and after considerable "backing and hauling" an agreement was reached in mid-February. By its terms some 320,000 datallers (about twice the number on the minimum) got an increase of 6 shillings per week, and the national weekly minimums were raised to £7 6s. 6d. and £6 7s. 6d. for underground and surface workers, respectively.

This brings to an end our review of the general wage increases negotiated at national level since Vesting Day. We turn now to an examination of the effect of these wage changes on geographical differentials.

Geographical wage differentials have been traditional in the industry, and one of the most difficult questions for the future is the extent to which they ought to be retained. The NUM, like most unions faced with this problem, would like to see geographical differentials abolished, but it knows perfectly well that this could only be done if rates could somehow be raised to the level of the highest-wage district. By 1954, the Coal Board had not announced any clear policy with respect to district differentials, though it was known to feel that any level-up to the highest-wage districts was out of the question. Indeed, in *Plan for Coal*, the Board explicitly assumed that district differentials would persist, though it did not commit itself on any detailed aspects of this issue. The Board hoped that disparities in wage levels within particular localities, brought about by anomalies in piece rates between one pit and another, would disappear; that disparities in wage rates for different classes of work would not undergo substantial changes; and that wages (on mid-1949 values) would rise only when productivity is rising, taking each coalfield separately. It is doubtful if all these differentials can be wholly eliminated. Chart 3 shows us the extent of such differentials for each of the industry's twenty wage districts as of the third quarter of 1952. It is easily seen that geographical differentials are most pronounced among faceworkers and least pronounced among surfacemen. The middle line, for "underground workers," includes the facemen, but its flatter slope reflects the fact that district differentials among day-wage underground workers are less pronounced than among facemen alone.

Chart 4 illustrates the *changes* in geographical differentials among underground workers that occurred during the first six years of nationalization. The broken line shows us how many percentage points above the industry's average earnings per manshift were the earnings in, say, South Derbyshire or Leicestershire in 1947 (to start at the left-hand end). These particular districts, it will be seen, were earning about 21 per cent more than the industry average right after nationalization. As we proceed along this broken line, the percentage-above-average falls until, with Durham, we come to the first of the nine districts where earnings were less than the industry-average. The tiny Bristol-Somerset district was more than 16 per cent below the average early in 1947.

The solid line gives us the same information for the third quarter of 1952. It is readily apparent that there were many districts closer to the average in 1952 than there had been in 1947 (some districts, it will be observed, had pulled away from rather than toward the average). The same phenomenon is apparent with respect to the below-average dis-

Chart 3

District Wage Differentials among Major Groups

(Per Cent by which Average Wages of Faceworkers, All Underground Workers, and Surface Workers in Each District Exceeded or Fell Short of the Average Wages of All Workers in the Industry in Each Group, Third Quarter, 1952)

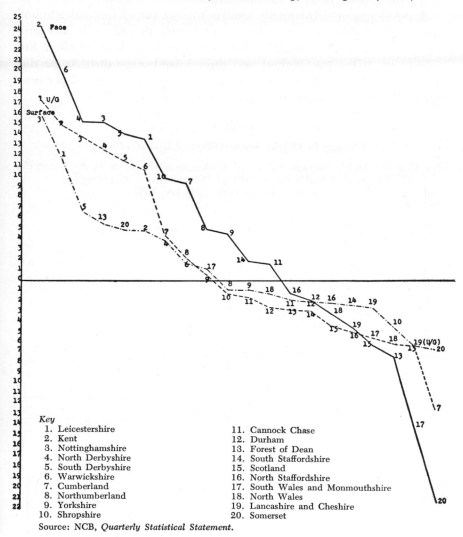

Key
1. Leicestershire
2. Kent
3. Nottinghamshire
4. North Derbyshire
5. South Derbyshire
6. Warwickshire
7. Cumberland
8. Northumberland
9. Yorkshire
10. Shropshire

11. Cannock Chase
12. Durham
13. Forest of Dean
14. South Staffordshire
15. Scotland
16. North Staffordshire
17. South Wales and Monmouthshire
18. North Wales
19. Lancashire and Cheshire
20. Somerset

Source: NCB, *Quarterly Statistical Statement.*

[147]

tricts: all the below-average districts in 1952 lay closer to the average than had been true in 1947. The fact that the solid, or 1952, line lies nearer the average line than does the broken 1947 line for both the above-average and below-average districts indicates that there has been a perceptible narrowing of differentials among the districts. But it is also obvious that geographical differentials had by no means been eliminated during the first six years of nationalization.

A development substantially similar to that found for underground men is observable among surfacemen, though the effect is less pronounced because their traditional differentials have been smaller. For surfacemen, if the two highest and three lowest-wage districts be disregarded, all the remaining fifteen districts lay within a range 5 per cent above or below the industry average as of the end of 1952.

Chart 4

Changes in District Wage Differentials, 1947–1952

(Per Cent by which Average Wages of Underground Workers in Each District Exceeded or Fell Short of Average Wages of All Underground Workers in the Industry, 1947 and 1952)

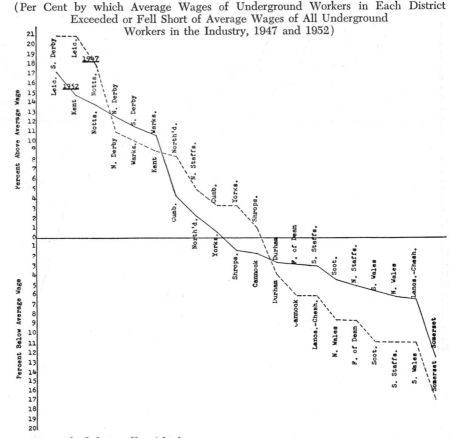

Average for Industry—Unweighted

Source: NCB, *Quarterly Statistical Statements,* Third Quarter, 1947 and 1952.

The narrowing of district differentials has come about almost exclusively as a result of successive increases in the national weekly minimum wage and not as a result of any direct attack on the problem. When Lord Greene rendered the 1942 Award which introduced the principle of the national weekly minimum, he estimated that only about 1 per cent of the underground workers and 5½ per cent of the surface workers in the industry would require an addition to their "earned wages" to be "made up" to the minimum. Since then the proportion of men on the minimum has increased startlingly, until in mid-1953 well over one-half of all day-wagemen were on the minimum. The following remarks of Arthur Horner at the 1951 Conference are instructive:

As a result of the national minima being raised, 66 per cent of the day-wage men underground are on the same rate of wages, and 63 per cent of the surfacemen are on the same rate of wages, doing the same job, and receiving the same pay in whatever part of the country they are working . . . If you take Northumberland, 77 per cent of their underground workers are on the national minimum. In South Derby only 9 per cent of the undeground workers are on the national minimum. In South Wales 78 per cent are on the national minimum, and in Scotland 40 per cent of the underground workers are on the national minimum, but 75 per cent of the surface workers there are on the national minimum.[14]

Any future adjustment of geographical differentials is bound to be greatly influenced by the relative numbers of men in the high-wage and low-wage districts. If there are large numbers of men in the low-wage districts, raising them up to the high-wage districts would be a costly process; this is a factor which weighs most heavily with the Coal Board. On the other hand, if there are substantial numbers of men in the high-wage districts, this defines the levels to which wages must be brought so far as the union is concerned.

In terms of numbers employed, the important low-wage districts (so far as underground men are concerned) are South Wales, Scotland, and Lancashire and Cheshire. It is the union leaders of these districts who seem to be able to summon the most sympathy for "the lower-paid workers" — particularly the Scottish and Welsh delegates. It is worth noting that one huge coalfield, Durham, ranked well below the average in 1947 but today stands modestly above the average: a development like this helps to explain why Will Pearson, the Communist General Secretary of the National Union of Scottish Mineworkers, and Sam Watson, the devoted Labor Party Secretary of the Durham Miners, are more than just political and national rivals. Mr. Pearson asks pointedly why Scottish miners should be asked to keep their wages below Durham's, Northumberland's, and Cumberland's so that the Scottish Division can show a profit, when the same standard is not applied to the three English coal-

[14] *NUM Conference Report*, 1951, p. 151.

fields "across the border" — three districts which have regularly shown *losses* ever since nationalization. If a loss can be sustained in England to permit "fair wages," why not in Scotland, he asks.

The high-wage districts (again taking underground earnings) are all in the Midlands: Nottinghamshire, Leicestershire, North and South Derbyshire, and Warwickshire. Chart 3 shows that the "spread" among these high-wage districts themselves has considerably narrowed since 1947.

How much of a differential is there — in terms of pounds, shillings, and pence — between the high-wage and low-wage districts? In late 1951 there was a difference of about 8s. per shift, which amounts to

Chart 5

District Differentials with Respect to Productivity and Wages: Faceworkers, Third Quarter 1952

(Districts ranked l. to r. by decreasing productivity)

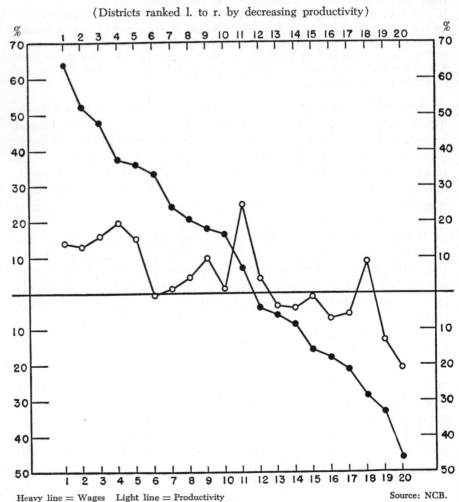

Heavy line = Wages   Light line = Productivity   Source: NCB.

£2 9s. for a five-day week. In percentage terms, the high-wage districts earn about 25 per cent more than the low-wage.

Wage Differentials and Productivity Differentials. The classic defense of wage differentials is that they correspond to differences in productivity among the districts. Interesting data on the extent to which this was true at the end of 1952 is shown in Chart 5, which compares differences in wages and output per manshift for the faceworkers in each district.

The heavy line represents a ranking of the twenty wage districts according to the per cent by which the productivity of their facemen exceeded or fell short of average productivity for all facemen in the industry. The light line plots the relative position of each of these same districts with respect to average earnings per manshift. The districts at any point in the horizontal axis are the same for both lines. Thus in the district where facemen's output is nearly 65 per cent above the industry's average, facemen's earnings are only about 15 per cent above the average earnings of all facemen in the industry. As the two lines are read along to the right together, it will at once be noted that there is no district correlation between the productivity ranking of particular districts and the relative wage position enjoyed by its facemen. All one can say is that in general the districts where face productivity is high yield facemen higher-than-average earnings; in districts where productivity is lower than average, earnings are usually (though not invariably — witness District No. 18) lower than average. The earnings differentials, it will be noted, are much less pronounced than the productivity differentials. The lack of a direct correlation between the wage position of a particular district and its productivity standing suggests that there could be some substantial reshuffling of wage levels among facemen in different districts without doing violence to an "ideal" relationship between productivity and wages; indeed, it could be argued that some reshuffling is necessary in order to improve on current relationships.

THE RELATIVE STANDING OF MINERS' WAGES

It is not easy to make a meaningful comparison between the wages of miners and other groups in the economy. This difficulty rarely discourages people from making such comparisons, comparisons invariably made on the basis of average weekly earnings of all workers in the industry.[15] Comparisons made on this basis ought to be taken with a grain of salt as far as British mining is concerned.

---

[15] A somewhat more sophisticated comparison was made by the Porter Tribunal in its Award of October 1952, denying any wage increase for the industry. After showing that the national minimum weekly wage had increased, between 1947 and 1952, at just about the same rate as the cost of living, the Tribunal went on to quote unchallenged Coal Board figures which showed that pieceworkers and day-wage mineworkers received higher hourly earnings than comparable grades in other heavy industries. See NCB, Memorandum of Agreements (VII), p. 459.

The greatest difficulty is the lack of accurate, full data, a lack which prevents any satisfactory comparison except on the basis of average weekly earnings, figures of which are published semiannually for one hundred industries by the Ministry of Labour.

If we compare the weekly adult male earnings of coalminers with those of men in all the hundred-odd other industries surveyed semi-annually by the Ministry, miners' cash earnings have indeed led those of all other industries since about 1948. The extent of the miners' lead, however, has not been dramatic: in September 1949, for example, miners enjoyed about a 4 per cent lead over the next highest industries, "motor vehicles and cycles" and "printers — newspapers and periodicals."

This kind of comparison, however, may be more misleading than helpful. In any industry it is normally true that the "average worker" does not receive the industry's "average wage"; but the extent to which an "average earnings" figure is misleading is probably much greater in British mining than in most other industries because of the great disparity in earnings between pieceworkers and daywagemen, who account for about three-quarters of the industry's wage earners. More than half of this 75 per cent earn no more than the national weekly minimum — which may be regarded as the "entry wage" for adults. Probably the most significant economic aspect of wage comparisons is to arrive at some measure of the potential recruiting-power of an industry's wage level, and for this purpose entry wages are much more meaningful than "average earnings." Unfortunately, precise figures are lacking on which this more realistic comparison might be made; however, the whole weight of interview evidence suggests that on this basis mining has not enjoyed any marked advantage over other key industries (such as engineering, shipbuilding, electrical manufacture, the motor trade, or iron and steel). There is even reason to suspect that in certain districts the entry wages of mining have been lower than those of competing industries for much of the postwar period.

Chart 6 gives some notion of the relative wage increases in mining since 1947, with increases recorded in "all other industries." It will be seen that during the six years from 1947 through the miners' wage increase of February 1953 miners did not even quite hold their own. This conclusion, however, must be interpreted with caution because of statistical limitations to the figures used. If Chart 6 is compared with Chart 7 on manpower fluctuations at page 186, a crude comparison between the industry's labor supply and its relative wage levels can be made. While no very close comparison between the two phenomena seems justified because of the number of influences at work, it can at least be said that the fall in mining manpower between 1949 and 1950 was not something which happened in spite of a vigorous use of wage-policy to draw men to mining. Perhaps the government ought to have used wage

Chart 6

Relative Changes in the Miners' Weekly Underground Minimum
and "All Wages" in the British Economy, 1946–1953

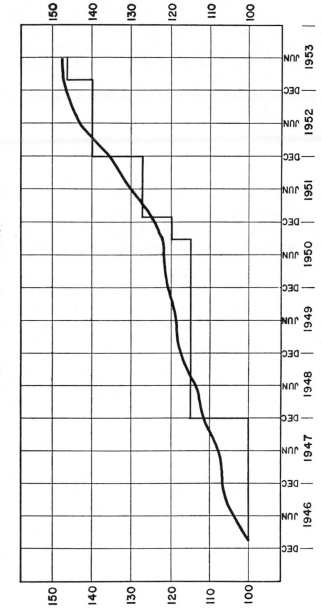

Heavy, smooth line = "All Wages"   Light, stepped line = Miners' u/g weekly minimum

Source: NUM–NCB Agreements and Professor Bowley's index as published in the London and Cambridge Economic Service. The term "All Wages" refers to the standard weekly rates, excluding overtime, of all wage-earners excluding miners.

policy more energetically by raising the weekly minimum in 1949 or 1950 to give mining a definite and continuing advantage. The difficulty with such a policy, of course, is that it would have clashed with the government's policy of "wage restraint" (begun in February 1948): it might well have proved politically impossible to raise miners' wages as a "special case" without inviting upward revisions in many other industries. But this political difficulty was probably not as real as the government's general reluctance to use the price system vigorously to allocate resources where they were required.

THE "NEW WAGE STRUCTURE" OF MARCH 1955

The agreement on a new wages structure early in 1955 after nine years of discussion was a major achievement. At bottom, it represents the completion of an industry-wide job evaluation program for the industry's 400,000 daywagemen: some 6000 job titles were reduced to about 360, which were then grouped into thirteen job classifications or grades. Before prices were agreed for each of these 13 grades, daywagemen at every pit in the country were allocated to an appropriate class in a grading program carried out by pit officials and branch officers. This process, which went forward with remarkably little friction, was completed early in 1955; at that point the Board and the Union could determine how much various alternative rates for each grade would cost the industry.

It is rarely possible to install a job evaluation system (i.e., to rationalize a wage structure) without increasing wage levels at the same time: the essence of the process is that some people get bigger raises than others (and some may get none at all, though this group cannot safely be a large one, else the plan is likely to be rejected). For this reason, the agreement of March 1955 contained not only a new set of national minimum rates for 13 new job classifications but a general wage increase as well: the national weekly minimums for underground and surface work (which nearly 80 per cent of the daywagemen then earned) were raised by about 8 per cent to £8 6s. 6d. and £7 6s. 6d.; respectively.[16] But the significant new element was the naming of national minimum rates for each of the 13 newly-negotiated job classifications. Thus how much of an increase each man got out of the 1955 agreement (and 95 per cent of the daywagemen were estimated to have received something) depended on (1) what he had been earning, (2) the class in which his job fell, and (3) the new national minimum rate for his job class. Because the new national rates are minimums only, they do not abolish completely geographical differentials among daywagemen: many of the 20 wage districts will con-

---

[16] The last minimums cited in the text were the £7 6s. 6d. and £6 6s. 6d. figures negotiated early in 1953. In a routine general increase not reviewed in the text, these figures had been raised to £7 15s. and £6 15s. in January 1954.

tinue to have district rates for all or some of the 13 job grades which exceed the nationally agreed minimums. The agreement thus affected geographical relationships among all daywagemen just the way the 1948 Craftsmen's Agreement had affected geographical differentials among craftsmen: such differentials were not wiped out but they were greatly reduced.

The new wage structure was not meant to apply to daywagemen alone; the 1955 agreement expressed the hope that a similar revision of differentials among the industry's 300,000 or more pieceworkers could proceed swiftly. But the wage problems of pieceworkers are much more complex than those of the daywagemen who have been the subject of this chapter. The next chapter concerns the main problems involved in the incentive wage schemes under which facemen usually work.

CHAPTER SIX

FACEWORKERS' PRICE LISTS

By and large, American mining conditions permit coalface workers to be paid by time rates, while British conditions, like those on the Continent, require that facemen be paid according to the amount of work they perform.[1] The old Miners' Federation had a standing resolution against piecework that it used to pass in "the bad old days," but today the NUM accepts piecework pending closer study of the conditions under which it might be abolished. As a practical matter, it is unlikely that piecework will be supplanted by day rates in any large number of pits except where cutter-loaders can be used (the wages of men working on these machines will be discussed in Chapter Ten). This chapter is devoted to a description of the common types of piecework contracts under which the overwhelming proportion of faceworkers are today paid — contracts which have evolved from long attempts to pay men in accordance with the amount of work they do. These contracts, usually known as pit price lists, are negotiated individually at each pit by the pit manager and branch officials.

The Transition from "Hand-Got" to Mechanized Working

Under the classical system of "hand-got" working which prevailed until conveyors and cutting machines came in during the 1920's and 1930's, the individual collier had to be a man of many skills, and there was a much smaller subdivision of labor than has since developed. The collier was customarily responsible not only for winning and getting the coal in his own working "place" (the section of a longwall face or a room where he worked with not more than two to five helpers) and loading this into tubs, but he was also responsible for setting his own

---

[1] A general discussion of the methods of paying facemen in continental mining is contained in the International Labor Organization's study, *Productivity in Coal Mines*, Coal Mines Committee, Fourth Session, Report III, Geneva, May 1951 (see pp. 84–115).

timber, doing his own ripping, extending the tub track in his own "gate," and other "deadwork."

Under this system of work (which has now all but died out in most districts), a price list specified either a tonnage rate for coal sufficiently high to cover all these operations, or a lower rate for getting and loading the coal alone, plus additional "allowances" for setting timber, ripping, plate-laying, and other "deadwork." Under such a system, each collier was in effect "working for himself" and his weekly earnings did not depend upon the work of other colliers working neighboring "places." When individual colliers filled directly into tubs, it was a simple matter to credit him directly with the weight of coal he produced. Colliers or hewers would normally have had one or two "putters" assigned to them — lads responsible for bringing empty tubs to the colliers' place and taking them out to the main haulage when filled. The collier would put a "token" or "mottie" on the tub so that it could be credited to his account at the bank; the putter, also on piecework, likewise put his token on the tubs he handled. If the particular contract required that the collier help the putter, he normally got paid for this piece of his total task. Both hewers and putters normally had a minimum rate to protect their earnings in bad places.

There inevitably grew up all sorts of informal work-practices by which men could squeeze something more out of their formal contracts than had been intended when they were negotiated. A faceworker in Northumberland, who was a putter under hand-got methods before World War II, described how the putters frequently used their idle moments to scramble off into an idle collier's "place" to "get themselves a tub of coals" — on which they would get their collier to put his token and the putter would then collect from the collier on pay day. When two putters worked in adjoining "bad places" that did not permit them to earn more than the minimum, one of them used to take credit for some of the other's tubs in order to earn something over the minimum (the second boy would of course be paid his minimum). This practice, called "frauding" or "snaffling" in Northumberland, was common there; it was illegal, of course, and the men were subject to company discipline if caught.

The particular form that face mechanization has taken in Britain since the first World War (machine-cutting and hand-filling onto conveyors on longwall faces) has greatly increased the degree of specialization and subdivision of labor. As described in Chapter One, there are now customarily found the following seven different classes of facemen: drillers or borers; cutters or machinemen; fillers or strippers; packers (sometimes called "coggers"); wastemen; conveyor-shifters, pan-turners or "flitters"; and face rippers.

This organization of the facework is not universal, but it is the most common. From the point of view of the amounts of money involved and of group psychology, the "filling price" is much the most important of the lot, and most of our attention will be given to various types of filling contracts. Before turning to this subject, however, we shall devote a few pages to a description of how the other six classes of pieceworkers are commonly paid.

Common Types of Non-Filling Contracts

Even today, considerable secrecy surrounds specific price lists: they are not given out freely to anyone who wishes a copy, though presumably anyone working under the contract is free to have a copy if he wants one. Each class of facemen normally works under its own price list — there are separate "contracts" for drillers, machinemen, packers, wastemen, conveyor-shifters, and face rippers (as well as for strippers). Most of the following examples come from Yorkshire, but many of them are representative of other Midlands districts and very similar provisions would be found in the remaining British coalfields.

DRILLERS

The relative simplicity of drilling prices is evident from the following award of a District Disputes Committee:

It is agreed that the contract price to be paid for boring shot holes in the *coal* with a rotary machine shall be, per dozen holes, 2s. 6d.; this payment to include the work done in dealing with cable or hose which supply power to the rotary boring machine . . . The above price is a base rate and subject to the percentage increase operative in the district . . . Shot holes in the *ripping* shall be put in by the workmen who are responsible for taking the ripping down, the Management to make available to the workmen power machines for boring the shot holes. — Primrose Hill Colliery, March 15, 1948 (italics mine).[2]

It will be noted that a distinction is made, as is frequent, between holes bored in the coalface itself and holes to be bored in the face lip ripped down nightly by the face rippers, who must do their own drilling. While the Coal Board normally supplies drillers and rippers with their initial bits, the workmen are commonly required to buy replacements at cost.

Since drilling in the coal is normally a one-man operation, the incentive operates directly on individuals: no problem of group incentive is involved.

CUTTERS OR "MACHINEMEN"

Machinemen normally work in small cutting teams composed of two or three men, one of whom frequently draws a flat-rate shift payment

[2] Award printed on pp. 36–37 of the *Findings of Disputes Committees,* vol. II (March 5, 1948, to Dec. 31, 1948), Yorkshire Area, NUM (178 pp.).

for being in charge (this bonus may not appear in the cutters' price list itself). Here is a cutting price for the Thorncliffe Seam at Brookhouse Colliery, Yorkshire:

1. To cutting at floor level to the full extent of the jib, setting the necessary roof supports and spragging the cut coal in full compliance with the posted timbering rules, picking and lubricating the machine, cutting the ends correctly and pulling the cables or hose on and off the face . . . per yard, 4½d.

2. Turning the cutting machine round, including removing and setting the necessary supports . . . each time, 6s. 0d.

3. Any work done for which a price is not stated in this list shall be paid for by a mutual agreement between the Management and the men.

The above prices are December 1911 basis and are subject to District Percentages.

The above prices, be it noted, represent total amounts paid for the job. This global amount must then be split up among the two or three men in the team who together share the work. The management has not much to say about the division of the total price: custom, perhaps modified by bargaining within the team itself, determines how the amount shall be shared out.

It will be seen that cutting is here, as normally, paid for on a straight yardage basis, with added payments for stipulated "deadwork" (such as turning the machine around).

While most cutting is paid for on a lineal basis, it is not uncommon to find cutting paid for on a tonnage basis. Thus cutters do not determine fully their own earnings: they are paid only for the tons that get credited to the face. This arrangement gives them an interest in the productive performance of the whole face and brings them into a financial dependence upon other classes of facemen — a situation cutters in many districts would not accept.

Sometimes contracts will be more specific as to how the "spragging" (placing wooden wedges under the cut coal to prevent its falling before being "fired") must be done, the care with which the "gummings" must be cleaned out and thrown onto the gob side of the conveyor-track, or as to whether the cutters shall receive added payment for attaching and dismantling a mechanical "gummer" to the cutting machine.

PACKERS AND WASTE DRAWERS

These two operations are often performed by the same men. A typical price — again, an all-in price that must somehow be "shared out" among the individuals in the team — is the following (Brookhouse Colliery, Yorkshire):

1. For building good solid goaf packs with a five-foot turnover (that is, the distance the face advances during one complete cycle) per yard, measured along the face . . . 3s. 0d.

Other depths of turnover, pro rata.

The above price of 3s. per yard to be paid where the seam is within the limits of 4 ft. to 5 ft. measured at the face side of the pack. When the section of the packing, measured as above, is a complete 6 in. above or below the limits specified above, then payment shall be on a pro rata basis.

2. For drawing off all face supports in the waste, per yard . . . 6d.

The waste measurement to be between the goaf side walls of the packs put on by the gate rippers.

3. For setting chocks, each . . . 6d.

    For withdrawing chocks, each . . . 6d.

4. Any work done for which a price is not stated in this list shall be paid for by mutual arrangement between the Management and the men.

The above prices are December, 1911, rates, and are subject to District Percentages.

CONVEYOR-SHIFTERS

Conveyor-shifters, like cuttermen, are normally paid so much per yard of length worked (conveyormen's yardage prices would refer to the length of the conveyor, not the face, though the two would of course be closely related). A price list from Maltby Main Colliery in Yorkshire is quoted below:

(1) For turning over face belts 26 in. wide with fully covered bottom belt structure . . . 9d. per yard, gross

(2) Moving gear heads, including tension — each time . . . 10s. gross

(3) Extending gate belt — each time . . . 8s. gross

These payments include: (a) That the gear heads are pulled over into the new positions, and properly fixed for loading onto the gate belts

(b) That conveyors are built to a straight line

(c) That all faulty buckles are changed (the Deputy in charge to determine the joints to be changed other than the obvious ones)

    Extra joints above two to be paid — single 1s. 6d. gross, double 3s. — gross

(d) That all rivetted buckles are properly rivetted and not put on with rivets bent over and that the joints are put on at right angles to the belt

(e) A section to be changed on each side per day

(f) That all work connected with the conveyors be done in a thorough and efficient manner and that the conveyors are left in running order for filling coal [3]

Since there are many different types of face conveyors, some of which are much harder to "flit" [4] than others, the price depends greatly on the type of conveyor. If the manager decides to install new types of conveyors on a face, he must usually expect to face a claim for a revision of the existing price.

Conveyor-movers usually share out the total price for the job on an

---

[3] NUM (Yorkshire Area), *Findings of Joint Sub-Committees Set up under the Conciliation Machinery*, March 19, 1947 to April 29, 1948, p. 3.

[4] Colloquially, the verbs "to shift" or "to flit" mean "to move" in many coalfields; the terms are often used when speaking of moving a residence.

equal basis (as in other small teams, there may be a chargehand who receives a shilling or two more than his mates, the money perhaps coming out of the contract). The flitters' contract often specifies the price they are to receive if they must fill off some coal remaining on the face after the strippers leave; it may be paid for at the strippers' rate and be charged against the stripping contract.

FACE RIPPERS

These are the men who, working in teams of four to six, attend to the daily extension of the junctions between the "gates" and the coalface. Since this area is likely to be disturbed, it is one where particular attention must be paid to safety. A ripping price from Yorkshire reads as follows:

1. For ripping at the face ripping lips, including the boring of all shot holes, the removal of all supports under the ripping lip and in the gate side pack holes, the setting of temporary supports and the disposal of all dirt into solid, well built packs to be paid at . . . per cubic ft. ripped, 2⅛d.

Any packing in excess of 5 yards on each side of the gate to be paid for at the rate specified in the goaf packing price list.

The Company to provide all explosives, drilling machines and boring drills.

2. For moving forward the cable tube, cover table and scaffold a payment shall be made each turnover . . . 4s. 0d.

3. For setting, strutting and covering in steel arches with either timber laggings or corrugated sheeting, each arch to be set on foot-blocks or stilts, the price shall be:

Heavy section steel arches 15 ft. wide by 10½ ft. high . . . 7s. 6d. each.
Heavy section steel arches 13 ft. wide by 9½ ft. high . . . 6s. 6d. each.

4. Any work done for which a price is not stated in this list shall be paid for by mutual arrangement between the Management and the men.

The above prices are December, 1911, basis rates and are subject to District Percentages.

Thus, rippers are frequently paid according to the cubic measure of the material they take down, plus flat rates for separate "pieces" of work. The actual measurement would consist of a measurement of the average thickness of material taken down, multiplied by the yardage by which the gate was extended and the sectional measure of the roadway.

Work Loads: The Size of Face Teams

The above sketch of how the main face groups other than fillers are paid has indicated that in most cases the contract consists of an "all-in" price for the particular job in question. The earnings of individual facemen thus depend upon how many of them are in the contract. This question naturally arises during the negotiation of any new price list and forms part of the bargain. The manager is always anxious to have work performed by as few men as possible, and he will normally seek to trade off higher earnings for fewer men; that is, if the men insist upon a higher manning of the job than he thinks necessary, he will usually ask that the extra man be paid for out of the contract. The argument is usually over the addition or elimination of only one man, as work standards in a district are fairly standardized and would not change by a large proportion where reasonably small work-teams of no more than ten or fifteen men are involved. In the end, it is the manager's refusal to pay more than a certain amount to get a particular job done and the men's refusal to take less than a certain individual amount out of a total contract price which combine to determine how many men there shall be in the contract.

In conclusion, it is worth noting that the problem of "work standards" on which depends a great part of the incentive of the "service" groups of facemen (that is, the nonstrippers) is different from that found in the incentive wage plans of many manufacturing industries: in the latter, small differences of individual effort can lead to small differences in earnings — the relationship between work and earnings is related by a "continuous function." This is not true of the piecework contracts held by the face groups so far described: the jobs are not variable as to their output because so much packing, no more and no less, has to be done every night; one conveyor has to be moved over and it is either moved over or not; and so on. Thus, bargaining over the prices for these jobs is more akin to bargaining between an employer and a subcontractor in building construction than to bargaining over piecework prices in shoemaking or metalworking. Indeed, many of these jobs have been handled on just such a basis in years gone by: colliery managers would negotiate a contract with a single individual, known in many districts as a "butty" and in some as a "puffler," who would take responsibility for driving a road, or taking care of the packing, or doing the face ripping. The "butty" would then recruit his own men from the pit's work force (or from his friends in the village) and make whatever arrangements with them he could. The "butty" then collected the week's pay for the job (an amount often kept secret from the men) and paid off his men personally. This system made many "butties" into working-class robber barons, and led to much inequality of treatment among different men and much petty racketeering; it has been effectively stamped out in recent years.

The Filling or Stripping Contract

We come now to the most interesting and important of the facemen's contracts: the price for filling off or stripping the face. The importance of this particular contract arises primarily from the large number of men involved. At bottom, the problem of negotiating a contract for filling coal is a problem in reconciling an individual's contribution to the work of a large team: the difficulty is that the more directly a contract appeals to individual performance, the lower is likely to be the performance of the whole group; and, conversely, the more the contract hinges on total group performance, the lower is likely to be the performance of each individual in the group.[5] "Filling" contracts fall into two broad groups: either the price is (1) tied directly to individual performance, or (2) relates to the total performance of a group of fillers, in which case it is customarily referred to as an "all-in" contract.

Pit managers pay fillers to produce "tons of saleable coal." However, it is not always possible to tell directly how many tons of coal a man or a group of men or a whole coalface produces in any period; this is because the amounts of coal contributed by any unit of labor is often mixed indistinguishably with that contributed by other units. This difficulty is not always present: wherever men fill directly into tubs which proceed individually to scales (almost invariably placed on the surface), the individual tubs can be credited to the accounts of individual men. But if this situation does not exist (and it does not at a majority of pits today), then an individual's contribution can only be imputed indirectly. There are two ways of making this imputation: (1) by a lineal measure of the yardage cleared by an individual, or (2) by a cubic measure of the volume of coal he extracts. Both measures offer a method for determining the proportion of a total tonnage contributed by one or more of the workers. The *lineal* basis for apportioning an all-in contract whose total earnings are determined on a tonnage basis means that earnings decrease if the seam "thins out" (and a seam can easily lose 10 per cent of its thickness as the face advances). This means that the men's pay would go down, even though they cleared their normal stints each day.

Let us trace through how these two key problems — the problems of incentive and of measurement — find expression in different types of stripping contracts.

INDIVIDUAL INCENTIVES

The most common method of individual incentive is to express the contract in terms of the number of yards of face which each individual stripper is responsible for filling off. These lineal measures, or "stints," may be the same for all men working on the face, or different men may take different lengths. As a rule, younger men will be able to fill off a

---

[5] The best account of the effect of modern longwall working on problems of group relationships at the coalface is by E. L. Trist and K. W. Bamforth, "Some Social and Psychological Consequences of the Longwall Method of Coal-getting," *Human Relations,* vol. iv, no. 1 (1951), pp. 3–38. The article contains many useful insights into the methods of wage payment of particular face groups and of the status-rankings on the face that underlie the wage structure.

longer stint than older men. Also, younger men are more often those whose family responsibilities are somewhat heavier than those of men in their late forties and fifties, and for these two reasons there is sometimes a coincidence of interest between both younger and older men in splitting up the face into variable stints which will yield earnings proportional to their length. Where variable stints are formally recognized, all fillers will not earn the same amount during a shift.

However, the arguments in favor of a variable stint do not always appeal to the men on a face even where there is a considerable variation in the age of the men involved. The older men are often found to take the view (and often the younger men readily concede its justice) that some day the younger men will wake up to find that they, too, have become "older men" and that they will then be grateful for help from younger men in the next generation just as they themselves helped out the older men when they were young. Where this psychology prevails, one often finds some form of "all-in" contract under which all share equally in the total earnings of the face, and this may be combined with an informal system of variable stints by which the younger men subsidize the older ones. Not infrequently, men will "buy" each other's stints according to how they feel on a particular day. It is impossible to generalize as to how this common divergence of interest between facemen of differing productivity will express itself in the formal price list. The influence of pit personalities at the time of the contract's negotiation and the subsequent force of custom has, of course, a great deal to do with it.

In addition to the problem of reconciling the sometimes conflicting interests of different facemen when they are paid on an individual basis, there is another shortcoming to such contracts, one which involves supervision. Under individual payment, each stripper comes to expect the deputies on the face (usually one or two) to give him prompt, individual attention so that he can get out as much work as possible. This makes life much more hectic for the deputies than when all strippers have an interest in the fair and orderly servicing of the whole face, getting supplies in, sending for the shot-firer, chasing tools, and so on. Furthermore, under individual contracts, men with places near the gates are in a favored position to grab supplies as they come onto the face, thus causing friction with others who may have felt they deserved what the best-placed men took. A yardage contract, therefore, often results in teamwork so poor that it negates the greater individual incentive it is designed to provide. Finally, yardage contracts sometimes involve deputies in so much "measuring up" work for recording individual performance that they are not able to devote as much time as they ought to their primary responsibility — safety.

"All-in" contracts include a great variety of arrangements. Fundamentally, the term is applicable to any contract which specifies a price for work done by a group of individuals, so that the share of each individual must somehow be determined. The two most common types are those which measure the total work performed by those in the contract either by (1) the total tonnage of coal which comes off the face, or (2) a cubic measurement of the volume of coal cleared from the face during, usually, a week. Our attention will be limited to longwall contracts, although "all-in" arrangements are found in the room-and-pillar mining on the northeast coast.

"Tonnage" vs. "Cubic Measure" Contracts. An "all-in" tonnage contract specifies a price per ton for all coal coming off the face, the coal being loaded off the face or gate conveyors into tubs which bear a mark indicating the face from which they have come. This makes possible the direct allocation of a specific number of tons to specific faces when the coal is weighed at bank, though not to specific individuals.

Under any system of tonnage payment the fillers have an interest in maximizing their gross tonnage. The manager has an equal interest in maximizing the net tonnage of salable coal which comes out of the pit; he is naturally anxious not to include in the tonnage for which fillers are to be paid any "dirt," which not only yields no revenue but entails additional disposal costs. Thus, there exists an inevitable conflict between the fillers and the manager over the question of "dirt"; the fillers know that dirt weighs as much as coal and all they want to do is maximize their tonnage, leaving it up to management to eliminate the dirt in calculating the tonnage of salable coal sent out of the pit. The difficulty is that the physical process of separating unsalable dirt from the salable coal has to occur after the weighing of the tubs. Consequently, some method has to be devised for determining the dirt content of the tubs at the time they are weighed, and the only way this can be done is by some rough-and-ready guesswork. This problem is usually handled by pit "dirt agreements" for inspection of the tubs once a month. These may be either verbal or, commonly, written agreements separate from the regular price lists. The price lists themselves for coal filling frequently contain clauses specifying that the payment shall be made only for "clean coal." But this does little more than establish management's right to protest against dirty coal.

At some Yorkshire pits there is a special man called a "muck clotcher," whose job it is, in coöperation with a man selected by the company, to overturn a specified number of tubs sometime during each shift and to determine on the basis of this sample the proportion of dirt in the coal being sent out of the pit for that day. This proportion would then be deducted from all tubs weighed during the day. The job was one often given to trade-union officials, since it is not a time-consuming one and leaves them free to attend to union business during the regular shift. The job, however, is gradually disappearing.

Despite the difficulty of satisfactorily coping with the dirt problem, either through prevention by attempting to incorporate incentives for the production of clean coal only or by penalizing or disciplining individuals if dirty coal has been filled, there are very considerable advantages in keeping the payment of coal fillers on a tonnage rather than a lineal or cubic measure. Under any contract where payment is related solely to the lineal distance or cubic measure of face cleared, the fillers' sole concern is to clear his place of coal: he has no incentive to see that as much coal as possible goes onto the conveyor and as little as possible is lost by being thrown into the "goaf" on the other side.

Nevertheless, there is one particular and increasingly common method of work for which payment by cubic measure is almost inescapable. This is where "concurrent loading" occurs — that is, where the coal coming off separate faces is brought together by different conveyors to a single loading point where the coal is then mixed indistinguishably as it is loaded into tubs. When the tubs are weighed at bank, it is impossible to tell how much tonnage has come from either face. This situation may be handled in either of two ways, both of which involve cubic measurement.

The first method is for each face to be measured up (usually weekly by the deputy) and then for each man to be paid in proportion to his share of the total volume of coal extracted. This is determined, of course, by the length of stint which each man customarily takes during his shift. Under this method, no account is taken of the tonnage which actually reaches the bank. The second method involves the process of measuring up, but includes a reconciliation of this cubic measure with the combined tonnage from both faces as recorded at bank. The procedure is to credit each face separately with that proportion of the total tonnage which its cubic measure bears to the total cubic measure recorded for both faces taken together. This latter method has the advantage of all methods of payment by weight: it gives the coal fillers an incentive to make sure that they send all the coal out of the pit. However, it sometimes gives rise to disputes between the men on separate faces as to the proper allocation of the global tonnage which they are sharing. The difference of opinion which arises is not usually over any question of fact as to the respective cubic measures excavated on each face but over the conscientiousness of the respective faces in keeping coal out of the "wastes": facemen who feel that those on another face with whom they must share have thrown more coal into the goaf sometimes grumble under the necessity of sharing a tonnage price on the basis of cubic measurement.

The number of concurrent loading schemes has been increasing with the reorganization of underground transport during the past few years, and it is not unusual to find a pit experiencing some difficulty in accepting as fair the methods of wage payment which this development makes inevitable. However, there are few if any instances where these difficulties have been so great as to force the abandonment of this advantageous method of organizing the underground transportation.

Warwickshire "All-In" Contracts. A type of "all-in" contract that differs significantly from those so far discussed has recently become quite common in the Warwickshire coalfield, though it is not confined to that district alone. The interest of this contract lies in the fact that it covers more than the single operation of coal filling, and that it frequently includes a special form of "retroactive incentive" which is not common in any other coalfield.

The general nature of such contracts is illustrated by the following extracts from a new price list negotiated for a machine-cut, hand-filled face at a pit near Coventry:

The next contract includes all workers on "Normal operations" [twelve are explicitly named] on the face, excluding cutters, conveyor movers, wayhead timberers, and congate rippers. These excluded grades are to work under separate contracts . . . All men to share equally in the contract . . . Men shall work such shifts as may be determined by the management to suit the cycle of operations . . . The class of work in the face area which a faceman may be asked to do may be varied from week to week. Men will be required to assist in other work which it is reasonable to expect them to do, or when their own particular task has been completed.

While this "all-in" contract included three of the five principal face operations, it still excluded the machinemen and conveyor movers, who have separate contracts. Men working under the above contract must be more versatile than men who work solely as "fillers," "packers," or "drillers." The principal advantage of the contract is that it allows, on a longwall face, more interchangeability of men among the various face jobs performed on the coaling shift, giving management greater flexibility in manning the necessary jobs and men more variety in their tasks — advantages normally found only in room-and-pillar working. A large number of coal-filling contracts in Warwickshire contain an unusual "retroactive incentive." This arrangement provides firstly for a basic rate per ton (say 3s. 6d. for filling off a specified tonnage — say, x tons). It then provides that 4s. shall be payable if a specified higher tonnage is reached (that is, x plus y tons) and an even higher rate, 4s. 6d., if a still higher tonnage (x plus y plus z tons) is reached. The significant point is that the six-penny steps do not apply only to the extra tonnage alone, but apply to the total tonnage filled off. Consequently, the men know that if they reach the level at which the bonus is brought into play, they receive the bonus on all the tons they have previously produced.

This uncommon type of incentive is seldom employed in any other area of the West Midlands Division and has apparently been developed locally in the Warwickshire coalfield since nationalization. It is well thought of by both Coal Board and trade-union officials in that district; but it met with objection from the Board's Divisional headquarters, where it was felt that the prices already quoted would lead to extremely high tonnage costs. However, the Area Labor Officer was able to forward statistics to Division showing significantly higher output and lower costs where such contracts had been installed, and, as the Area Labor Officer put it, "We didn't hear anything more from them."

"SECTIONALIZED" FACES

In order to get around the lack of teamwork which often accompanies payment on an individual basis and the lack of individual incentive which accompanies an "all-in" contract, the Coal Board in 1949 sought to extend the use of an intermediate type of contract which had been used in some districts (e.g., Yorkshire) before nationalization.

To understand the problem which "sectionalizing" the face tries to overcome, it is important to realize the great weakness of the common type of "all-in" filling contract where all strippers share equally in whatever the face earns during the week: everybody expects everybody else to do the work. This is overstating the tendency; but that there will often be some people who "do not pull their weight" is a very common complaint, and not only from management. "Not pulling one's weight" can cover a multitude of sins, from not exerting oneself unduly when at work to not being overly conscientious about getting to work — in which case the rest of the men in the contract have to take up the slack.

If this happens, the absentee does not get the same payment at the end of the week as those who have worked the full week: the face's total earnings would be divided by the number of manshifts worked on the face during the week and each man would receive this amount times the number of shifts he worked. This is manageable on an occasional basis, but not as a steady diet — which is what it can become if there is a real "rotter" in the group.

Apart from the problem of the slacker, an "all-in" contract shared equally by all (that is, where everyone takes the same stint) tends to make the better men work at something below their full capacity. Not only does it not pay them to do more than the minimum, but it invites criticism and ostracism by the rest of the group. The result is that the equal-shares contract is often a "low productivity" contract, and this is management's main reason for disliking it. The device of "sectionalizing" faces into "panels" or "breaks" is an attempt to get away from the short-comings of the equal-share type of "all-in" contract without going all the way to an individual incentive which has the disadvantages already noted.

"Sectionalizing" means dividing a face of 25 to 40 strippers into a number of "sections" or "panels" which stand as independent accounting units for purposes of wage payment. Each section is worked by a "team" of 4 to 6 fillers who share equally in what they are able to earn as a team. Teams of different size or capacity on a face would all be paid on the same basis (so much per lineal or cubic measure cleared) but they do not receive the same amount. Instead of keeping separate measurements for each individual, the deputy has only to keep a few separate records of work done by each team (this necessity would of course involve an increase in his work if the face had changed over from an "all-in" contract).

The aim of the sectionalized contract is to combine teamwork and individual incentive by designing work-groups sufficiently small so that these two forces reinforce instead of canceling out each other. Ordinarily the separate sections would be formed by "natural" groupings of men who "decide they'd like to take a contract together." Such groups may be found to discipline "dodgers" more effectively than when the latter can lose themselves in the much larger group of an "all-in" contract that includes everyone on the face. Indeed, the advantages of small work groups over large ones are so great that any change (in technology as much as in methods of wage payment) that encourages the breaking up of large groups is usually welcomed by management and most union leaders.

The apparent attractiveness of sectionalized working does not mean that the Coal Board has always found it easy to persuade particular pits to accept it. For example, where men are used to "sharing out the whole contract" for a face on an equal basis, they often regard sectionalization as a divisive threat that will raise some men above others; they may also suspect that the Coal Board hopes to get men to do more work as a result of a change in the method of payment — which of course is true. A major aim of sectionalizing faces is to get away from the tendency for "everyone to work at the pace of the slowest man." Where group ties are strong on a face, the strippers may not entertain a proposal to abandon an "all-in" contract; there have been several pits in Yorkshire where sectionalization has been turned down by the branch committee, although union officials at higher levels do not challenge the Board's preference for this method of payment. It is probably easier to introduce sectionalized contracts on a face that has been accustomed to some kind of individual contracts yielding differential earnings.

On the most important question of all — the effect of this change on productivity — there is not much information. Casual references to the subject by people in Yorkshire indicate that while sectionalization is an improvement over the "all-in" contract, it has not shown any very dramatic results.

Relative Earnings of the Main Grades of Facemen

There is considerable divergence of interest among the major classes of facemen, each of which bargains separately and seeks to maximize its share of the total wages fund represented by the manager's notion of what he can afford to pay, in toto, for coal from a given face. If the packers, say, are able to get a good price out of the manager, this (1) makes the other face teams all the more anxious to improve their own prices "to restore differentials," and (2) it makes the manager all the more anxious to see that they do not get it. One of the main advantages of the Warwickshire multi-team contract described above is that it reduces the amount of "sectional bargaining" among the face teams by bringing several of them under a single contract. This jockeying for position among face groups can make life difficult for those union officials who feel that a responsible wage policy requires considerable stability to the wage structure. The rivalry and jealousy among face groups has led more than one union official to conclude that "the men aren't ready for a Socialist wage policy yet." Changing conditions on a face can affect earnings, perhaps differently for different groups, and they can also lead to the revision of one or more price lists. Hence the relative earnings of the different classes are not rigidly fixed, nor are they always the same in different coalfields.

In a great many districts today, however, the machinemen are the aristocrats of the pit: they earn the "big money." Frequently the conveyor-shifters are not far behind. The comparative newness of these two functions as compared with drilling, filling, and ripping largely explains their advantage: when machine mining and conveyors came into widespread use after the first World War, managers had to offer men attractive prices to undertake this unfamiliar work. A second advantage enjoyed by these two groups (especially by the cuttermen) is their smallness: when a manager bargains with two or three cutters, why should he waste time haggling over a halfpenny or so a yard when it will not significantly affect his total costs? When he argues with thirty to forty fillers, a halfpenny can make a tremendous difference. Finally, machinemen occupy a position of strategic advantage in bargaining, because there are not many of them employed in most pits and it would take time to train replacements (this is less true of conveyor-shifters and much less true of strippers).

The combination of sectional bargaining and changing natural conditions in the several seams normally worked in a pit make the relative earnings of the different classes highly dynamic. Rippers in one seam, for example, may go along for a few months earning less than rippers in another seam — and then the relative position of the seams may change. In any particular seam (though not necessarily on the same face), drillers might rank second among the face groups in April and third or fourth six months later. Nevertheless, there is a status system on any face, and no group will allow itself to depart too markedly from its accustomed ranking without taking steps to restore its position.

Facemen earn from 75 to 100 per cent more than daywagemen. If you ask a union leader why pieceworkers deserve this much of an advantage over datallers, his answer will usually cite two reasons: (1) their work is "twice as hard" as the work done by daywagemen, and (2) pieceworkers are more subject to accidents, injuries, and industrial diseases than are men who work "outbye." This latter reason means that their earnings are more liable to interruptions during their working career on the face and that they run the risk of permanent loss of earning capacity through loss of health or even of their lives. Most daywagemen would certainly acknowledge the justice of these claims, but this does not mean that there are no limits to the differentials which daywagemen will allow without becoming aggrieved. During the relative improvement of pieceworkers' earnings as compared with daywagemen's during the immediate postwar period, many pieceworkers themselves acknowledged that they had about stretched the differential to the limit of what they felt they deserved.

The Pricing of Work

Bargaining between pit managers and "deputations" of pieceworkers over price lists has traditionally been conducted with somewhat more energy than in most other piecerate industries. Some people believe that wrangling over "prices" has been the main cause of coal's poor labor relations; others think that the difficulties over "prices" are only the symbolic expression, in a "cash nexus," of deeper tensions. Apart from the sociological complexities of pit life, there have been at least two very important external influences which go far to explain the atmosphere of price-list bargaining. These factors have been the exceedingly competitive nature of the industry and the variability of the conditions under which coal must be produced. Competition has kept managers cost-conscious; the variability of mining conditions has made it difficult to compare costs or labor prices from one pit (or part of a pit) to another. These difficulties have not kept either managers or union spokesmen from relying heavily in their arguments on prices and practices at nearby pits; but competition and variability have prevented these com-

parisons from exercising as strong a coercive effect as they do in many
other industries. Without strong coercive standards, the bargaining process inevitably becomes personalized and settlements are likely to depend
on primitive trials of wit and strength. One of the important achieve-
ments of the Coal Board during its first years has been to remove piece
rates from the arena of cutthroat competition, to develop more objective
standards for determining what the price ought to be, and thereby to
depersonalize the bargaining process. The millennium may not be at hand,
but many would agree that it is closer.

There has been an inevitable tendency in the past for men to earn
somewhat more money in a "good" seam of coal than in a "bad" one.
A good seam is determined by two main factors, by no means always
found together: (1) coal that is easily won, so that physical productivity
is potentially high, and (2) coal that is of a good quality, so that it commands a high price in the product market (in other words, its revenue
productivity is high, in the language of economics). Another important
determinant of physical productivity per manshift lies not in the inherent potential productivity of the seam itself but in how long men will
be at the coal once they are down the pit: an easily won seam that lies
a mile and a half from the pit bottom may not show any greater productivity per manshift than a more difficult seam that lies nearer the
shaft. The farther out men have to travel (a fatiguing as well as a time-
wasting exercise) the lower is the tonnage price the manager will want
to pay and the higher the men's asking price.

So far as bargaining over a price for coal filling is concerned, these
two factors tend to "buck" each other: ease of getting and filling argues
for a low price per ton, while high quality argues for a high price. Since
it is customary the world over to give capital, rather than labor, a first
claim on the rent accruing from consumer preferences in the product
market, it is by no means true that a highly productive seam of high
quality coal would yield the men who work in that seam a high tonnage
price: they can earn high wages from the productivity factor while the
employer earns high profits from the revenue factor (plus low total
costs). If it were not customary for capital to take most of the economic
rent, then high physical productivity and high revenue productivity
would not buck each other but would reinforce each other — which
would introduce much more inequality among wages and more uniformity among profits than people would consider fair.

Indeed, it is this subjective notion of what constitutes a fair wage for
a given class of work that determines what tonnage price will emerge
from bargaining over a price list. It is most realistic to think of the
negotiators' subjective evaluations of jobs (any face operations) as em-
bracing a range within which it is feasible to bargain for an advantage
of interest: but the pit manager will realize that he cannot expect men

to work for "unreasonably" low wages, and the men will realize that they cannot expect to earn "unreasonably" high wages. The tonnage price must be set to yield earnings that fall within this range of acceptability. Naturally, the negotiators' subjective "range of acceptable earnings" is not independent of time and place: what were acceptable earnings for strippers in the highly productive Nottinghamshire coalfield in 1947 would not be acceptable in 1954; nor would acceptable earnings in South Wales in 1954 be within the range of acceptability in Nottinghamshire in 1954.

Another set of considerations to be taken into account is the specific elements in the content of a particular class of work or special factors in the working conditions (such as excessive dust or the presence of water). The custom of giving explicit recognition to specific job elements is perhaps more important in nonfilling than in filling contracts, though the same principle holds in the latter case. For example, a stripping price will depend partly on the kind and amount of timbering that strippers must perform; if the manager or the law requires a change in either the kind or amount of timbering (there may be a change over from wood to steel; or new government regulations, as in 1947, may require that props be set closer together), the men are likely to ask for an additional payment in their contract. Usually these considerations are taken into account implicitly during the original negotiation of price lists; but the memory that this was the case may grow dim with time and there may be agitation to revise the price list to take account of elements that the men do not think they are being paid for. There are a great many identifiable job elements which may call for special or extra payment, such as payment where two men are required to push tubs, payment for boring bits, payment for dusty conditions, and so on.

Working within the foregoing framework of economic pressures, price-list negotiators at any pit begin by trying to agree on how their situation compares with similar situations at nearby pits. Any major seam in a district will be worked simultaneously by many neighboring pits and the costs and prices in these pits would play a big role in pit negotiations on one of these seams. It does not follow, of course, that there will be an exactly uniform price for coal filling, cutting, and so on in the Brockwell seam in Durham at all pits in the county where that seam is worked. The knowledge of prices for the Brockwell will constitute key benchmarks in each side's level of expectations; but whether the men will get something more than nearby pits have or the manager will force them to take something less will depend on bargaining attitudes, ability, and strength.

A somewhat primitive but well-established custom in County Durham is the following: when the manager has a new "room" to open out or a new drift to drive, he customarily posts a "To let" sign at the pithead, notifying the men that a contract job is available and inviting them to bid for it. "Sets" (teams) of men then submit bids, usually bidding something over the going "seam price" for similar work in the same seam. The manager is free to take whatever bid he likes (apparently without appeal by the men) — and he may not always take the lowest bid if he thinks another "set" will be more productive. If no one bids for the job, it is put into the next quarterly "cavil" and all work-teams participate in a "draw" for it.

Today the pit managers are not under nearly as strong pressure to "keep down costs at all costs" as they were before the war. But to a great extent they are still judged by their pit's costs and it is not true to say that pit managers no longer have any incentive to keep their costs down. The change has been one of degree and of emphasis.

The Shift in Bargaining Strategy since Nationalization

The basic change that has come over the industry's approach to price lists since nationalization has been a shift on the employer's side from putting the main emphasis on a tonnage price designed to yield a particular labor cost to one that will yield a particular level of earnings. This shift has probably not been formally endorsed by the National Coal Board, but the change in practice is frequently testified to in conversations with both Board and Union officials. The approach since nationalization has been to agree on "earnings norms" within a particular wage district to which the prices for various grades of work should be keyed. These standards are based on average earnings for a particular class of work (stripping, stone-driving, packing, conveyor-shifting, and so on) for the immediate district; in certain low-wage districts the norms seem to have been set somewhat higher than actual average earnings. These standards have not, apparently, been formally negotiated in each district; but they must certainly represent informal bargaining since the figures adopted have a coercive effect on the formal negotiation of pit price lists. The existence of such informal standards in the background of any negotiation has doubtless softened the bargaining process somewhat, but it has scarcely altered the essential nature of bargaining as a contest for economic advantage.

The use of informal norms as a guide for the negotiation of price lists does not represent a principle wholly new to the industry. The use of such norms merely extends to many districts which have never used them the wage-setting principles which have long governed the negotiation of price lists in Durham, Northumberland, and, to a lesser extent, Scotland. In Durham and Northumberland, price lists were regulated for many years by a concept known as the "county average." There, district agreements specified that when the piecework earnings under any price list fell either above or below average earnings for the class of work involved over each separate county as a whole, either party was permitted to seek a revision of the price list (adjustments were not automatic). This rule operated to keep the range of earnings among separate pits much narrower than in most of the other British coalfields.

In Northumberland, "average county earnings" are still used by the union as an informal criterion for reopening price lists but the device is not a formal part of the wage-regulation machinery of the district. The Divisional Coal Board was reportedly not happy about the practice, since every adjustment of below-average prices has the effect of raising the average, thus encouraging more contracts to be reopened.[6]

Scotland likewise had for years a somewhat looser control over the negotiation of price lists through a device known as the "common" or "field" price. "Theoretically, the 'common price' [was] an agreed district minimum, and if the average earnings given by any one cutting price [were] either much above or below this figure, there [was] supposed to be, *a priori*, a case for an alteration in that cutting price . . . the system does not seem to have had any effect on the rates paid to men on day wages." [7]

We may say, therefore, that the Coal Board has extended to most, if not to all, of the English coal fields a system very similar to that which had long prevailed in Scotland. Indeed, the use of specific figures for norms in the different districts has produced a bargaining guide very close to that provided by the county average system of Durham and Northumberland, the main difference being that there is no formal or explicit recognition given to the norm.

The reason the Coal Board has moved toward the guiding principle described above lies chiefly in the hope that it will facilitate the negotiation of price lists with fewer work stoppages and fewer frayed relationships than have been customary in the past. In addition, it is perhaps fair to say that the Board, as compared with the former owners, is more ready to agree with the general trade-union principle that marked variations in earnings for similar work within particular districts are not easy to defend. Being a monopoly employer, the Board is much better able to afford this magnanimity than were the owners of competing firms. Finally, one very practical explanation of why the Board has moved in this direction may be that the first Board member in charge of labor relations, Ebby Edwards, was for many years General Secretary of the Northumberland Miners' Association, and he may well have persuaded

[6] A useful description of the county average system employed in Durham and Northumberland since the early 1870's will be found in Jevons' *British Coal Trade*, pp. 358–364. Cf. also Rowe, *Wages in the Coal Industry*, pp. 42–44. Rowe notes that "these are the only districts where any detailed system of control over the price lists and time rates at each pit has been evolved."

[7] Rowe, *Wages in the Coal Industry*, pp. 44–45.

the Board to adopt a successful policy of which he had had direct experience.

Since the above principle tends to yield fairly similar earnings under
the very dissimilar natural conditions normally found in the different
pits of any district, its effect is to do away with the long-standing principle that seams of low productivity must be expected to yield lower
earnings. Consequently, one would now expect to find an even greater
dissimilarity in the labor costs per ton produced under different price
lists than has been true in the past. In other words, the new tendency is
for the burden of unequal natural conditions to be transferred from the
labor market to the product market, though the Board's price policy
does not make this apparent to consumers.

Price lists have traditionally been based on normal working conditions for the seam in question. If abnormal conditions are subsequently
encountered (faults, water, bad floor, and so on), it has always been
customary for the men to claim an extra allowance to enable them to
maintain their earnings at a level somewhere near that which the contract
was intended to yield. The amount of the allowance, by its nature, has
to be negotiated and this has always provided a second arena for the
running battle between the manager and his pieceworkers over their
wages and his costs. Such allowances are still in widespread use and
would seem an inevitable supplementary element in any system of
piecerates when used under variable conditions.

Most pit managers personally keep "little black books" showing the
amount of allowances they are carrying on each face. If these special
costs climb too high, they bear down on the deputies to prevent them
from being too generous. Allowances have also permitted dishonest
deputies to pull the wool over their men's eyes: they could put in to
the manager or undermanager for allowances and then pass on only
part of this to the men themselves. This was not common, but it was not
unknown and not the least important of the post-nationalization reforms has been the "cleaning up" of certain pit practices which went on
without the knowledge of higher pit management or the branch officials.

Traditionally, it has been customary in all or almost all coalfields to
have pieceworkers bear the cost of the essential "tools of their trade,"
though there has been great variety in the treatment of particular tools
and among different classes of men. In most cases, the cost of these items
was deducted from men's wages. The aim of this arrangement was of
course to make men careful and economical in the use of their equipment. Nevertheless, men had argued from time immemorial that the cost
of these items should be borne by the employer, who was often suspected
of charging the men something more than cost.

A resolution passed at the union's 1947 Conference asked the Executive Committee to see if the Board would not bear these costs, a change

that would amount to an increase in take-home pay. The union's national office approached the Board on this subject late in 1948. The Board did not wish to commit itself to a principle without some knowledge of how much money would be involved, but it proved impossible to get a clear picture of costs without careful local investigations. The union was invited to participate in local surveys, but the matter has not been of sufficient importance to compel attention from the busy Area officials of the union. Consequently little progress has been made towards a national agreement on the cost of tools, a subject which continues to be ruled by differing local customs. Articles of protective clothing must almost always be provided by the men themselves, though at some pits the Board furnishes these at cost so the men are saved the trouble and extra expense of buying through retail stores. The Board was required, by a new law effective in 1949, to provide safety lamps without cost wherever such lamps were required (at the few remaining naked light pits, miners often still have to buy their own lamps and the carbide used as fuel).

The NUM Secretary agreed to this procedure, but to date the cost data has not been collected from the Divisional Boards. The position with respect to the cost of powder stands on a somewhat more formal footing than tools and protective clothing. The Board and the Union agreed as early as 1947 that the cost of powder ought properly to be considered as part of the industry's wage structure and, as Arthur Horner then explained to delegates, "Prices have been determined at certain levels because powder is either provided free or must be found by the workmen involved in the case." [8] Thus there has been a conscious postponement of any deliberate policy with respect to powder until such time as the Board and the Union agree on the principles which will underlie their "new wage structure."

The Attempted Shift in Bargaining Tactics

In addition to the shift in bargaining "strategy" we have just noted, at least some Divisions of the Coal Board have also suggested to their own negotiators that they adopt new "tactics" for the actual conduct of negotiations, particularly when confronted by claims for revisions in existing price lists. The Yorkshire Division, for example, sent the following circular to all its Areas early in 1948:

It would be advisable to insist that all applications for a revision of price lists be in writing, showing not only full details of the proposed new rates, but the reason for the revision. In the absence of such written application, there is a tendency of the workmen to put out "feelers" in the form of verbal applications for unreasonably high prices. Such applications would be con-

[8] *NUM Conference Report,* 1947, p. 45.

siderably modified at the outset if it was insisted that they be in writing, and much preliminary negotiation would be eliminated. Further, in those cases where the matter of a revision of price lists was pursued to the Disputes Sub-Committee staff, such written applications would prove of considerable assistance to the NCB representatives, because the terms of reference for the Sub-Committee could be based on the actual written application of the workmen. In other words, the question before the Sub-Committee would not be HOW MUCH OF WHAT THE WORKMEN REQUEST CAN WE GRANT, but rather (and more properly), IS A REVISION OF PRICE LISTS JUSTIFIED? [9]

An earlier letter had urged upon each Area Labor Officer the view that he "had a responsibility for offering the workmen's representatives what he thought the job was worth at the outset," and the Labor Officers were requested "to consult their Area General Managers to urge Unit Managers to offer what they felt to be fair and responsible from the first deputation." [10]

The first of these advices (if the union could be made to respect it) seems admirable. But the second — while it reflects commendable good faith — scarcely seems to square with the realities of bargaining, which require each party to discount the sincerity of the other's offers in the early stages of negotiations. After all, the Coal Board's bargaining tactics have to take cognizance of union tactics, and it cannot be assumed that branch secretaries or miners' agents will often be on such good terms with Board officials that each party can afford to lay his best price on the table at the first meeting. Nationalization has not killed the old trade-union habit of trying to force up their members' earnings, though the union may be more willing to arbitrate and less enthusiastic about striking. Within this limit, bargaining still proceeds accordng to the natural laws of the bargainer's art. The following description of bargaining over a new price list was given by a candid and able branch secretary at a pit in the West Midlands Division and it is useful in reminding us that nationalization has not repealed the laws of human nature; the method of payment being described is one which expresses the rate in terms of so much per yard:

When we are opening out a new face, the men are usually first given a trial period on a fairly low day rate. I have to try to make sure that the men take it easy on the stint so that we can eventually get a high rate for a low stint when we get around to bargaining for a contract. In other words, I have to educate my men on how to work when we're negotiating, because sometimes the manager "can get round them" and persuade them to do more for the price he has in mind — that is, he offers them the price they want for the stint they know in their hearts they can do, and they sometimes tend to settle too quickly. For example, if the men clean up the face an hour or two before the end of the shift, the management will certainly note this and throw it in my face during the negotiations. But if all goes well and we end

[9] Divisional Labor Circular, Northeastern Division, Feb. 27, 1948.
[10] Divisional Labor Circular, Northeastern Division, Sept. 25, 1948.

up with a high rate for a low stint, then — but not too quickly! — they can begin to earn some extra money and then, if the management sees that we have beat him on the stint and thinks of renegotiating the contract, he knows just as well as I do that we would counter with requests to reopen a lot of other price lists in the pit — and this would lead to so much instability that he just lets the situation ride.

'Twas always so and nationalization has not made it otherwise, though perhaps the degree of unreasonableness on both sides has been somewhat narrowed. Can a free society expect more?

CHAPTER SEVEN

THE PROBLEM OF LABOR SUPPLY

In this chapter we shall be concerned only with manpower, the number of men in the industry. It should be obvious that this figure does not determine by itself the "amount of labor" which goes into coal production. A dozen or more subsidiary questions (such as the length of the work-day and work-week, the number of holidays per year, the rate of absenteeism, the number of strikes, and so on) also play a vital part in the amount of labor available to the industry. Still another dimension to the notion of labor input is determined by the effectiveness with which manpower is utilized when it is available; this depends on employee morale, good machinery maintenance, and a smooth flow of working supplies.

It is because the manpower situation of the industry has been so tight during and since the war that these many subsidiary questions, separately and collectively, have become questions of national concern. And because no amount of attention to problems such as absenteeism and strikes has led to any marked increase in output, the fundamental problem of labor supply has remained one of great and continuing urgency.

The problem had its origin very early in the second World War when about 85,000 of the industry's 775,000 employees used the opportunities presented by full employment to get out of the industry. By the second quarter of 1941 there were only 690,000 men left; and it was only through the use of emergency controls that the wartime labor force was thereafter kept between 700,000 and 710,000 men.[1] This number was the bare minimum necessary to produce the amount of coal Britain then needed — an amount somewhat below the levels of a prosperous peacetime period, thanks to the wartime loss of export markets.

Despite the relative success in maintaining manpower through the war period, a qualitative deterioration in the labor supply inevitably occurred, so that output per man-year steadily fell from 296 tons in 1941

[1] A description of the wartime manpower controls can be found in Harold Wilson, *New Deal for Coal*, and W. H. B. Court, *Coal*.

to 246 tons at the low point in 1945. At the end of the war, the average age of the industry's labor force was too high and there were many unwilling and relatively inexperienced workers, frozen to mining by manpower controls, who wanted to get out at the first opportunity. There were several thousand ex-miners in the armed forces (though none had been taken after 1941), but these were far from enough to restore employment to prewar levels even if all of them elected to return to the pits.

The existence of a labor shortage, it should be noted, has been common to all the coal-producing countries of Europe during the postwar period, a situation quite different from that in the United States. This situation, which contrasts dramatically with the heavy unemployment of the prewar period, has accompanied full employment, which for the first time in two decades has given miners effective freedom in the choice of employment. The existence of this freedom has tested relative wage levels, and other employment considerations, in a way that they had not been tested for many years previously: What kinds of rewards did coalmining have to offer in order to compete successfully with other industries for the number of men available? How important are nonwage factors in building and maintaining a labor supply? How could the status of pit work be endowed with a dignity in keeping with the industry's importance?

At the outset, we need to emphasize one characteristic of recruitment in British mining that is crucial to an understanding of the problem. Coal is an industry whose labor supply is, to an extraordinary degree, recruited internally — that is, most of the people who enter the pits come from "mining families." This means two things. It means that the industry is relatively powerless to attract labor from nonmining areas (and even from nonmining families within mining areas). It means also that mining must be able to attract young boys at the very beginning of their careers, normally as they leave school. This is so because of the empirical fact that once a boy, even though he be from a mining family, has settled down in a nonmining job, it is very unlikely that he will ever return to mining unless he has worked in the pits at some time early in his career. The ability of the industry to compete with other industries for adult British labor is thus limited, in the main, to that relatively small group of men who have worked in the pits at one time or another. Given similar wages and other benefits, nonmining industries can attract adults from mining much more easily than mining can attract men from other industries. This is attributable chiefly to the peculiar nature of working conditions in mining — physical and psychological conditions to which men find it very difficult to adjust unless they have been born to them — and also to the inferior social status which mining has long enjoyed in the general public mind.

For this reason the Coal Board has had to pay particular attention to juvenile recruitment, since the Board's inability to develop alternative sources of manpower on any large scale has confirmed the view, widely held by those within the industry, that the main job of the industry is to reëstablish juvenile recruitment as the chief source of fresh manpower over the long run. The main contribution of the alternative sources tapped since the war has been to protect the industry's total manpower on a short-run basis until the basic reforms necessary to the reëstablishment of juvenile recruitment could be carried out. By 1952, it began to look as though the Board were succeeding in attracting sufficient numbers of boys to the industry.

Nationalization and the Supply of Labor

Whether or not nationalization would, as many people had come to believe, restore recruitment and reduce wastage to satisfactory levels depended on the kinds of answers which the Coal Board and the government gave to three general types of questions. We shall discuss these questions briefly before going on to analyze the manpower figures themselves.

PROBLEMS OF "DOCTRINE"

The war ended with the industry's precarious manpower situation protected only by direct manpower controls: would the Labor Government dismantle these controls summarily as incompatible with the new status of labor promised by the election of a Labor Government? If the government's position of responsibility advised taking a more cautious approach, would there be impatient protests from the rank and file and difficulties in the government's relationship with the NUM? In practice, the government took the cautious approach; this did not invite protests from the NUM, though controls did become less enforceable as time went on.

With foreign labor available for British mines, what would be the attitude of the Labor Government (and of the NUM) toward the introduction of such labor? Would the government,[2] and the NUM, take the

---

[2] The word "government" instead of "Coal Board" is used advisedly because when collective bargaining on very important issues reaches the final stage of decision, the Coal Board is almost forced to consult the Prime Minister and the Cabinet as to what is the best offer it should make. As long as a distinction is effectively maintained between the big, substantive issues that affect the country's interests and lesser ones that do not, it is more desirable that the last word should lie with the government than with the management of the industry (whose judgment of what the country can afford is not as broad as that of the government). The process of collective bargaining in critical industries of the United States does not seem to differ greatly in its dependence upon government approval or disapproval than is true of the nationalized industries of Britain.

line that the country would have to accept whatever wage level in coal might be necessary to man the industry exclusively with Britons? The use of foreign labor did develop into a major problem, and the Board, after trying, was eventually forced to abandon this source of recruits.

In January 1946, the NUM Executive Committee approved a Miner's Charter that enumerated twelve improvements in wages and working conditions which it regarded as due the nation's mineworkers.[3] Among these demands were the five-day week, two weeks' vacation with pay each year, and a retirement plan. The adoption of these demands would obviously reduce the amount of labor available to the industry in the short run and would thus reduce output. Would the government be found agreeing with the NUM's bargaining "line" that only through the adoption of these reforms would the industry be made attractive enough to solve the problems of recruitment and wastage? (Progress had been made in implementing all these demands by the middle of 1952, but the effect of any single reform, and even their total effect, remains ambiguous.) ·

PROBLEMS OF "ECONOMICS"

When an industry is trying to recruit people who feel that mining is a relatively disagreeable industry (those people "at the margin" for whom mining has a high disutility), the wage offer must be correspondingly high to get these people, and it will have to be offered not just to them but to those already in the industry.[4] Wage increases in an amount sufficient to solve the manpower problem would inevitably impose much higher costs on the country's coal consumers and might result in making coal so much of a "preferred industry" (employment-wise) that a Labor Government would incur unmanageable political problems in its relations with the consuming public, especially the nonmining unions whose members do not receive miners' wages but must still buy coal.

Since nationalization, the "law of supply and demand" has certainly been operative in the industry's labor market. This law tells us that if demand exceeds supply, the employer will bid up the price of labor by offering "higher wages" in the hope of persuading more people to offer themselves for employment.

[3] The text of the Charter is included as Appendix III.

[4] It is conceivable that there is no level of wages, within the range Britain can stand, which will be sufficiently attractive to enough men to man the industry satisfactorily. In the long run, the level of mine wages "Britain can stand" is a matter for testing in international trade (primarily in the coal trade, but also indirectly through other exports made with coal); in the short run, before any such test can be conducted, this upper limit is set more by habits of thought about "traditional" or "reasonable" prices for coal than by any objective knowledge of where the limit lies. Readjusting British habits of thought about "reasonable" coal prices has so far been a more important aspect of the problem than the danger of "pricing coal out of the market."

A major difficulty the Coal Board has been up against is that this system of "trial and error" has not operated with sufficient certainty or with sufficient speed to protect the industry's labor supply.

It was not at all clear, for example, whether higher money wages would suffice to attract enough men or whether these would have to be combined with such other forms of "wage increases" as a pension scheme, the five-day week, employer-financed educational opportunities, longer vacations, and so on down a long list of the many forms that "wages" can take. Nationalization had meant that there would eventually be a general improvement in the terms of employment; but no one could have any very confident prediction as to which improvements would yield quick returns in terms of labor supply. The one form of real wage whose favorable effects seemed most certain happened to be the one that seemed least possible politically — the allocation of more houses to mining communities (see pages 191–192 and 261–273).

PROBLEMS OF "TECHNIQUES"

Finally, there is an important relationship between nationalization and the ability of the coal industry to introduce those modern personnel skills which are ordinarily thought to be part of good management and which aid recruitment and help curb wastage. Coalmining is an industry which has not made much use of these techniques in the past — it has not needed them, simply because the external pressures on the work force were so strong that recruitment was automatic and wastage was low, perhaps too low. Part of the industry's recruiting job must lie in "raising the status of pit labor"; this includes such steps as getting out attractive movies about mining, publishing appealing vocational literature, developing good local relationships with the education authorities, organizing the kinds of welfare activities that interest boys and other kinds that interest men, planning an effective training program, and so on. It also implies the ability to develop the use of a "personnel approach" on the job — partly by the introduction of such specialists as Training Officers but more importantly by educating pit management to think in new terms, to adopt new practices, and to keep new kinds of records. At best, the development of such new techniques and attitudes can proceed only slowly — so far, too slowly to have yet been of much help in protecting the industry's labor supply.

Recruitment and Wastage: 1946–1953

Some men and boys join the industry, and some leave it, on every day of the year. The crudest of the available manpower figures — changes in total employment — show which of these opposing tendencies has been strongest during any particular period. Chart 7 shows the

average number of men attached to the industry during each week from
the beginning of 1946 until the end of 1952.

There are three different statistical concepts used to measure the
volume of employment in the industry: the number of men listed on
the books of all collieries in the industry; the foregoing number minus
all men who have not been at work for six months or more (for reasons
of sickness, accident, working at another job, or any unexplained absence
of six months); this figure is known as the "standardized" number of
wage earners on colliery books; finally, there is a yet smaller total
known as the number of "effective" wage earners on colliery books,
which includes only those men who have been at work at least once
during a given week; in any week the number of "effectives" is usually
only about 90 per cent of the number of "standardized" wage earners,
though the "effective" figure is considerably more erratic than the number of "standardized" wage earners. We shall use the "standardized"
figure except where noted otherwise.

It is obvious that there was a substantial influx during 1947–1948,
an even heavier exodus during 1949–50, a second build-up of manpower
in 1951–52, and a moderate loss after mid-1953. This over-all weekly
figure of coalmining employment has probably been one of the half
dozen most closely watched statistics in postwar Britain, for it must be
remembered that whenever mining manpower is below about 700,000
the whole country faces the possibility of a major economic disaster if
certain things should occur; and no one, inside or outside the industry,
can feel at all comfortable until manpower exceeds 720,000.

With the labor supply on a precarious basis ever since nationalization, the direction of the week-to-week trend has had a great influence
on management's thinking and morale. It is not difficult to imagine the
gloom which spread in 1950 when, after three years of greater reforms
than the miners had received during the past fifty years, manpower fell
to a lower level than in 1900.

It is true that the Board's long-run plans call for a labor force of
about 618,000 men by the mid-1960's; but the 1949–50 loss was coming
much earlier than the Board could afford and was proceeding at a rate
which, if continued, would have left the industry with less than 400,000
men by 1965. When the tide finally turned in 1951, uncertainty replaced
despair, for no one could be sure that a new reversal would not set in,
as, indeed, moderate ones did in the middle of 1951 and 1953.

In any event, the first seven years of nationalization had seen the
industry stem what threatened at one time to be an almost fatal erosion
of its labor supply. By 1952, the problem of labor supply had ceased to
be general throughout the industry: the most troublesome problem now
was local shortages in certain key areas (about a dozen of the Board's

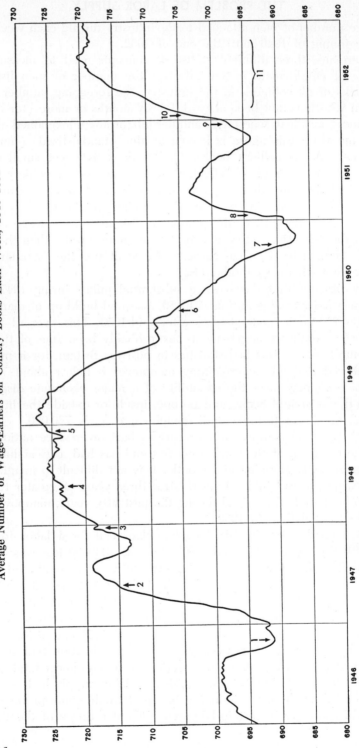

Chart 7

Average Number of Wage-Earners on Colliery Books Each Week, 1946–1952

Key: See bottom of p. 187. In late spring 1955, manpower was just under 710,000.

Source: NCB, *Annual Reports*.

fifty-two areas were short of men in 1954). Since district variations in
money wages are no longer permissible to take care of local supply and
demand relationships, the Board has placed primary reliance on special
housing allocations to attract men to the labor-short areas. While the
industry's manpower experience has been affected somewhat by the
general level of employment in the economy, mining does seem to have
been made sufficiently attractive so that it is able to recruit and hold
a labor supply in a free labor market under conditions of full employment. For two years, 1949 and 1950, the possibility of this had seemed
doubtful, to say the least.

We must now go behind the net figure of total manpower fluctuations
to examine separately the sources of recruitment and wastage. The main
conclusions will be somewhat disappointing, for the largest influences
on manpower changes are found in the statistical categories about which
least is known — the residual or "balancing" categories. It is important
to understand how little is known about the sources of recruitment and,
especially, wastage; for the administrative difficulties experienced by
the Board in securing informative figures about why men leave the
industry can mean only that far-reaching policy decisions have had to be
made on the basis of information that is very incomplete.

The Board's statistics on recruitment and (especially) wastage have
improved since nationalization, particularly since the effort to record
reasons for wastage was begun in 1950. However, better and more complete statistics require better and more record-keeping at the pit-level.
Pit officials skilled in exit-interviewing are scarce, and the Board cannot
ask for more elaborate record-keeping without risking criticism from pit
officials, who already have many reports to make out.

Table 10 shows us broadly the four main sources of new manpower
for the twelve years, 1942–1953, inclusive: juveniles, newly employed
adults, reëmployed adults and juveniles, and transfers from one pit to
another. Somewhat finer breakdowns are available with respect to the
second and third. Gross and net recruitment can be derived from these
figures for all but the last four years.

"Gross" recruitment includes all persons signing on at all pits regardless of where they came from (even from another pit); "net" recruitment

---

Key for Chart 7

1. Five-day Week agreed to "in principle."
2. Five-day Week put into effect.
3. Weekly U/G minimum raised from £5 to £5/15 (15 per cent); juvenile rates also increased.
4. Mineworkers' Supplementary Injury Scheme effective.
5. Redundancy Pay Agreement signed.
6. Fatal Accident benefits become payable.
7. Weekly U/G minimum raised from £5/15 to £6 (4.3 per cent); juvenile rates also increased.
8. Weekly U/G minimum raised from £6 to £6/7 (5.8 per cent); juvenile rates also increased
   and pension plan agreed to "in principle."
9. Weekly U/G minimum increased to £7/0/6 (10.5 per cent); juvenile rates also increased.
10. Pension plan comes into effect.
11. First substantial postwar unemployment (chiefly in Midlands soft-goods trades).

TABLE 10

Recruitment, 1942–1953

	1942	1943	1944	1945	1946	1947	1948	1949	1950	1951	1952	1953
1. Juveniles newly employed	12,900	12,130	10,400	9,400	12,700	14,600	11,000	14,200	15,100	19,600	27,100	19,900
2. Adults newly employed												
a. From residential training centers	—	—	31,300	17,000	4,800	15,700	18,000	6,100	16,100	20,400	22,200	11,700
b. From other industries	—	—	1,700	1,400	6,200	17,300	16,700	13,500				
Total: New Adults			33,000	18,400	11,000	33,000	34,700	19,600	16,100	20,400	22,200	11,700
3. Reëmployed												
a. Ex-miners from Forces	14,100	3,600	6,400	11,600	27,900	12,200	4,400	1,100	1,200	32,800	28,100	20,000
b. Ex-miners from other industries	—	—	6,900	8,100	19,800	28,700	18,800	12,800	18,500			
c. Compensation and long-term sickness cases	9,800	10,000	9,200	9,500	8,400	7,000	5,300	4,900	—			
d. Others	13,100	13,400	—	—	1,400	4,900	4,800	4,300	4,400			
4. Transferred from other pits	31,000	28,000	27,000	29,300	29,300	72,100	76,600	76,300	n.a.[a]	n.a.	n.a.	n.a.
5. Gross recruitment	80,900	67,100	93,000	86,300	110,600	172,400	155,600	133,200	n.a.	n.a.	n.a.	n.a.
Minus: returned compensation and long-term sickness cases and transfers from other pits	40,700	37,900	36,200	38,900	37,700	79,200	81,900	81,100	n.a.	n.a.	n.a.	n.a.
6. Net recruitment	40,200	29,200	56,800	47,400	72,900	93,200	73,700	52,100	55,300	72,800	77,400	51,600
7. Net recruitment as a per cent of gross recruitment	50	43	55	55	65	54	48	39	n.a.	n.a.	n.a.	n.a.

Source: Figures for 1942–1949 taken from Table 29, Ministry of Fuel and Power, *Statistical Digest, 1948 and 1949*; remaining figures taken from NCB, *Annual Reports*. Figures rounded to nearest hundred. Blank spaces indicate either that the figures are not available or were not recorded.
[a] n.a. = "not available."

includes only those signing on from outside the industry, though not necessarily for the first time.

Notice first the large difference between gross and net recruitment for the years this figure is available. Though some of the difference between the two is accounted for by the return to work of men who have been on the sick list for more than six months, three-quarters of the difference arises from transfers or turnover within the industry, a process that increased, as one would expect, right after the war.

The second point to notice is the natural change in the relative importance of the various sources since the immediate postwar period: ex-miners from the armed forces and from other industries fell off, as both categories contained a large nonrecurring source. The category "ex-miners from other industries," however, has included not only the ex-miners who went off to war plants early in the war, before controls were imposed, but also the very important recurring source of ex-miners who may or may not be persuaded to return to the pits.

We have narrowed attention to newly employed juveniles, newly employed adults, and the reëmployment of ex-miners who leave other industries. Since the last-named group involves individuals who have once worked in the industry, our interest finally centers on juvenile recruitment and the ability of coalmining to draw men away from other industries. Unfortunately, not very much can be said about coal's ability to attract adults for the first time: by comparing row 2-b in Table 10 with the sum of 3-b and 3-d we find that more men have returned to mining from other industries than have turned to it for the first time. A large proportion of newly employed adults seem to be of a relatively unsatisfactory type who do not adapt well and frequently leave the industry after a few months.[5]

[5] This statement must be read with some caution, as it is based on narrow though highly placed interview evidence. There is great room for more research (statistical and nonstatistical) on where new recruits come from, how long they "stay recruited," and why they leave, when and if they do.

A study of the "reputation of coalmining in the eyes of men outside the mining industry" was conducted by The Social Survey early in 1948 and was published, confidentially, in August of that year. Some random quotations from this study follow: "The miner himself is generally regarded to be a decent hard working man, doing a rotten job. It appears that his social standing has improved considerably in recent years . . . It is apparent that nearly half the men think it would be unusual in some way if any of the men they worked with were to enter the industry . . . The younger men, that is, the men aged 16–34 — the men who could be taken by the mining industry — were somewhat less favourably disposed toward mining than the older men . . . Of five vital industries and services coalmining is the least popular . . . In general the best chances of recruitment seem to exist in mining areas and perhaps among unskilled operatives . . . Lastly, so far as can be estimated from a not very exact comparison between the views of parents in 1946 and in 1948, the reputation of the mining industry has not advanced to such a degree that more than a tiny proportion of parents are in favour of their children entering the industry." The main complaints men had were not about wages and hours or security of employment but rather about the alleged sombreness of mining

TABLE 11
Wastage, 1946–1953

	1946	1947	1948	1949	1950	1951	1952	1953
I. Natural wastage								
1. Deaths	2,900	3,100	3,200	3,200	3,300	3,500	3,300	3,300
2. Retirement	6,500	3,500	4,000	4,400	4,800	2,500	6,500	8,600
3. Excess of new compensation and medical cases over those returned	24,800	10,600	10,200	8,600	7,000	10,500	8,600	11,200
4. Total natural wastage	34,200	17,200	17,400	16,200	15,100	16,500	18,400	23,100
II. Other wastage								
5. Released from National service obligation	3,100	13,000	7,000	18,300	—	—	—	—
6. Dismissals	28,700	16,700	17,800 [a]	9,600 [b]	5,900 [b]	4,200 [b]	4,600 [b]	4,000
7. Balance	11,000	21,000	23,800	24,300	54,900	42,800	32,600	35,000
8. Total "other wastage"	42,800	50,700	48,600	52,200	60,800	47,000	36,600	39,000
9. "Balance" as a % of "other wastage"	26%	41%	49%	47%	90%	91%	87%	90%
10. Total net wastage	77,000	67,900	66,000	68,400	75,900	63,500	55,000	62,100
11. Natural wastage as a proportion of total net wastage	44%	25%	26%	24%	20%	26%	33%	37%

Source: NCB, *Annual Reports;* figures rounded to nearest hundred.
[a] "Dismissals, releases," etc.
[b] "Dismissals (other than redundancy)."

Table 11 will give us an elementary understanding of the pattern of wastage from the industry for the eight years, 1946–1953. Unpreventable wastage (wastage from death, retirement, and medical causes) is seen to account for roughly a quarter of the total leaving the industry. The remaining or "other wastage" was broken down into three classes for the years 1946–1949, and into two classes thereafter. Manpower controls ended at the end of 1949 and it was no longer necessary to account for the release of people who formerly needed government permission. The significant point about these other wastage figures is the high proportion represented by the "balance" figure — a figure which has averaged nearly 90 per cent of all the nonnatural wastage since 1949. A large part of this very large residual category must include great numbers of men whom the industry might hope to retain if only the Coal Board (or the union or the government, for that matter) knew how to make the industry attractive enough. To some extent, this is not a fair statement, for the Board and the government have been confident that something could have been done to decrease wastage if the economy could afford the cost — chiefly the cost of allocating more houses to mining areas.

In the spring of 1952, after five years of acute housing shortages in nearly every coalfield, the Board set up its own Housing Association to expedite the construction of a special allocation of 20,000 extra houses allotted to it by the government. These houses, built by private builders, were sited where labor was needed most urgently (85 per cent were built in Yorkshire and the Midlands). By the spring of 1955, nearly all these houses had been built and the program was generally felt to have been a marked success. Three-quarters of these special NCB houses in any community were earmarked for local miners, the other quarter being reserved for transfers from other coalfields. This arrangement, for example, enabled the Board to transfer 3200 men into the West Midlands Division from declining coalfields in 1953–1954—a shift which was barely sufficient to offset very heavy internal wastage from the Division's normal labor force. But the success of this special and somewhat tardy housing program is costly: the Board will carry a £50 annual subsidy on each of these 20,000 houses.

The housing problem has been exceedingly difficult since the war and it is not easy to be dogmatic as to whether the Board ought to have taken this step in 1947 instead of 1952. Consider the government's position in allocating scarce housing resources during the postwar period. How many should the bombed-out areas get? The critical export industries? The police who must be enlarged to stem the postwar increase in crime? Added to these purely economic choices has been the political difficulty of showing favoritism to particular groups.

---

towns and the arduousness of the working conditions. See *Men and Mining*, by Geoffrey Thomas of The Social Survey, Central Office of Information, N.S. 113, Aug. 1948, 23 pp.

Even when the Coal Board has been able to persuade the Ministry of Health to increase allocations to mining areas (as it had been able to do before 1952 on a very modest scale), this has meant only that the local authorities in those areas had more houses to rent. It did not necessarily mean that these extra houses would be let to miners, for local authorities have set their own rules in allocating houses to people on their long waiting lists. The Board could thus only hope that more houses in mining areas would lead, in a general way, to more houses for miners. Much of the new building in mining areas simply relieves the "doubling-up" that exists among natives of the village — the new houses are decidedly not available for tempting men in London or Birmingham or Leeds to move to a new job in a mining town in preference to townspeople who have patiently waited their turn on their local list. A considerable number of the "beds" and "lodgings" which are vacated by people lucky enough to get new houses are "taken off the market" and are not available to "strangers." These are some of the considerations which led the Board into the unremunerative business of building its own houses.

But even if houses had been available earlier, there would still have remained an embarrassingly high rate of additional preventable wastage that is not well understood and hence cannot be prevented. The Coal Board naturally has somewhat more complete unpublished wastage data than the summary figures presented in Table 11. However, the Board did not attempt to quantify the various reasons for wastage until early in 1950, when it began to collect from each colliery a weekly report on the "movement of manpower" based on pit interviews with each man who had "given his notice." In the autumn of 1950, a joint subcommittee of the National Consultative Council (the top-level NCB-NUM "labor-management committee") made a special report on manpower which listed eight main steps it thought would do most to reduce wastage. (The report noted that the Coal Board had already acted on most of these points!)

The eight factors listed were these, (1) *housing* — "the most effective means to increasing recruitment, reducing wastage and facilitating the re-deployment of the labour force"; (2) *dislike of hostels* and the need for improving living conditions in them; (3) *better reception of new recruits* — a special responsibility of the Training Officers; (4) *systematic interviewing of intending leavers* — a problem of getting line management to follow sound personnel procedure; (5) *stopping the call-up of army reservists* employed in mining and discouraging volunteers through interviews; (6) *disappointment among some recruits about upgrading opportunities* — sometimes they were blocked by too long a waiting list, often by the fact that there was no one available to take their present jobs if they were allowed to go forward; (7) *improving amenities in colliery villages,* and especially, an acceleration of the building of *pithead baths;* (8) a postponement of the proposal to lift *manpower controls on foreign workers* who had been in the country for three years or more as of January 1, 1951 (the Government did not accept this recommendation — which came only three weeks before the foreigners had been promised free choice of occupation).[6]

One additional fact about wastage that is only too well understood is that it has been concentrated much more heavily in the age groups under 40 than in those above this figure — the same desirable age groups among whom recruitment has been discouragingly low until the more favorable experience of 1952. The difficulty of getting and retaining men in the age-groups under 40 led to a serious rise in the average age of employed mineworkers, a process that had been going on ever since the early 1930's. Not until 1952 was the slow advance of the miners' average age finally reversed (from 40.5 to 40.2 years), a reflection of the extraordinarily encouraging recruitment of 27,000 boys that year and a halt to the loss of men in the important 20 to 30 age group. In 1953, the average age once again turned upward.

In 1931, the average age of British miners was 34.6 years and the subsequent rise has not nearly been accounted for by the general aging of the British population. As of mid-1950, coalmining had a lower proportion of its labor force in the "under 20" and "20–39" groups, and a higher proportion in the 40–64 group, than agriculture, fishing, engineering, shipbuilding, and the electrical trades, textiles, building construction, and many other lighter industries.[7] The result has been a shortage of men between 20 and 35 to man the key coalface jobs so that older men, whose productivity is lower, have had to be retained on such jobs longer than is desirable.

TRANSFERS AND THE TURNOVER RATE

The number of interpit transfers and the turnover rate are figures which afford some indication of the stability of the labor force. Unfortunately the Coal Board's annual reports do not include information on transfers or gross wastage, so we must rely instead on the figures supplied every three or four years by the Ministry of Fuel and Power in its *Statistical Abstract,* covering the years 1942–1949, inclusive.

For the eight years 1942–1949, the scale of interpit transfers was large enough to make gross wastage (men terminating at any pit) almost double net wastage (men leaving the industry); the Board would very much like to see the number of transfers reduced in the interest of

---

[6] National Consultative Council, Manpower Committee, *Report of the Council,* issued by the NCB press office, Dec. 14, 1950 (6 pp., mimeo.).

[7] See the article, "Age Analysis of Employed Persons," *Ministry of Labour Gazette,* June 1951, pp. 223–229, especially the table on p. 224.

stability, but the problem cannot be said to have been critically serious even during the unsettled postwar years. As for the turnover rate (gross wastage divided by average employment for the year), this ran at about 10 per cent in 1942–1944, rose to about 15 per cent in 1945–46, and rose further to 20 per cent for 1947–1949. The latter figure means that one out of five employees left his job during the year. To people familiar with turnover rates, a 20 per cent figure may not seem alarmingly high; it has, however, been considerably above what coal has been accustomed to in previous years.

The Use of Foreign Labor

During the early years of nationalization, when the manpower crisis threatened to become desperate, the Coal Board and the government turned to temporary and extraordinary policies designed to plug the manpower leakage until a more permanent reconstruction could be achieved — a reconstruction based on lower wastage and much higher juvenile recruitment. Among these extraordinary stopgap measures were such policies as a prohibition against quitting the industry (this manpower control was lifted for British citizens as of January 1, 1950, and for foreign labor one year later), special demobilization privileges for miners serving in the armed forces, draft exemption for juveniles entering the pits, and the introduction of foreign workers into the industry. The last of these several policies is of the most interest and is the only one we shall review in any detail.

The foreign labor issue developed many more pressures within the union, the government, and the Coal Board than did manpower controls. The abolition of the wartime controls was recognized by all parties as "only a matter of time"; their continuation was not widely regarded as either an effective or acceptable method of protecting the industry's manpower: the controls (which after 1945 did not affect hiring — only wastage) did not so much determine whether or not people could leave the industry as they did the tactics by which people left.

By and large, the policy of introducing foreign labor was not a great success, though it was certainly not a total failure, and to know the reasons why it was not will sharpen our understanding of the industry's labor problems.

The Employment of Poles and EVW's, 1946–1949. The suggestion that certain groups of foreigners might provide additional manpower for British pits arose in the middle of 1946. During 1945 manpower had fallen from 712,000 to 695,000 and juvenile recruitment had fallen to the very low rate of 9500 boys. This net wastage had occurred despite the return of some 12,000 men from the armed forces. It was against this background that the Ministry of Fuel and Power suggested in

June 1946 that some 1000 Polish ex-miners — then stationed in Britain with a Corps of the Polish Army — be employed in the industry. The proposal was put up to the NUM's Executive Committee which decided to defer any recommendation for acceptance of this foreign labor until more information was available as to the exact numbers likely to become available, the possible reaction of British mineworkers, and the specific guarantees that would be set up to protect the interests of the union and of British miners. The union's 1946 Conference, held in July just after the original proposal had been made, approved this delaying action of the Executive Committee.[8] The union, of course, wished to make certain that these foreign workers would become union members and it also wished to avoid being in the position of recommending to its members that Polish miners be accepted when it had complaints that at certain pits (undoubtedly only isolated ones, but enough to raise fears) some older men and some wartime "green" labor were being discharged to make room for men returning from the forces. This attitude was the initial expression of that mixture of commendable economic caution and the irrational suspicion of "foreigners" which characterizes British mining communities, and perhaps, in lesser degree, Britain generally.

After the NUM Conference decision, the Ministry again approached the NUM executive committee with a view to the immediate placement of two hundred Poles who had had previous underground mining experience. The union informed the minister that it did not believe it wise to seek to introduce Polish labor "and thus cause a diversion." The union expressed concern that there were still a number of its own members being refused employment on the ground that they had suffered accidents or disease. "It was made clear that the Union could not acquiesce in the employment of outside labour whilst our own members are debarred from working in the industry." [9] The union was obviously using the Ministry's proposal as a bargaining weapon to bring pressure on the employers to rehire men receiving workmen's compensation. This is one instance of the tendentious character of workmen's compensation which the Industrial Injuries Act (1946) was designed to eliminate. It is perhaps also true that this issue was really an excuse for the union to avoid facing up squarely to an issue that it felt might be fraught with difficulties.

Nevertheless, by December the industry's manpower had sunk to the all-time low of 692,000 and this led the Ministry to make a further approach to the union on the question of Polish labor. According to the Executive Committee's own report, considerable pressure of opinion was brought to bear on them ("the serious economic position of the country,"

[8] The point is reported in the debate on pp. 72–79 of the *NUM Conference Report*, 1946.
[9] *NUM Conference Report*, 1947, p. 223.

"the great need to employ every available person," etc.), and on January 16, 1947, they agreed in principle to the employment of Polish workmen, subject to three conditions which were embodied in the NUM-NCB agreement of January 31, 1947:

It is agreed that the Union will raise no objection to Polish workers being employed as civilians in British coalmines on terms and conditions identical with those applicable to British workers provided that:

(i) No Polish worker shall be placed in employment at any colliery without the agreement of the Local Branch of the Union.

(ii) Polish workers who enter the industry shall join the Union and any who fail to do so shall be dismissed.

(iii) In the event of redundancy Polish workers shall be the first to go. They shall be transferred or dismissed wherever their continued employment at a particular pit would prevent the continued employment or re-employment of a British mineworker who is capable of and willing to do the work on which the Polish worker is employed and who might otherwise be displaced by redundancy at that pit or another pit.

This national employment agreement has been supplemented at pit-level by many local agreements giving British miners preference in up-grading. Also, the NUM originally insisted upon participating in the actual screening of the Poles conducted by the Coal Board, but the number of volunteers and the very low proportion of undesirables found in the early experience made the administrative effort involved too burdensome for the union; it therefore decided to leave all screening up to the Coal Board.

The ink was no sooner dry on the agreement on Polish labor than the Minister of Labour (as distinct from the earlier approach on Poles by the Minister of Fuel and Power) asked the union to consider the employment of foreigners other than Poles, provided the same safeguards were observed. The union agreed to accept such additional recruits "should it be found that the manpower requirements of the industry could not be met from available British manpower and such Poles as were willing to take up employment in the industry." Thus was the basis laid for the subsequent employment of "European Volunteer Workers" (or EVW's, as they came to be called), persons recruited from the camps of displaced persons on the Continent. Poles and EVW's were recruited to mining on the understanding that they would remain in the industry for at least three years. The government abolished this final manpower control (which applied not only to the 17,000-odd foreigners who entered mining but to some 60,000 additional D.P.'s who took up other employment in Britain) on July 1, 1951.

The great majority of foreign recruits did not know English well and had never worked in a mine. There thus had to be mounted a very considerable program of language education and preliminary pit training, the latter being provided at the Adult Residential Training Centers (a late wartime development) which also catered to British recruits without previous mining experience. The big difficulty that developed was not training; it was the reluctance of many miners' lodges to authorize the employment of foreign workers at their pit. Ironically, local resistance proved strongest in some coalfields where the manpower shortage was most acute. In Yorkshire, for example, about one-half of all the miners' lodges voted not to accept foreign workers. One difficulty in the arrangement which gave each branch a veto power over the employment of foreigners was that it encouraged each branch to hope that others would assume this responsibility. This result was probably not foreseen by the Executive Committee of the NUM when it entered into the original agreement, which was simply modeled on the traditional union policy that in matters such as this the union was accustomed to respect local autonomy as part of trade-union democracy. Indeed, the NCB gave National and Area officials of the NUM public credit for overcoming local opposition to foreigners in many branches: there was simply more opposition than union officials could overcome.

Nevertheless, the foreign-labor program was certainly not a failure. In 1947 over 6000 Poles and nearly 1000 EVW's were placed at pits and in 1948 about 8600 additional EVW's were added. In 1949, some 2300 more EVW's joined the industry — but by the middle of the year, the number of new placement opportunities fell off sharply as the saturation point of permitted openings was approached. The exhaustion of placements permitted by local branches, however, was by no means the only reason why the program was tapered off. Housing was short in many coalfields where more foreign workers might have been sent. Although many foreign workers lived in "digs" (lodgings) in colliery villages, the great majority were billeted in temporary miners' hostels, several of which had been built during the war in nearly every coalfield. These barracks accommodations fairly well suited the young, single men who were the typical foreign workers.

Additional recruitment among EVW's was stopped during the last half of 1949: the supply of Poles had been pretty well exhausted by the middle of 1948. In three years these sources had provided nearly 18,000 young men for the industry; this amounted to about one recruit in every eleven entering the industry and equaled nearly half the number of juvenile recruits for the whole three-year period (1947–1949). Qualitatively, many of these foreign workers (particularly the Poles) have proved very useful workers. Most of them were relatively young and many of them were bigger than native Britons; these characteristics sometimes made them appear very desirable to pit managers and as a "threat" to native miners. However, this is a delicate subject about which

it is hard to generalize: the degree of acceptance of foreign workmen
has varied considerably from district to district and even from pit to
pit.

   The Unsuccessful Attempt to Import Italians, 1950–1952. As noted,
the recruiting of foreign labor came to a halt in the last half of 1949, chiefly
because of difficulty in finding pits where such labor was acceptable to
the miners' lodge. We have also seen that the slippage in total manpower
was proceeding inexorably all through 1949 and 1950: there seemed no
hope of keeping the industry manned with British labor, at least not in
the immediate future. Some further stopgap seemed imperative, and
in June of 1950 the government suggested to the NUM that it might be
possible to employ five to ten thousand unemployed Italians. It was not
until the signing of the wage increase agreement of January 1951 (after
several key areas had already voted against taking any Italians), that
the NUM Executive Committee agreed to try to persuade their branches
to accept this new class of foreign labor. A year after the initial suggestion, the first contingent of twenty-four Italians arrived in Yorkshire
for training. Within a few weeks Yorkshire branches completed voting
on the question of whether or not to accept Italians: 87 of the Area's
114 branches turned down the proposal, despite the pleas of Area
officials.

   By the late spring of 1952, rank-and-file opposition to the further
employment of Italians, an opposition which defied the advice and
pleading of both Coal Board and union officials, forced an embarrassing
end to the whole scheme. The 2200 Italians imported on two-year contracts had been sent to Yorkshire, South Wales, and Scotland, but it
was in Yorkshire that the fatal difficulty arose. There, at Bullcroft
Colliery (the largest pit in Britain), a group of haulage hands struck in
March against working with a group they did not like. Nothing Coal
Board or union officials could do could make the Italians acceptable to
the determined Bullcroft haulage workers, and rather than risk a series
of strikes by forcing a showdown on "who was boss," the Board (to the
great relief of the union) decided to stop recruiting Italians in Italy
and not to try to place any more of the Italians already brought to
Britain who were still undergoing training. Of the thousand men affected, about a third found jobs in Belgian mines, another third found
nonmining jobs in Britain, and nearly a third were repatriated.

   The Italian Government raised the question as to whether the Coal
Board would not be liable for fulfillment of the men's two-year contracts.
This question was finally settled by the payment of cash settlements of
£60 to the men repatriated, £40 to the men sent to Belgium, and £25
to those who found nonmining jobs in Britain. Including the training and
transport costs, the Board lost in the neighborhood of £200,000.

   The fiasco caused considerable embarrassment to the British Government in its relations with Italy, to the Board in its domestic public relations, and to the NUM as an agency for social justice, international friendship, and working-class solidarity. At the NUM's 1952 Conference, both Sir Will Lawther and Arthur Horner made strong public statements condemning the action of the union's Bullcroft members. Neither the national nor the Yorkshire Area officials made any attempt to "pussyfoot" this incident; however, the union understandably made no attempt to punish its offending members.

The Problem of Juvenile Recruitment

We have already emphasized the heavy dependence of coalmining on internal recruiting of its labor supply; the most important aspect of internal recruitment is the ability of the industry to attract and hold boys under eighteen. If the industry should prove unable to attract enough individuals from this group, the long-run labor supply would appear hopeless. This point is, of course, only too well realized by the Board and the government, and a special effort has been made to reëstablish this long-run source of labor, a source which had seriously broken down before nationalization.

Before the second World War there was no concern about any inadequacy in the labor supply and little attention was paid to the fact that a decreasing number of boys were coming forward to take up mining as a career. True, it was vaguely realized that fewer boys were entering the industry, but no good quantitative measure of the shortage of boys was available until separate statistics on juvenile recruitment were begun in 1942. There then developed a widespread realization that the industry's labor force was no longer reproducing itself. In order to maintain the 1942 number of employees under twenty-one, some 18,000 boys would have to be recruited every year. At that time, the industry was attracting only about 12,000 boys annually, and this low rate threatened to fall still further.[10]

To prevent this, the government appointed a special committee (the Forster Committee) to analyze the reasons for the decline in juvenile recruitment; its findings pointed to six critical factors which were not ranked as to relative importance. These were: the poor employment record of the industry, which was marked by recurrent crises, prolonged and massive unemployment, and widespread short-time working in the 1930's; the unfavorable wage levels for both adults and boys; the widespread feeling that the working conditions imposed by natural conditions

[10] These figures are taken from the *Manpower Memorandum* prepared in the Mines Department, Feb. 12, 1942, reprinted in the MFGB's *1942 Conference Report*, pp. 215–222. That Memorandum notes that 18 years earlier, in 1924, the industry had received 30,000 juveniles — though the total labor force was then much larger (about 1.2 million).

made mining inherently uncongenial; the discouragement boys received at home and at school, and the disappearance of the family ties on which recruitment and training had traditionally been based; the rapid development of local transport facilities in the 1920's and 1930's, plus the growth of light industries, both of which greatly enlarged the juvenile labor market; and, finally, the decline in the birth rate in colliery districts during the years following the 1926 strike.[11]

The industry's labor force since the war has not differed greatly from that on which the 1942 estimate of juvenile recruitment was based, and we may take 18,000 as the datum figure of the number of boys the industry would like to secure at least until about 1955. During the first four years of nationalization, juvenile recruitment averaged only about 13,500, or three-quarters of the minimum needed. But after 1950 the number of boys joining the industry rose significantly, as Table 12 shows. The 1952 experience, which was matched by the Board's over-all experience with recruitment, was exceptional in that it was undoubtedly achieved only with the help of the heaviest unemployment in the British economy since the war. This increase in juvenile recruitment has been the most encouraging single development in the labor supply picture since nationalization and gives strong evidence that the Board's many reforms are beginning to make themselves felt among the mining communities. The measure of the 1951–1953 record can be appreciated only by fitting it into the general problem of juvenile recruitment in postwar Britain.

The key factor in determining the potential supply of boys for the mining industry in any year is the birth rate in mining communities fourteen or fifteen years earlier. Since 1947 the age at which boys might leave school has been raised from fourteen to fifteen. This defines the group to which the Board can hope to appeal in any year. The fall in the British birth rate which began in the mid-1920's and which lasted up to the second World War meant that there were fewer boys and girls entering the labor market each year during the 1940's and 1950's. Whereas 420,000 boys reached the age of eighteen in 1939, only 320,000 did so in 1950. The combination of the drop in the birth rate and the raising of the school-leaving age from fourteen to fifteen in 1947 was expected to make the number of young persons in industry in 1955 only half the number present in 1945.[12] These underlying population trends mean that mining today faces a degree of competition for boys which the industry has never before known.

[11] The conclusions of the Forster Committee, whose recommendations were chiefly responsible for the extensive reforms in the wartime training and wages of juveniles, are summarized in PEP's *The British Fuel and Power Industries* (London, 1947), p. 75, and in Harold Wilson's *New Deal for Coal*, pp. 154–155.

[12] *Training and Employment of Young Workers*, pamphlet published by the Central Youth Employment Executive of the Ministry of Labour and National Service, Sept. 1949, p. 1.

TABLE 12

New Entrants under Eighteen [a]

Division	1946	1947	1948	1949	1950	1951	1952	1953	1952–53 ave. / 1946–47 ave.
1. Scottish	1,920	1,800	1,190	2,050	2,320	2,840	3,460	2,700	+ 65%
2. Northern	3,420	3,630	2,770						+ 65%
a. North'd & Cumb'd	(Already Satisfactory)			1,140	1,200	1,420	1,670	1,350	
b. Durham				2,430	2,530	2,730	3,040	2,320	
				} 3,570	} 3,730	} 4,150	} 4,710	} 3,670	
3. Northeastern	2,050	2,350	1,940	2,430	2,560	3,430	5,070	3,780	+100%
4. Northwestern	1,220	1,960	1,330	1,590	1,480	2,150	3,910	2,420	+100%
5. East Midlands	1,370	1,690	1,660	1,740	1,670	2,360	3,440	2,520	+ 95%
6. West Midlands	900	1,160	890	990	940	1,590	2,710	1,830	+120%
7. Southwestern	1,690	1,920	1,320	1,670	2,240	2,850	3,540	2,770	+ 75%
8. Southeastern	120	140	120	130	180	230	290	185	+ 82%
9. Total: Great Britain	12,690	14,650	11,000	14,160	15,120	19,590	27,110	19,870	+ 72%

Source: NCB, *Annual Reports*. Figures rounded to nearest decile.
[a] Does not include juvenile reëntrants, though this number would not significantly change the above figures.

As already emphasized, family attitudes have a great deal to do with this industry's ability to compete successfully for boys. Useful insight into this aspect of the problem faced by the Coal Board is given by some of the findings of the 1946 Social Survey into family and juvenile attitudes in selected mining districts.[13] This survey found that while one-third of the boys from mining families had entered mining, only 7 per cent of the boys from nonmining families had done so; that despite the foregoing fact, only 6 per cent of the schoolboys from mining families wanted to enter the pits — which suggested that a high proportion of pit boys were entering mining against their will or only as a last resort; that most mining parents (but only half the nonmining parents in the same colliery districts) had seriously considered mining for their children — a finding which suggested that parental attitudes towards the industry were not, even before the massive reforms of nationalization, hopelessly and irrevocably set against mining for their sons; that the most hopeful source of recruitment would continue to be from mining families, especially from among boys still in school before they accepted other work; and that the status of mining needed to be raised to that of a trade, to which most parents and boys looked.

An additional index of the acceptance of mining by boys is the proportion of them willing to accept underground employment. During the 1930's and 1940's, a growing proportion declined underground work: whereas in 1930 some three-quarters of all boys under twenty were normally employed underground, by 1940 this figure had fallen to two-thirds and to less than 60 per cent in 1948. By the end of 1952, however, the Board had been able to reverse this trend, and the figure climbed to 66 per cent and to 70 in 1953. This fact is important, for among boys who remain on the surface wastage is relatively high, and boys who leave the industry before going underground are not likely to grow into men who can be tempted back into mining by the high wages at the face.

The Board has made a great effort to improve its personnel practices in the handling of boys, to widen chances for advancement by the generous provision of training opportunities, to instill greater pride in key occupations by the establishment of apprenticeship programs, and, by instituting numerous reforms that affect adult employees, to convince parents that mining is an occupation that they can urge their sons to

[13] "The Recruitment of Boys to the Mining Industry," by Geoffrey Thomas, being "An inquiry carried out in Six Coalfields for the Directorate of Recruitment, Ministry of Fuel and Power, and the National Coal Board, August–October 1946." *The Social Survey*, N.S. 77/2/3, April 1947, 77 pp., mimeo. The first 18 pages summarize the very extensive statistical tables which make up the remainder of the study. The sample, taken entirely from mining communities, consisted of 566 parents in nonmining families, 602 fathers and 1425 mothers in mining families, 550 schoolboys, and 884 working boys.

enter rather than avoid. The recruitment record of 1951–1953 strongly suggests that these measures have begun to bear fruit. In the light of the prewar and wartime history and of the over-all decline in the juvenile labor force, the Board deserves credit for a major achievement. In order to learn something more about juvenile recruitment than over-all figures can teach, a study of juvenile recruitment in three areas of the West Midlands division was made by the author in the middle of 1950. This Division, with Yorkshire, has experienced the most trouble in getting boys; it is also one where the coal industry must compete directly with other industries for its labor supply. Some notion of the geography of the Areas studied is afforded by the accompanying map, but the significant aspects of the demand for and supply of labor, particularly juvenile labor, are extremely local. The main conclusions and impressions relevant to each of the three Areas are summarized below.

Cannock Chase: Though not quite 15 miles from Birmingham, Cannock and the dozen mining villages that surround it form a labor market that is largely self-contained, at least as far as boys are concerned. The town does have some small light industry, but from one-half to two-thirds of the job openings for boys lie in the coal industry. Public officials concerned with juvenile placement reported that the coal industry seemed to be getting a reasonable number of boys; although they acknowledged the parental objections to mining, they also noted that since nationalization coal had developed training programs that were far more complete than those characteristic of any other local industries. This attention, coupled with coal's undoubted local advantage with respect to wages and hours, was gradually undoing the prejudice against mining which had grown up over the preceding generation. Cannock pits, however, drew almost exclusively on local boys for recruits — none at all from Birmingham and practically none from nearby Walsall. Town-bred boys, brought up on expectations formed by other industries in their own localities, practically never enter mining even though the mine is only a thirty-minute bus ride away, pays significantly higher wages, and offers considerably more leisure.

Warwickshire: The dozen pits comprising this Area are strung out north and south fifteen miles or so to the east of Birmingham. The latter city is too far distant for all but a handful of boys to commute daily (a few large Birmingham firms do send buses out to the villages in the Warwickshire coalfield, as they do to Cannock, and a few boys use them). At the southern terminus of the field, however, lies Coventry, the Detroit of England; this city is able to attract large numbers of boys from the mining villages at the southern end of the coalfield. A high Area official explained his difficulty in getting boys as follows:

We're mainly up against the automobile industry, both in Coventry and Birmingham. We offer the lads a shift of 7½ hours, plus about half an hour's

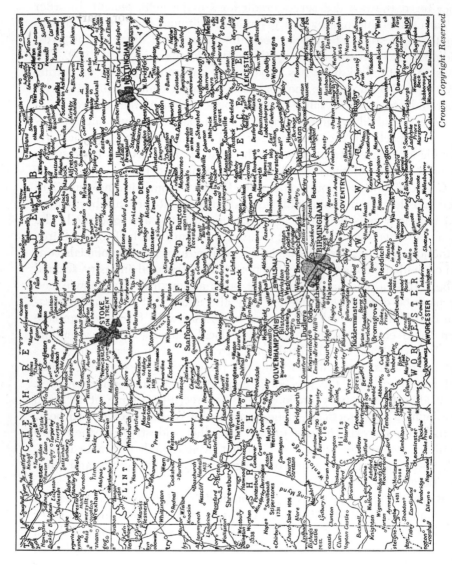

Fig. 3. Cities and Towns in the West Midlands Division of the NCB

Scale: 1 in. = 10 mi.

[204]

travel to and from work, on the average; the auto industry has an 8-hour day and the lads have to travel an hour each way to and from work. But the lads today just prefer 10 hours of daylight employment to 8½ underground. A youth of 17–18–19 can, it is true, earn 5s. to 10s. more a week in mining, but that's not enough to keep a young single man in the industry.

Coalmining, in fact, did not in 1950 appear to hold the wage advantage which the above-quoted official assumed. True, entry wages may have been slightly higher in coal, but it is not hard to get on piecework in the automobile plants and when boys do so their earnings exceed those of coal by a substantial margin.

Statistics showed clearly that much higher numbers of boys were entering mining at the northern end of the Warwickshire field (at the "country pits" north of Nuneaton) than at the southern end. These northern pits are located in what is predominantly rural country, spotted with many small "mining villages" built long years ago. In addition to the geographical barrier that would deter boys in these villages from seeking work in Birmingham or Coventry, there would also exist a set of expectations and social mores much more favorable to pit employment than is true as one moves southward beyond Nuneaton towards Coventry.

A further fact of some importance relates to the number of schoolboys entering mining in the village of Atherstone during the two years 1948 and 1949. This town in the northern end of the field is an important center for the manufacture of men's hats. In 1949, the number of schoolboys entering mining was nearly double the 1948 number even though there were only 92 boys graduating as against 258 in 1948. School authorities ascribed this shift to the development of a local depression in hat-making in 1949. This suggests that even in a country district, where many boys normally enter mining, many do so only in default of alternative opportunities and not because of any positive pull exerted by the rewards of mine employment.

North Staffordshire: This Area is one where the Board hopes to expand mine employment by about 16 per cent between 1950 and 1965. In 1950, however, the Area was getting less than half the number of boys needed merely to replace its present labor supply; in the early 1950's, the Area would like to have been getting three or four times as many boys as it was in fact attracting. The North Staffordshire pits are all located in or right on the edge of Stoke-on-Trent, the center of the British pottery industry. The dominance of "potting" is indicated by the fact that it provided nearly 60,000 jobs (many of them for women) in 1948, while the Area's twenty coalmines employed 18 to 20,000. Building, rubber, and the engineering trade are also important employers.

In 1950, the pottery industry appeared to offer boys 4s. to 5s. a week more than coalmining for a normal workweek, and in many potting firms boys had a chance for overtime work that was not available in coal. The combination of plentiful job opportunities within a convenient travel radius and higher potting wages was widely taken to explain coal's difficulty in recruiting boys. The two top officials of the Kemball Training Center, through which pass all boys recruited to the industry in North Staffordshire, both thought that boys' wages would have to be raised from 10s to £1 a week in order to secure a definite wage advantage for coal in its competition with the potteries. Incidentally, neither in North Staffordshire nor Warwickshire was there any evidence that pit officials were paying any kind of informal premiums above the national scale in order to try to keep boys from leaving the pits.

The fact that mining did not seem to have a competitive wage advantage in parts of the Warwickshire coalfield and in North Staffordshire suggests that boys' wages ought to be raised in those districts. However, such a remedy would require a departure from the centrally negotiated national rates secured by the union, and neither the Coal Board nor the union had seriously considered the possibility of district rates for juveniles, let alone for adults, in the interest of "meeting competition." Either the system of national bargaining and national rates could be maintained at the cost of not getting enough boys in those local labor markets where alternative jobs are plentiful, or nationally bargained rates could be modified to permit adjustments tailored to the needs of local labor markets. The only way to satisfy both objectives would be to raise the national rates to the point where they were competitive in the highest-wage districts; this would mean that boys' wages in Durham and Northumberland (where there has been no trouble recruiting boys) would become considerably higher than they need to be. Such a course would also disturb traditional relationships between juvenile and adult wage levels and would very probably invite a substantial general wage increase. In point of fact, national bargaining and nationally uniform wage increases for juveniles since nationalization have not done away completely with the district differentials that had long been customary. The low-wage districts have, indeed, been raised up considerably, but even in 1953 about a half dozen of the industry's twenty-two wage districts had juvenile scales that ran above the national minimum. The districts with high juvenile rates have not been denied the general increases won for boys in national bargaining and have thus maintained their advantage.

There remains a host of subjects that are important to a rounded understanding of juvenile recruitment which must nevertheless lie beyond our interests. For example, the industry has developed an elaborate training scheme which is now characteristic of the whole industry and which is part of a combined effort to improve safety and to ease the boy's induction to the industry. There is still considerable debate as to whether the scheme adopted has been properly conceived and — a separate question — whether it is being effectively conducted. To many boys, training (involving as it does both classroom and training-pit requirements) looks too much like "more school." Another problem is the controversy over whether boys and young men can reasonably be expected to accept shift work. At many pits boys work on the night shifts, as they do not in many other British industries. In many cases an alert training officer could do much to reduce wastage attributable to dislike of night work. A related question is the early hour at which most day shifts start in mining. Boys in their late teens who like late evenings may often be discouraged by mining's early-rising requirement, a requirement that persists more from force of habit than from anything else. Other important questions are the role played by the miners' present exemption from the draft[14] and by the potential recruiting advantage that might be gained from giving concessionary coal to boys entering the industry from nonmining families, or from defining an "adult" as anyone over eighteen — a change that has been discussed and which would give boys an earlier opportunity for "good money." Indeed, it is in the handling of boys on the job that the Board especially hopes for a positive contribution from improved personnel practices and changed attitudes among managers, training officers, and the line management down the pit.

[14] Boys working in agriculture were likewise exempt from military service until Feb. 1951. When the exemption was lifted for agricultural workers at that time, there was a noticeable shift of boys from agriculture to mining in some districts (e.g., the West Riding of Yorkshire). However, there is no magic in military exemption as a recruiting device for coal independent of relative wages: after the rise in military pay in the fall of 1950, the pits of Northumberland and Durham lost alarming numbers (on the order of 1 per cent of their total manpower) to the forces. The subsequent national wage increases of Nov. 1950 and Jan. 1951 appeared to halt this leakage on the Northeast Coast.

CHAPTER EIGHT

ABSENTEEISM

Although absenteeism has been somewhat lower since nationalization than it was during the second World War, the Coal Board has not been able to reduce it below a level that is almost double the prewar rate of 6 to 7 per cent. A persistent absenteeism rate of 12 to 13 per cent has caused a considerable amount of embarrassment for both the Coal Board and the NUM, for it is frequently interpreted as a symbol of managerial incompetence and worker irresponsibility. Is it fair to conclude that postwar absenteeism rates demonstrate that the miners are incapable of that "new morality" which Sir William Beveridge has called essential if full employment is to be tolerable in free societies? [1] The absenteeism rate has as much practical significance as it does moral and political significance: when absenteeism runs at the rate of 11 to 13 per cent a year, a decrease of just one percentage point would add about a million tons of coal to the industry's output.

It will be recalled from Chapter One that absenteeism is most acute, in both its size and its consequences, among coalface workers. If the "fillers" do not complete their work, it must usually be finished by the next group of men who come onto the face during the cycle, that is, by the conveyor-shifters. (There are usually local agreements at pits which determine what proportion of a face must lie unfilled before the conveyor-shifters can refuse to fill it off.)

If the conveyor-shifters do not clear the face, the cycle is broken and output is reduced not just on that cycle but on the next one as well, since the cuttermen cannot do their work until the face is finally cleared. Such interruptions to the normal cycle cause considerable transfers of men to unfamiliar work until their own face can be put back on schedule.[2]

Absenteeism among haulage workers underground can also have serious effects. The usual source of replacements for haulage jobs when haulage hands are off work is to pull a pieceworker off development

---

[1] See Sir William Beveridge, *Full Employment in a Free Society* (London, 1945).
[2] A good discussion of the qualitative aspects of absenteeism will be found in Zweig, *Men in the Pits* (1949), pp. 56–70.

work or off another producing face (replacements are scarcely ever sent down from surface jobs). This either slows development work or raises the danger that the face from which the replacement was pulled may not get filled off.

It is only among surface workers that absenteeism is not much of a problem, either in the sense of magnitude or consequences. Absenteeism is thus highest precisely at the point where it has the worst effect and lowest where it has the least effect.

About all that pit managers can do to discount the effects of absenteeism is to inflate their manning schedules: if they could operate a face with twenty-five strippers who lost very few shifts every week, they must try to assign thirty men, say, to that face if absenteeism is running high. In a qualitative sense, then, it might be said that high absenteeism tempts management into labor hoarding, though the manpower shortage has made this intention difficult to execute. The hoarding of labor to "cover" absentees has probably not been nearly as serious a drag on output as the daily crises which require overmen and undermanagers to shuffle their men around from job to job at the start of each shift, thus delaying the day's start on the jobs to which men are temporarily sent and preventing the accomplishment of jobs to which they are regularly assigned.

Some of the Characteristics of Absenteeism

The absenteeism rate is defined by the following ratio:

$$\frac{\text{Manshifts not worked}}{\text{Manshifts worked plus those not worked}}$$

Such a figure can, of course, be calculated on many different bases: on a daily, weekly, monthly, or annual basis; for specific classes of men, such as facemen, daywagemen, surfacemen, deputies, and shot-firers, or all workers; for any desired geographical unit, such as a single pit, a Coal Board Area or Division, or the industry as a whole. All figures naturally must begin with records compiled at individual pits and, while the Coal Board has attempted to make sure that the records are compiled on a uniform basis throughout the industry, there is some reason to doubt that it has been wholly successful.

Since absenteeism is, in part, a measure of "shifts not worked," it might be thought that a rise in the rate of absenteeism over time would necessarily mean that people were working fewer shifts. This does not always follow, however, since absenteeism is calculated on the basis of shifts lost as a proportion of shifts possible, which may fluctuate with the length of the workweek. Harold Wilson reminded critics of the miners during World War II that although absenteeism more than doubled between 1938 and 1944, the number of shifts actually worked increased.[3] The same thing occurred again between 1950 and 1951: absenteeism increased slightly, but so did the number of shifts actually worked, presumably due to an increase in Saturday working.

There is considerable difference between the amount of absenteeism experienced on different days of the week; this difference is regular in the sense that it is predictable and is undoubtedly characteristic of nearly all pits. There is also an observable seasonal ebb and flow to absenteeism, as well as specific weeks in the year when absenteeism is invariably very low; these are the so-called "Bull Weeks" when men have a particularly strong incentive to maximize their earnings — the weeks that precede the weeks in which holidays fall. There is, as indicated above, a great difference in the amount of absenteeism incurred by the main functional groups in a pit — (1) faceworkers, (2) others underground, and (3) surfacemen. These functional differences correspond to different earnings levels and confront us directly with the problem of whether or not absenteeism always tends to increase as wages increase. The Board's divisions have shown considerable variation in experience. Some have consistently had low absenteeism, others have struggled with a consistently high level.

Let us say something about each of these aspects of the problem.

DIFFERENT DAYS OF THE WEEK

In coalmining, as in other British industries, absenteeism is regularly much higher on certain days of the week than on others.[4] An examination of statistics in several Areas reveals that absenteeism is invariably higher on Mondays and Saturdays than it is over the week, averaging all shifts from the Monday afternoon through the Friday day shift. Among facemen, Monday absenteeism is customarily about 50 per cent higher than the weekly average. Among "others underground," Monday is about 25 per cent worse than the weekly average. For surface workers, the Monday rate is only slightly worse than the weekly average. Taking all three classes of workmen together, the Monday rate runs roughly 20–25 per cent over the weekly rate.

Saturday is considerably worse than Monday, this being especially true for facemen and "others underground" and not particularly true for surface workers. This poor Saturday attendance has two main effects:

---

[3] *New Deal for Coal,* pp. 54–55.

[4] An excellent, detailed study of absenteeism in five Midlands metal-working firms (covering 57 plants) is contained in *Absence under Full Employment,* by Hilde Behrend (Monograph A3, Studies in Economics and Society, University of Birmingham, 1951, 138 pp., mimeo.). This and a companion study on labor turnover were designed in part to secure data on key aspects of labor utilization under conditions of full employment; the studies found both absenteeism and turnover to be much higher during the postwar period of full employment than they had been prewar.

(1) it disrupts the cycle of operations, so that Monday's work requires a lot of "tidying up" from the disorganization left by Saturday (Monday often is bad enough in its own right), and (2) there is often a serious imbalance in the manning of pits on Saturdays, with the normal ratio of underground and surface workers to facemen greatly distorted. This, most managers believe, leads to very high production costs for Saturday work. Except during "Bull" weeks, 40–60 per cent of the facemen are usually off on Saturdays; the figure is only slightly lower for "others underground." However, surface workers generally attend well on Saturdays. This is because they are the lowest paid workers and "the only way we can 'make a go of it' is to get the time and one-half for the Saturday work."

This explains why there is sometimes conflict of interest between the facemen and the surfacemen (and it must be realized that the branches are predominantly "facemen's unions") when a decision is being sought as to whether the pit will work a voluntary Saturday shift or not. The facemen generally do not like Saturday working, because, even though it is "voluntary," there is a lot of feeling generated against those who do not volunteer; the surfacemen cannot afford to forego this extra day at time and one-half as easily as the facemen can. In many cases, facemen may have voted for Saturday working with no intention of working themselves but simply because they know that it would give surfacemen a chance to "make out."

During the "Bull" weeks, especially the most important Christmas "Bull" period — really two or three weeks — absenteeism drops sharply on Saturday and is just about the same through the week. But Monday absenteeism remains high — considerably higher than on Saturday at Christmastime. This would indicate that some workers were choosing to take Monday off in the knowledge that they can work Saturday at overtime rates. It also indicates that the effect of taxation is not so severe that men are never willing to work overtime (particularly facemen, who would be most affected by the higher rates of tax). Clearly, it pays them to work overtime at Christmas, anyway. During holiday weeks, not only does weekend absenteeism shoot up far above normal weeks but the "weekends begin earlier" — and Friday noon and night absences are high. If weeks following holidays were shown, Monday absenteeism would of course also be much higher than normal, as every colliery manager expects.

"BULL" WEEKS

As indicated above, there is a characteristic dip in absenteeism during every "Bull" week, of which there are about six each year. These predictable spurts of output occur in the weeks for which men will be paid just before a holiday or before their vacation; since payday follows by one week the week for which men are being paid, "Bull" weeks usually occur two weeks before the event which explains them. Sometimes, there is more than one "Bull" week before a holiday, as men bear down to accumulate spending money; this is naturally most pronounced in the Christmas season when expenditures are abnormally high. While absenteeism decreases markedly during "Bull" weeks, it by no means disappears altogether (even "voluntary" absenteeism remains at 3 to 5 per cent). We shall return to this point in our later discussion about the difficulty of establishing norms of "reasonable" absenteeism.

SEASONALITY

There exists a fairly regular seasonal pattern to absenteeism, a pattern which makes it more difficult to think of reducing absenteeism to some dead level that would remain unchanged throughout the year. The normal pattern of seasonality is as follows: a seasonal low (not necessarily the lowest of the lows) is reached in November; a modest increase then occurs for December, January, and February. The rate then falls from March through May, roughly. From the late-spring low, absenteeism normally rises during June, July, and August, reaching the annual peak during the month in which almost all miners take their vacation (August). During the fall months, from September through November, attendance improves; these are the critical months so far as the country's stock position is concerned and if any particular "season" can be called the "Bull" season, it is these fall months when individual and national needs luckily coincide. This seasonal pattern is based on an examination of the national figures for total absenteeism. Figures are also published for "voluntary" and "involuntary" absenteeism; an examination of these separate figures shows a somewhat greater regularity for involuntary than for voluntary absenteeism; during the past seven years, voluntary absenteeism has shown occasional sharp increases (followed by decreases) which involuntary absenteeism has not.

DIFFERENCES AMONG THE DIVISIONS

Just as we saw that some Divisions have much better strike records than others, so too with the level of absenteeism—though the Divisions arrange themselves somewhat differently on the two counts. The most notable differences are that Scotland, which has had a lamentable strike record, has maintained an admirable absenteeism record; on the other hand, the West Midlands, which has had an admirable strike record, has had a lamentable absenteeism rate.

Table 13 gives the figures for total absenteeism for each Division, 1946–1953; the right-hand half of the table shows the rank of the Division each year, the highest rank implying the lowest absenteeism. The ranking of the Divisions remains remarkably constant from year to year,

TABLE 13

Over-all Absenteeism by Divisions, 1946–1953

Division	Absenteeism Rates								Divisional Ranking							
	1946	1947	1948	1949	1950	1951	1952	1953ª	1946	1947	1948	1949	1950	1951	1952	1953
1. Scottish	12.57	9.48	9.64	10.58	10.63	10.19	10.25	9.25	1	1	2	3	3	3	3	3
2. Northern	12.78	9.68	8.47						2	2	1					
N & C				9.31	9.42	9.27	8.64	8.91				1	1	1	1	1
Durham				9.79	9.96	9.45	9.09	8.96				2	2	2	2	2
3. Northeastern	19.08	15.15	14.38	14.80	14.34	14.64	15.19	14.37	6	6	6	8	9	9	9	9
4. Northwestern	17.28	13.25	11.78	12.28	11.93	14.52	13.53	11.70	4	4	4	5	4	8	6	5
5. E. Midlands	18.07	14.36	12.96	13.24	12.27	12.03	11.81	12.84	5	5	5	6	6	4	4	7
6. W. Midlands	19.62	15.69	14.84	14.83	13.38	13.71	13.54	14.26	7	7	8	9	8	7	7	8
7. Southwestern	14.16	11.04	10.57	12.24	12.11	12.37	12.02	11.29	3	3	3	4	5	5	5	4
8. Southeastern	21.05	16.43	13.08	13.27	12.30	12.39	14.28	12.81	8	8	7	7	7	6	8	6

ª Absolute figures not comparable with earlier years; the 1953 *Report* contains absenteeism figures based on a five-day week, a basis not used uniformly in all divisions in earlier years.

Source: NCB, *Annual Reports* (1947–1953).

a phenomenon that suggests the existence of real differences in the underlying influences at work, whether they lie with temptations to poor attendance or the effectiveness of control measures.

The three northern Divisions have a markedly superior record; at the other end of the scale lie the Northeastern, West Midlands, and small Southeastern Divisions, which quite consistently have shown roughly 40 per cent more absenteeism than the three northern Divisions. One often hears offhand explanations for these differences (for example, lower mine wages in the north, fewer opportunities for supplementing family income from female employment); but such "explanations" scarcely scratch the surface: adequate explanations would be highly complex and to date the necessary research that might provide sure answers has not been undertaken.

Postwar Absenteeism

Now that we understand something of the detailed ebb and flow of absenteeism, we can proceed to examine the course of the over-all absenteeism rate since nationalization. Chart 8 shows us the course of total absenteeism (voluntary and involuntary) from 1943 through 1952. The first year in which separate figures were published for voluntary absenteeism was 1943. One should bear in mind that during the late 1930's (and indeed all during the 1920's and 1930's) total absenteeism ran at what now seems the very low rate of 6 to 8 per cent. The figure rose quickly very early in the war until it reached the 12 per cent figure in 1943 with which Chart 8 opens. The high point was reached in 1945: the large amount of "green labor" then employed and the relaxation of pressure attendant upon the ending of the war may account for that year's record.

The figures leave no doubt that absenteeism has come down from the very high levels to which it had gone by the end of the war; indeed, the reduction was about 25 per cent — from a rate of around 16 to about 12 per cent. The year in which this reduction was mostly secured was 1947, the first year of nationalization and the year in which good attendance was first rewarded with a substantial bonus. Since 1947, the over-all rate has fluctuated inconclusively, with the 1952 rate not greatly different from the 1947 rate.

The point of greatest interest, perhaps, is not the bald trend of the annual figures but the question of how to interpret them: has the Coal Board done a tolerably good job in controlling absenteeism or has it failed? How conscientiously have the miners "accepted their responsibilities" under nationalization and full employment?

It is extremely difficult to be positive in suggesting standards by which the reasonableness of absenteeism rates ought to be judged. Al-

most any definite standard is open to criticism of one kind or another;
let us look at some of the standards which have been suggested or implied in the course of the political debate that has raged around this
subject.

Some extremists have felt that there ought not to be any "voluntary"
absenteeism. The unreasonableness of any such standard becomes apparent when one stops to reflect on the way in which a relatively few ab-

Chart 8

Voluntary, Involuntary, and Total Absenteeism per Year, 1943–1952

(Per Cent)

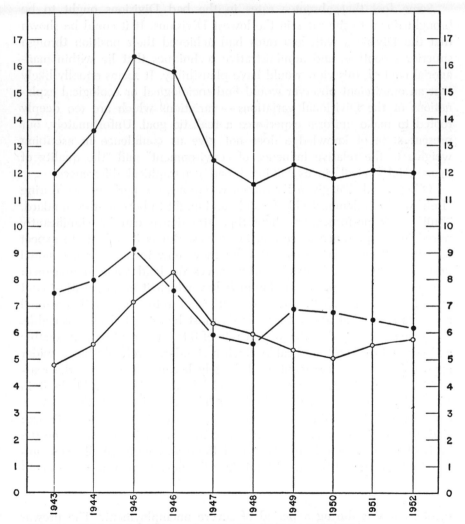

Top line = Total; black-dot line = voluntary; white-dot line = involuntary

Source: Ministry of Fuel and Power, *Statistical Digest, 1948 and 1949;* NCB, *Annual Reports;*
Coal Figures.

sences affect the absenteeism rate: a man who loses one shift a month incurs an absenteeism rate of 4–5 per cent. Long absences by a very few members of a large group also have a marked effect on the group's total rate: if, for example, one man in a group of 100 men should be off continuously for a month in which all could have worked 20 shifts, he alone gives the group a 1 per cent rate. Given the nature of coal-mining and the ambiguous meaning of a "voluntary" absence (it may sometimes be involuntary even though formally unexcused), a rate of 3, 4, or 5 per cent cannot be called intolerable. Actually, the voluntary rate has run 5 to 7 per cent.

Some feel that absentee rates in the bad Divisions ought to be brought down to the rates in the lowest Divisions. If it could be shown that the Divisions with low rates had achieved their position through superior incentives and administrative techniques that lie within managerial control, this view would have plausibility. It seems equally likely that an omniscient observer would find sociological or ecological explanations of the Divisional variations — variations which are too deeply rooted to make uniform experience a realistic goal. Unfortunately, our present state of knowledge does not give us confidence in ascribing weights to the relative influence of "environment" and "the quality of managerial control" in explaining broad geographical differences.

Others hold that given Divisional variations, the performance during "Bull" weeks is demonstrably feasible and ought to become the standard. "Bull" week performances show that attendance can be significantly raised in the presence of certain incentives. But is it realistic to expect men to perform continuously at what they show to be their occasional best? However, "Bull" week performances do hold some significance: they suggest an absolute minimum below which it is unrealistic to set our sights. More realistically, the "Bull" week rates suggest that annual averages somewhat above these levels must be accepted as reasonable.

Some critics say that the prewar rate of 6 to 8 per cent was reasonable and ought to be still applicable. Such a standard probably has a wider appeal than any so far mentioned, largely because it assumes the reasonableness of "getting back to normal." If the miners could do it in 1938, why can they not do it in 1952? The difficulty with this view is twofold: (1) the validity of the prewar figures is widely questioned; since the problem was not a pressing one then, reporting was done on the basis of a small sample of firms, and there was undoubtedly less uniformity in the definitions used than since the development of government control during and after the war; (2) more important, how reasonable is it to expect attendance to be as high during a period of full employment as it was during a period of severe unemployment? The prewar years certainly constituted no Golden Age for the mining industry and

it seems a doubtful justice to base full employment standards on performance under conditions of massive unemployment.

Finally, some feel that people ought to attend work in coalmining at
least as faithfully as they do in other trades. This is not often given the
dignity of a serious standard, for most people recognize that coalmining
gives rise to special reasons for absences not present in most other
industries. Still, people inside and outside coal often do make implicit
comparisons between the miners' attendance and those of "factory
hands." But little information is available about absenteeism in British
industry outside coalmining — which is the only industry for which regular statistics are published in government sources. This puts the coal
industry at a real disadvantage in its public relations: its own figures are
so well known, and all other industries' so unknown, that the public is
tempted to compare coal's 12 per cent with zero rather than with, say,
6 to 9 per cent.[5]

It is worth remarking that for 1948 and 1949 absenteeism in British
mines was slightly lower than in every European country except the
Saar (where the rate was so low that one is forced to wonder about their
comparability). Letters to twelve coal associations in eastern United
States in 1951 yielded three impressions (no absenteeism figures are
regularly published for the American coal industry): (1) during World
War II, total absenteeism in the reporting districts was at least as high
as in Great Britain (that is, 13 to 20 per cent); (2) four of the districts
report that since the war absenteeism has dropped to 4 to 6 per cent
(over-all), but three reported a rate of 10 to 15 per cent in mid-1951
which they did not feel exceptional; finally, (3) several correspondents
noted that short-time working in their district had undoubtedly contributed to a reduction in the rate during the postwar years. If one makes
allowance for (1) the more difficult working conditions in British pits,
and (2) the greater amount of short-time in the American industry since
the war, British absenteeism does not seem extraordinarily high.

A "rule of reason" suggests that if the over-all, annual rate could be
brought down to about 10 per cent under conditions of full employment,

[5] In the study cited in fn. 4, p. 210, Miss Behrend reports total absenteeism rates
of 4.6 per cent and 3.9 per cent for men in 51 Midlands metal-working plants
in 1947 and 1948, respectively. There was, of course, a considerable range among
the 51 factories studied (from 7.1 per cent to 1.3 per cent in 1947). In citing these
figures, we do not mean to imply that absenteeism in coal ought to be expected to
drop to these levels — there are *a priori* reasons for accepting a somewhat higher
rate in coal. With respect to the plants she studied, Miss Behrend suggests that
the "practically attainable minimum" rate might be the rate normally found on
Fridays, when, as in coal, absences tend to be fewest (perhaps because it is payday, though other subjective forces based on the rhythm of weekly living may reinforce this economic force). She does not deal with the problem presented by the
probable difference between the Friday rate in 1948 and the Friday rate of the late
1930's.

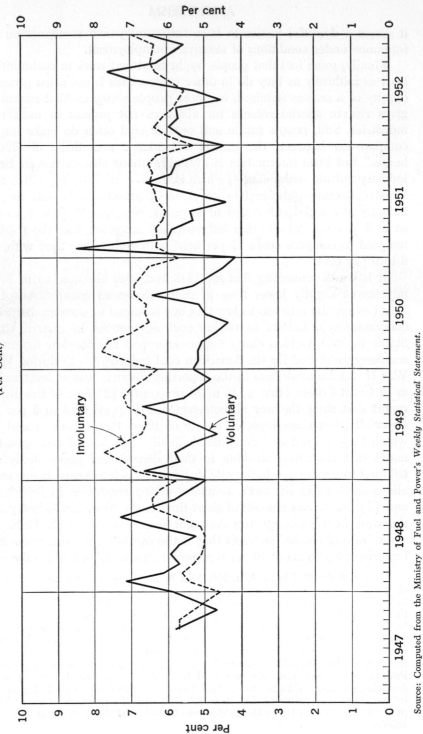

Chart 9

Voluntary and Involuntary Absenteeism by Months: August 1947–December 1952
(Per Cent)

Source: Computed from the Ministry of Fuel and Power's *Weekly Statistical Statement.*

such a rate would represent about all that even the best handling of the problem could yield. Within this figure, we should certainly have to accept seasonality, "Bull" weeks, Divisional differences, and daily and occupational differences — though the latter three might reasonably have their present wide ranges reduced somewhat. Even a 10 per cent rate may be unrealistically low as an industry-wide rate for coal under conditions of full employment: certainly many more years at 11.5 to 12.5 per cent will put a stamp of "normalcy" at that level. The difficulty, of course, is that the persistence of any rate for a long period under similar employment conditions implies that "whatever is, is right." This is no more defensible than the implication of many Coal Board critics that "whatever was, is right."

State Welfare Benefits

When National Health Insurance (entitling people to sick benefits) was introduced in July 1948, nobody knew how seriously this new right would affect industrial attendance. Critics feared that cash payments made to workers who were sick for three or more days would operate as a subsidy for workers desiring more leisure, thus encouraging malingering. Trade-Union and Labor Party members bet on a higher estimate of "human nature," accepted the calculated risk involved, and reasoned that even if attendance suffered somewhat, this would reflect a triumph of justice instead of malingering. What do the coal industry's statistics tell us about the trend of absenteeism following July 5, 1948, the date of the scheme's introduction?

If health insurance encouraged malingering, this should be reflected in a rise not only in involuntary absenteeism but possibly in total absenteeism. A second statistical measure may also have something to say on this matter: this is the weekly figure of wage-earners off work all week. Did either or both of these series rise after midsummer of 1948?

THE TREND OF INVOLUNTARY ABSENTEEISM

The course of voluntary and involuntary absenteeism for each month from August 1947 through December 1952 is shown in Chart 9 (annual but not monthly figures are available from 1943). The annual figures for 1943–1948 do not reveal any "normal" relationship between voluntary and involuntary absences: for 1943–1945, involuntary was slightly above voluntary; for 1946–1948, voluntary was slightly higher. Since the difference between the two rates is not usually great, small changes have little significance. From Chart 9 it can be seen that the two curves run close to each other until the end of 1948, when involuntary absenteeism climbed to a level clearly higher than the 1947–48 levels and voluntary absences fell. There is a clear separation of the curves during 1949 and

1950, but they are seen to return to a much closer relationship early in 1951. If the curves had continued apart after 1950, the obvious inference would have been that the introduction of the health scheme in mid-1948 led to a rise in involuntary absences. While this was probably true for a short period, the fall in involuntary and the rise in voluntary absences after 1950 rob our conclusions of much definiteness. Nevertheless, it is significant that total absenteeism did not rise perceptibly after the introduction of the health scheme. If malingering were widespread (nobody denies that there is occasional malingering), the over-all standard of attendance would have fallen; indeed, this would be one of the most significant tests of the extent of malingering. Instead, there seems to have been only some temporary substitution of involuntary for voluntary days off. Many pits have "sick funds," financed and controlled by a committee of workmen elected at the pit. In 1950, several pits reported that the drain on these funds had risen sharply after the introduction of the health scheme, reflecting the spread of the habit of securing a doctor's note under the stimulus of the new right to state sick-benefit.

The second significant figure for testing the effect of the health scheme is the trend in the number of men who remain off work for a full week: did health insurance not only encourage more men to "get sick" but also discourage them from "getting well"? If this were the case, presumably the figures of men staying off work for a full week would rise after July 1948.[6]

National figures are of some, but not great, help; consequently we shall cite evidence drawn from a detailed examination of the weekly return for one of the Board's fifty Areas (Castleford, Yorkshire). These figures left no doubt that there was a significant rise in the number of men taking full weeks off after July 1948: during late May, June, and early July for 1949, such absences were running 55 to 60 per cent higher than in the same months of 1948. By 1950, in the same weeks, the figure was 30 to 35 per cent higher than in 1948 (though down from the 1949 rate). There were also counterseasonal trends in the figures for 1949 which confirmed the impression given by the absolute numbers. No national figure on the number of men off for whole weeks is published. However, the Board's Annual Reports have included, since 1948, a figure of "effective" to "standardized" wage-earners. Standardized wage-earners refers to the gross number of men listed on a pit's employment records adjusted slightly "for those absent over very long periods, mainly for reasons of injury, sickness, etc." "Effective wage-earners" is the standardized figure "less those absent for the whole of any week from what-

---

[6] The Board's 1950 *Report* stated (p. 12) that "nearly two-thirds" of all absenteeism was accounted for by absences "of one complete week or longer, and most of these were caused by sickness or accident." But the Board pointed out that these relatively lengthy absences were not as upsetting to operations as were the short, voluntary absences which accounted for the other one-third of all shifts lost.

ever cause." If the proportion of effective to standardized falls, it means that more people have been accustomed to taking off full weeks. This figure did in fact fall from 1948 to 1949 (from 91.32 per cent to 89.77 per cent); however, it rose during the next three years, reaching 90.4 in 1952, before dropping sharply to 88.5 in 1953. There does not, therefore, seem to have been any progressive deterioration in the numbers of full weeks lost since the health scheme was adopted in 1948.

What is the meaning of these figures? Although not conclusive, they suggest just about what one would have expected: that after the introduction of sick benefits in July 1948, it took workers a short while to become familiar with their new benefits; but, once they had done so, more people took off full weeks than had done so previously. But does this imply malingering on a large scale? Does it, in short, represent "justice under the welfare state" or "malingering because of the welfare state"? Were people exercising rights which they had formerly been denied (that is, the right to remain off work until fully recovered) or had the health benefits led to a deterioration of personal responsibility toward their employment? The answer must be that, in the absence of extreme changes in behavior patterns, judgments remain highly subjective. People who "believe in the health scheme" will interpret evidence such as that just given as indicating a new standard of justice; people who do not believe in it will read the evidence less generously.

MALINGERING THROUGH SICKNESS BENEFITS

Few miners' leaders would deny that there are some men in their districts who take advantage of the state's sick benefits. Nevertheless, as one prominent miners' leader remarked with some relief: "Really, you know, there has been much less abuse of the scheme than we feared." Assuming this judgment to be valid, the problem of malingering becomes one of identifying and dealing with specific individual cases rather than one of recasting the health scheme to alter the general behavior of large numbers. The return of the Conservatives in October 1951 led to reform proposals that included only a charge for medical prescriptions — not for the privilege of visiting one's doctor. Thus no new barrier to securing "doctors' notes" was raised. The government has been much more concerned about the effect of the scheme on the demand for prescriptions than on the supply of labor to industry, though it may be politically easier to show concern for the cost of prescriptions.

One complaint some pit managers make is that the introduction of health benefits in July 1948 affected the pattern of absenteeism: before the scheme, absentees took their days pretty much at random; since the health scheme, at least some men have learned to take their "time off" in blocks of full weeks every few weeks. The reason the health scheme might produce this change in behavior is that sick benefits are paid for the first three days of absence only if the individual subsequently incurs nine additional days of illness during the succeeding quarter; hence in order to "make it pay" to be sick three days men may find it necessary to be sick nine more days (any combination of three or more days and nine or less will yield the same income). Despite the logic of such behavior, its extent is most difficult to prove statistically, and no pits could offer records to substantiate the occasional belief that the health scheme had produced such behavior.

MALINGERING THROUGH ACCIDENT BENEFITS

Men injured on the job become entitled to weekly injury benefits that are considerably higher than the sick benefits payable in case of illness that is not employment-connected. The private coal-industry scheme simply increases the main benefits paid by the state; both these so-called supplemental injury benefits are dispersed by the Coal Board's payroll offices at the pits, but the industry itself does not participate in the administrative process which determines a claimant's eligibility; the Board accepts liability upon notification from the Ministry of National Insurance that the man has been adjudged eligible for the regular state benefits under the Industrial Injuries Act.

It is not as easy for a would-be malingerer to feign an accident on the job as it is to feign sickness in his home. Charges of malingering on accidents, therefore, are found not in the initiation of claims but in their expiration: many managers suspect certain men of prolonging their injury-period beyond that which is medically justified.

They also feel that some men take advantage of the scheme by claiming the right to return to light work for an indefinite period when they could perfectly well resume their former jobs; if put on light jobs, they become entitled to a benefit which makes up part of the difference between their former pay and their new, lower rate (if, as is usually the case, the light job carries a lower rate; typically, a man on piecework may return to a daywage job off the face and, through his supplementary benefit, receive earnings not much lower than a faceman's earnings). The net result is that a man is relieved, to some extent, from economic pressure to return to the work on which the management would often like to place him.

There is not yet enough evidence at hand to pass judgment on the extent of malingering of the kind just noted. As in the case of the health scheme, few pits have marshaled any data that reveal their experience on a quantitative basis. But that a problem exists, albeit of unknown magnitude, is beyond doubt. The great difference between the prewar administration of private workmen's compensation, now superseded by the Industrial Injuries Act, and the present arrangement is that the employer is no longer represented by his own medical witness before the tribunals which determine eligibility: indeed, it was with the desire to remove accident claims from the arena of interest and litigation that the Industrial Injuries Act was enacted. Under such a purpose, it would seem inevitable that the employer's natural interest in returning men to work quickly and to their former jobs would receive less recognition than it once did.

Causes and Distribution of Absenteeism among Various Groups

People in the industry are perfectly familiar with the wide range of reasons which give rise to absenteeism, even voluntary absenteeism; but practically no one can evaluate accurately the quantitative importance of the many separate causes at work.[7] The reasons men "play" (to use the colloquialism for "not working") would include such things as high individual wages, high family earnings where wives or daughters work, a relatively low standard of living, the necessity for putting in a garden in favorable spring weather or for reaping in the fall, the widespread love of sport which takes many men from work on days of race meetings or football games, the proneness of miners to such "invisible" illnesses as rheumatism and muscle strains, the distaste of younger men for night-shift work, the adverse tradition of short-time working in most districts in the 1930's, the lack of stamina among some of the older men, and so on. Any weighting of these reasons is not only difficult; it would have to be varied from district to district and indeed from pit to pit. It is obvious, too, that many causes can be at work simultaneously — a man who wakes up a bit late, with a stiff leg, and hears it raining outside, is not likely to go to work if he has been earning good money lately, has a low standard of life, can look forward to a tax rebate the following week if he stays in bed, and recalls that he and his mates are currently working in a wet seam where he is drenched to the skin half an hour after starting his job, especially if his wife and eldest daughter are working.

A pure rate of absenteeism tells us nothing about whether the absenteeism consists of many absences by a few individuals, a few absences by many, or some intermediate frequency distribution. This question of whether absenteeism is intensive or extensive has great significance for

[7] The Coal Board has undertaken a limited amount of research on the subject and the findings of one study whose results have come to my attention are of great interest (this is a study conducted by the East Midlands Division entitled *A Further Enquiry into the Causes and Incidence of Voluntary Absenteeism* (summer, 1950, 35 pp., mimeo.). There was established, in 1948 or 1949, at the Board's London headquarters, a "Joint Committee on the Investigation of Attendance," which was presided over by the Board's chief scientific member, Sir Charles Ellis. The Committee conducted one statistical field survey of the problem in 1949 and has tested the routine figures on absenteeism for correlation with other figures. The Committee's findings were mimeographed for internal use.

the problem of control: if it represents a lot of absences among a few men — by an identifiable minority — then both the Board and, especially, the Union are in a much stronger position to punish offenders. If, on the other hand, absenteeism is spread over a large proportion of the work force, then not only is the control problem made much harder but one would also have to take more seriously the indiscriminate indictments of the miners which occasionally find their way into print.

There can be no doubt that almost all men take an occasional unexcused absence — behavior everyone takes for granted and readily "excuses." But there seems equally little doubt that the "problem" of absenteeism is primarily one created by a fairly small minority of men who take off an undue amount of time. This minority, like that which engages in unofficial strikes, is referred to as men "who are not playing the game," a phrase heard frequently in union circles. The phrase came into use during World War II before either the government or trade-union officers were completely convinced that it was only a minority who were "not playing the game." The gradual accumulation of statistical evidence that this "minority" assumption was in fact true has led to its freer and more confident use.

This conclusion emerges quite firmly from most talks with union and production officials and it is confirmed by the rather meager statistics which have been collected on the subject. Probably the most thorough analysis of the subject is a study conducted by the Labor Department of the Board's East Midlands Division in February 1950.

This investigation carries no references to similar studies of the problem in any other Divisions; it does, however, refer to a statistical investigation of absenteeism among coalface workers by a research group at the Board's London headquarters, presumably the study organized by Sir Charles Ellis in the late spring of 1948.

The East Midlands study consisted of a careful statistical analysis of both voluntary and involuntary absenteeism at fifteen selected pits in the Division's four wage districts, plus interviews with 5133 absentees. We shall summarize the study's findings with respect to (1) the distribution of absenteeism among the work-force, and (2) the main causes of absenteeism.

The findings confirm the view that a high proportion of total absenteeism is accounted for by a low proportion of the men employed. They concluded that 17 per cent of the men who voluntarily lost any shifts during the period of the inquiry accounted for 49 per cent of all the shifts voluntarily lost. This 17 per cent of the voluntary absentees constituted only 6 per cent of the men employed. This means that about one-twentieth of the men accounted for about half the voluntary absenteeism. The foregoing figures referred to all men employed; an almost identical conclusion was reached with respect to faceworkers: 15 per cent of the facemen who lost shifts without excuse accounted for 44 per cent of all unexcused shifts lost. This 15 per cent represented 8 per cent of all facemen employed. Slightly more than one-third of all men employed voluntarily absented themselves at some time or other during the inquiry. The figure for faceworkers was 55 per cent. Another way of putting this is to say that about two-thirds of all workers, and nearly half the facemen, did not lose any voluntary shift during the period under review.

These conclusions give strong support to the widely held belief that the problem of absenteeism is mainly a "minority" problem.[8] But they tell us nothing about that minority. However, the study presents further evidence by classifying (and cross-classifying) voluntary absentees according to class of work, age, and shift worked.

Two aspects of voluntary absenteeism which are commonly assumed to be true are borne out by the East Midlands findings: that men in their twenties take off a disproportionately large amount of time, compared with older men — the percentage figures decrease with almost every advance in age; and faceworkers take off considerably more time than "others underground," who in turn take off considerably more than surface workers. Indeed, faceworkers in this sample took off almost exactly twice as much time as "others underground" and more than three times as much as surface workers. Do faceworkers take off more time because they tend to be the younger men or does the younger group show a higher rate because it is heavily weighted with faceworkers? Undoubtedly the two influences reinforce each other. It is certainly true that surfacemen, whose absenteeism is the lowest of the three customary classifications, are heavily weighted with older men and enjoy a wage level roughly half that of facemen.

Further tables in the East Midlands study leave no doubt that, for that sample, voluntary absenteeism increased significantly as one moves from the day shift to the progressively less convenient afternoon and night shifts. Night shift absences were, on an over-all basis, 46 per cent greater than for the day shift, with the afternoon shift about midway between.

The study also analyzed the causes of voluntary absenteeism according to five characteristics of the absentees: age, grade of work, and three marital statuses (single, married men with children, married men without children). The principal findings were that the main causes were alleged (that is, uncertified) sickness; lying-in; illness in the family; high income ("irresponsible attitude other than attendance at sports events"); domestic reasons other than births, bereavements, and illness in family; and men not formally scheduled as incapable of five days' continuous

---

[8] Additional confirmation of this view is given in connection with figures on the proportion of men who fail to qualify for the attendance bonus (see below, p. 227 ff).

work but considered to be so incapacitated. Just over half (50.55 per cent) of all voluntary absenteeism was put down to alleged sickness. When "lying-in" and "illness in family" are added to this, almost three-quarters of the voluntary absenteeism was accounted for. The foregoing causes operated with marked similarity of importance (though not of intensity) among coalface workers, others underground, and surface workers. The main exceptions were that lying-in was less common with surface workers (whose shifts commonly begin an hour or two later than that of underground men), as was high income. There was a noticeable similarity of causes by age groups: alleged sickness, lying-in, and illness in family ranked 1-2-3 for all age groups except for men "over 50." The analysis by marital status confirmed what one would expect: that single men showed the highest percentages of "lying-in" and "high income — irresponsible attitude other than attendance at sports events" and that married men with children showed the highest percentages for "illness in family" and "domestic reasons other than births, bereavements, and illness in family."

Certain of the reasons given for absence are "excusable" while others are much less so. In any interview, the absentee's natural bias in almost all cases would be to hide socially inexcusable reasons for nonattendance by giving the interviewer a more acceptable reason, of which sickness is the most convenient. "Real" reasons for absences could not be checked at all after the fact; to check them on the day of occurrence would require much more fully developed personnel work by the NCB, plus a willingness to risk sharp criticism from the mining communities.

Control Measures

A huge effort has been made to reduce absenteeism to "the lowest possible level." What steps have been tried and what problems have they involved?

There are two classical prescriptions which have long been used to hold men to high standards of behavior, "carrots" and "sticks." The provision of "carrots," or incentives to good behavior, may involve either the establishment of automatic, self-administering rewards (such as wage incentives or attendance bonuses) or the awarding of discretionary benefits such as merit increases or promotion, which require positive action by management. The wielding of "sticks," on the other hand, involves heavier reliance on disciplinary measures which are not automatic but which must be administered individually after poor performance has been demonstrated. From top management's point of view, the administration of penalties *ex post facto* is (in coal, at any rate) much more difficult than the establishment of self-administering rewards, as it requires a standard of judicial administration of which operating management may be incapable or which it may be unwilling to invoke for fear of certain undesirable consequences, such as trade-union disapproval or unofficial strikes. Indeed, we need to ask ourselves two somewhat more specific questions regarding the administration of discipline.

How ought responsibility for dealing with absenteeism to be fixed? Should it be left wholly in the hands of pit management? Should it be shared somehow with local trade-union officials? Should the trade unions themselves be given primary responsibility for dealing with it? Should the problem be made a responsibility of some outside agency, such as the Ministry of Fuel and Power or of the Courts? Or, finally, should responsibility be divided among all these eligible agencies in accordance with some defined procedural sequence?

And secondly, regardless of what agencies are made responsible for their application, what specific methods ought to be invoked to control the problem? Should there be individual interviews? Personnel counseling? Automatic fines? Fines for "certain cases"? Warnings — by letter or by interview? Downgrading of offenders? Bonding devices, to permit the recoupment of fines upon subsequent good behavior? Legal prosecutions through the courts? Or is reliance on publicity and propaganda sufficient — or is it even helpful?

We shall examine the attempt to reduce absenteeism through the introduction of an attendance bonus in 1947 and through the various supplementary administrative measures used in recent years.

THE ATTENDANCE BONUS

When the Five-Day Week Agreement became effective on May 5, 1947, one of the conditions insisted upon by the Coal Board was that men would receive six days' pay for five days' work only if they did in fact work a full five days. It is difficult to think, *a priori*, of a stronger incentive to good attendance than this self-administering monetary bonus. How well has it worked? The agreement recognized a limited number of acceptable excuses (including illnesses certified by a doctor's note) which would not disqualify a man from receiving his bonus. However, men who worked less than five days were eligible for only a proportional bonus, a stated proportion of their actual earnings (roughly 20 per cent).

There was a second major "condition" attached to the introduction of the Five-Day Week Agreement which may be mentioned here. This was the stipulation that the Union and its members would agree to a "reassessment of tasks" wherever such reassessment was adjudged reasonable. This condition was sought by the Board in the hope of increasing output and lowering costs by eliminating the custom of "early lousing," which it was called in some districts. "Early lousing" means that when facemen had completed their normal "stint," they were free to go home; the Board was anxious to retain them at work for a full shift and to do so men might be asked to "fill off" seven yards, say, instead of six. The proposal had no rate-busting aspect, as men were to be paid for the extra tonnage at their regular contract rate.

It can be appreciated that the job of reassessing tasks in a pit where those tasks have the sanction of long custom presents a very delicate problem; local union officers would often find themselves siding with the pit manager in negotiations over how large the "stint" ought to be; and managers themselves might often prefer to "let sleeping dogs lie" rather than invite sullen or overt resistance from their men. Thus while the Coal Board and the NUM's top officers negotiated this provision in all good faith, it is not surprising that the job of actually getting tasks reassessed has gone disappointingly out where the coal is got. More often than not, people at the pit level have simply taken no action — for which no penalty was provided in the national agreement. That this had been the over-all result between the spring of 1947 and the fall of 1950 (three and a half years) is evident from the inclusion in the wage agreement of October 1950 of a clause pledging NUM coöperation in the reassessment of tasks — something the Board had secured, it might be thought, more than three years previously.

There are two statistical series that can help us judge the effectiveness of the attendance bonus: (1) the usual absenteeism figures and (2) figures on the number of men who do not qualify for their bonus.[9] The former series is more significant, as the whole purpose of the device is to reduce absenteeism; but the "loss of bonus" figures are helpful in telling us something about the distribution of absenteeism among individuals. The "loss of bonus" figures refer to individuals while absenteeism refers to manshifts; the following example will make clear why one must not equate the two. Assume a coalface manned by twenty men, each of whom might work five shifts per week, giving a total of 100 possible manshifts for this group for any week. If five of these men each lose one voluntary shift in the week, voluntary absenteeism is 5/100 or 5 per cent; but the proportion of men failing to qualify for bonus is 5/20 or 25 per cent. If, on the other hand, one man missed all five shifts in the week, the absenteeism rate would remain at 5 per cent — but the loss of bonus figure would be much lower than previously (it likewise would be 5 per cent).

A high "loss of bonus" figure indicates that a lot of men are taking off at least one voluntary shift per week; it gives no information on how many additional shifts they may be losing. If the voluntary absenteeism rate for the same period is low, this indicates that few if any men are "playing" for more than one shift. If both rates are high, this suggests

---

[9] The latter figure is not published by the Coal Board but is available at most Area and Divisional offices.

that a considerable number of men are losing more than one shift (but the suggestion may be widely misleading in the absence of an actual frequency distribution of shifts lost by individuals).

These remarks become relevant in reading Table 14. This shows the figures for voluntary and involuntary absenteeism and for "loss of bonus" for one NCB Area during seven selected weeks in 1950. The most relevant comparison is between the "loss of bonus" figure and voluntary

TABLE 14

Voluntary and Involuntary Absenteeism and Loss of Bonus
Area No. 8 (Castleford), Yorkshire, Selected Weeks
During the First Half of 1950 (Per Cent) [a]

Week Ending	Absenteeism		Individuals Not Qualifying for Bonus
	Voluntary	Involuntary	
Jan. 7			
a. Face	12.3	8.5	26.9
b. Elsewhere u/g	10.2	6.9	23.7
c. Surface	6.9	5.6	18.3
d. Total	10.2	7.2	23.7
Jan. 28			
a. Face	7.0	8.3	19.7
b. Elsewhere u/g	5.2	7.3	14.9
c. Surface	3.0	5.5	9.5
d. Total	5.4	7.2	15.5
Feb. 18			
a. Face	6.9	8.8	19.4
b. Elsewhere u/g	5.3	8.0	15.2
c. Surface	3.2	6.4	9.8
d. Total	5.4	7.9	15.6
Mar. 11			
a. Face	8.2	8.8	24.3
b. Elsewhere u/g	6.4	8.2	18.6
c. Surface	3.6	6.6	10.2
d. Total	6.4	8.4	18.9
April 1 (Bull Week)			
a. Face	5.4	8.5	15.9
b. Elsewhere u/g	4.3	7.6	12.9
c. Surface	2.4	6.0	7.2
d. Total	4.2	7.6	12.8
May 20 (Bull Week)			
a. Face	4.4	7.4	13.0
b. Elsewhere u/g	3.5	6.5	10.9
c. Surface	2.3	5.3	7.3
d. Total	3.5	6.6	10.9
June 3 (Holiday Week)			
a. Face	10.4	7.9	20.9
b. Elsewhere u/g	6.5	6.6	12.7
c. Surface	3.1	4.3	7.2
d. Total	7.1	6.5	14.7

Source: Area Statistical Officer.
[a] Area 8 then included 15 pits employing in total about 18,000 to 19,000 men.

TABLE 15

Percentage of Men Not Qualifying for Five-Day Week Bonus, Warwickshire Area (15 Pits) for Selected Weeks in 1949 and 1950

Week Ending:	1/8	3/12	5/21	5/28 ᵃ	(6/4) ᵃ	7/23 ᵇ	(7/30)	8/13	10/22	12/17 ᵃ	(12/24)	12/31	3/11/50
Area, by occupation													
a. Face	22.5	24.4	23.4	15.5	27.6	11.6	32.7	26.2	24.7	12.5	33.2	27.7	21.4
b. Roads	20.5	19.8	16.2	10.8	18.0	9.4	25.7	17.2	17.0	9.4	23.8	18.6	17.4
c. Others u/g	15.3	17.3	13.8	10.0	12.8	8.1	13.1	11.0	15.1	11.1	16.9	14.8	13.8
d. Total u/g	20.9	21.7	19.3	12.9	22.0	10.3	27.7	20.8	20.4	11.0	27.6	22.5	19.0
e. Surface	16.3	11.9	9.2	6.3	8.4	5.4	10.0	9.5	10.8	6.6	11.1	9.0	9.6
f. Over-all	18.6	19.3	16.8	11.2	18.6	9.1	23.2	17.5	17.9	9.9	23.4	18.9	16.6
Over-all rates for individual pits ᶜ													
1. Alvecote	11.9	14.3	7.7	8.5	16.7	7.4	17.9	9.2	15.3	7.0	16.4	16.0	12.4
2. Pooley Hall	22.9	19.6	12.4	11.4	16.3	7.2	23.6	14.9	16.5	7.9	21.7	22.0	16.3
3. Amington	12.5	12.1	10.3	4.7	10.0	5.8	17.2	11.8	12.1	5.9	15.9	12.5	13.2
4. Birch Coppiece	18.9	20.1	14.8	12.9	18.5	11.6	22.0	22.3	18.4	12.0	24.0	20.6	16.6
5. Kingsbury Deep	16.0	14.0	11.7	6.1	13.8	5.8	14.9	13.1	14.2	8.1	18.3	21.5	11.3
6. Kingsbury Dexter	15.7	13.1	11.3	7.1	12.5	5.1	13.8	9.4	14.0	6.0	15.7	17.4	9.3
7. Baddesley	16.6	18.7	21.2	8.3	19.1	9.3	25.0	21.8	18.9	10.8	27.3	28.4	15.3
8. Ansley Hall	17.8	26.4	17.6	12.6	22.2	8.6	21.3	15.9	18.8	8.6	27.3	15.9	19.8
9. Arley	20.8	22.7	20.7	15.5	26.7	11.7	34.7	17.6	22.6	9.2	28.9	19.1	22.1
10. Haunchwood	19.8	20.8	19.9	13.1	22.4	11.4	26.6	19.3	19.0	12.3	26.0	23.0	18.7
11. Newdigate	21.5	20.6	18.4	13.9	21.1	9.4	24.4	17.5	20.0	8.8	23.7	16.0	18.0
12. Coventry	18.2	19.5	18.1	11.9	17.6	8.9	21.7	20.0	15.6	11.8	20.9	16.2	15.9
13. Griff No. 4	21.7	22.3	24.8	15.0	23.0	11.9	25.5	20.3	28.1	13.5	29.7	18.0	26.2
14. Griff Clara	18.1	22.8	16.3	11.3	18.0	8.3	25.0	16.7	17.7	9.3	25.0	19.6	16.6
15. Binley	19.4	18.5	13.4	9.5	12.5	7.2	23.1	16.6	14.6	10.0	22.5	11.3	13.6

Source: Table supplied by the Area General Manager.

ᵃ Indicates a Bull Week.

ᵇ Indicates a week containing a holiday.

ᶜ Pits with the highest and lowest figures for each week have been italicized. Highest percentages most frequently: Griff No. 4 (6 weeks) and Arley (3 weeks). Lowest percentages most frequently: Amington (5 weeks), Kingsbury Dexter (4 weeks), and Alvecote (3 weeks).

absenteeism, since some of the men off work involuntarily will (if they have worked all other possible shifts in the week) receive their bonus. The chief significance of the table lies in the absolute level of the loss of bonus figures themselves: they show that in most weeks more than four out of every five men did not lose a single voluntary shift.

We are now ready to look at Table 15, which shows the proportion of men not qualifying for bonus at each of the 15 pits in the Warwick-shire Area for 13 selected weeks in 1949 and 1950. The table also gives the figures by main class-of-work for the Area as a whole. Several points are worth noting: Looking first at row *f*, it is apparent that the proportion of men not qualifying in weeks that are neither "Bull" weeks nor holiday weeks was just under one-fifth. In other words, over 80 per cent of the men do normally qualify for bonus. During "Bull" weeks (when men usually need extra money for their holiday or to tide them over the holiday week when they may plan to lose their bonus) only about one man in ten fails to qualify for bonus. During the three holiday weeks shown, slightly more than one-fifth of the men lost their bonus.

By looking separately at rows *a*, *b*, *c*, and *e*, one sees that in every week save one (12/17) the proportion of men losing bonuses fell as one comes "outbye." The rate for facemen is invariably 2½ to 3 times higher than the surfacemen's rate, with men working on roadways losing it less frequently than facemen and those on other underground jobs losing it, in turn, less frequently than roadmen. Thus loss of bonus is related positively to earnings (and arduousness of work).

A look at the over-all rates for individual pits shows that, while no pit consistently had either the highest or lowest proportion for the weeks shown, there are good pits and bad pits. Amington has the lowest figure for 5 of the 13 weeks, followed by Kingsbury Dexter (4 weeks) and Alvecote (3 weeks); these three pits accounted for the lowest rate in 12 of the 13 weeks. These three pits are at the northern end of the coalfield, away from Coventry; they are "country pits" and the men live mainly in semi-rural villages. The pit which most frequently has the highest percentage, Griff No. 4, is close to Nuneaton (and not far from Coventry) and is known to have working conditions among the worst in the Area.

One thing the weekly figures do not tell us is whether the same individuals forfeit their bonus each week or whether there is a high turnover within the fairly stable percentage. Interview evidence suggests that there is a "hard core" of men who can be counted on not to qualify for their bonus, week after week; but of course all men must lose their bonus occasionally.

LOSS OF BONUS AND MONETARY INCENTIVES

If it be correct that there is a hard core of around a sixth or a fifth of the total labor force in mining who can afford to lose two days' pay per week (one-third of their potential earnings), this situation is symptomatic of a general British attitude toward higher earnings that differs from characteristic American attitudes. One quite often hears it said that many British miners (and presumably other industrial groups would show the same tendency, though often to a smaller degree) have a relatively fixed standard of living, so that the incentive to work beyond an amount necessary to satisfy these fairly limited goals is relatively weak. Some of the factors which lend credibility to this view are the less extensive use of installment-plan buying in Britain as compared with America (though "hire-purchase" has grown considerably since the war); the lower pressure of British advertising, compared with American; the relatively long period during which it was difficult to purchase many goods in shops, so that older people have learned to "do with less" and a generation of younger people has come along without strongly motivated buying habits. The wartime shortage of consumers' goods was felt to explain some of the voluntary absenteeism after the war and it was often expected that the return of goods to the shops would reduce absenteeism. However, an experiment was tried in one mining community whereby special allocations to that community were made to test the effect on absenteeism. In that case, more goods did not mean less absenteeism. This was true generally after the wartime shortages began to melt: the hoped-for decline in absenteeism did not materialize. Also, the relatively rapid advance of mining wages during and since the second World War, combined with much more regular work, may have advanced miners' incomes more rapidly than their tastes; the housing shortage has prevented many younger couples from incurring the spending obligations which would lie on them if they kept house independently; and last, the relative isolation of many colliery villages provides both fewer pressures and fewer opportunities to spend money. Perhaps these explanations do not entirely account for either high absenteeism or why a significant proportion of men forfeit their bonus every week; but they are the reasons one is quoted in seeking explanations. Although Americans often underestimate the prevalence of fixed living standards among certain groups (and perhaps whole areas) of the United States, it is certainly a common impression of Americans who live in Britain (and of Britons who have lived in America) that, generally speaking, there is a greater gap between the wants of American wage-earners and their ability to satisfy those wants than is true in Great Britain. It would be folly to push this observation too far, but it is not misleading to suggest that there are cultural factors in British mining which may blunt the effectiveness of "automatic" monetary incentives. In one sense, the "value of money" is determined not only by what it will buy but by what people want to buy with it.

He would be rash who would discourse confidently on just how much effect the attendance bonus has had on absenteeism, though there is little doubt that it has been a help. It can, indeed, be said that since the bonus went into effect, attendance has in every year been considerably lower than it was during 1944 to 1946. But this improvement may well be due to other causes — for example, the qualitative improvement in the labor force or the persistent administrative measures which we have yet to examine. If we study the fairly short period of time defined by the few months before the bonus came into effect on May 1, 1947, and the few succeeding months, we find the following: absenteeism was running at a rate of 15.03 for March and April and fell dramatically to 9.13 for May and June. Part of this dramatic improvement, however, "may well have been due to the existence of the five-day week itself which reduced the number of days on which a man could be an absentee." [10] More important is the fact that the improvement did not hold: by the end of 1947, the absenteeism figures had begun to drift upward again (though they did not go above 11 per cent until 1948) and it had become clear that the bonus incentive was not, by itself, sufficient to control the problem.[11]

Perhaps it was inevitable that pressure should develop in many NUM Areas to alter the Five-Day Week Agreement so that men who lose a shift would lose not their whole bonus but only one-fifth of it. Two separate reasons lie behind this desire — a desire which found unanimous endorsement at the union's 1951 Conference. The first reason is that the original Agreement did not make any provision for men who lost shifts because of such reasons as attending funerals or weddings of their immediate family, serving on local government bodies, or attendance at medical boards. Being penalized for shifts lost because of these compelling personal reasons seemed unfair to many and ever since the Agreement was signed there has been pressure to make allowance for such personal reasons. In the fall of 1952 the Board agreed to a somewhat more liberal list of excusable absences.

The second reason behind the pressure is simply that many people do not like to live under the conditional nature of the bonus. "Why should we lose 100 per cent of our bonus when we lose only 20 per cent of our working time?" "Other workmen don't lose more than a day's wage when they're off for a day — why should miners?" Thus a feeling has grown up that the bonus is deserved as a right and when it is withheld, a man is being unfairly penalized.

[10] *NCB Report,* 1948, p. 14.

[11] The insufficiency (though not the "failure") of the attendance bonus had become apparent by August of 1947, when the NCB submitted to the NUM four proposals to reduce absenteeism. The union's National Executive Committee decided to "indicate our willingness to discuss the adoption of methods for dealing with habitual absentees as well as the introduction of penalties to deal with absenteeism . . ." (*Minutes* of the N.E.C. meeting of Aug. 12, 1947).

Until 1951, the National Executive Committee of the NUM opposed the removal of the conditional nature of the bonus, though it agreed that the list of excused absences should be broadened. In 1951, however, the Executive Committee gave in to pressure from the districts and accepted a conference resolution committing it to seek an unconditional (a daily) bonus. The rationale was that removing all conditions was "a simpler solution than trying to prove the man was ill, when it is oftentimes questioned." The Miners' Executive was able to cite "collieries in South Wales which since 1950 have been paying not six for five, but one and one-fifth shifts for each shift worked, and that is the position we want to achieve." [12] Prior to nationalization, many pits had paid a substantial daily attendance bonus.

One does not have to entertain a very low estimate of human nature to believe that the conversion of the weekly attendance bonus into a daily "bonus" (it would amount to a 20 per cent increase in the daily minimum) would completely remove whatever incentive to better attendance the Agreement now holds.

It will not, of course, hurt the Union's general bargaining position to have this demand to trade off against something else it may want; but it would be a great mistake to assume that the Union's present attitude on the bonus was developed purely to secure bargaining leverage. If the Board and the Union had worked out a more realistic "rule of reason" with respect to excusable absences, the Union might well never have moved to its present extreme position.

Penalties

The industry and the government tried several different approaches to the control of absenteeism during the war, none of which succeeded in halting the upward drift in the absenteeism rate. The first and least successful approach was the negotiation of a one-shilling per shift attendance bonus in the spring of 1941 (men who worked all week got an extra shilling for each shift worked). This approach to absenteeism lasted less than three months, for the definition of excusable absences was so tightly drawn that the whole idea became discredited overnight and the conditional one-shilling bonus became a straight wage increase. The coalowners and the union (which did not deny the existence of the problem) then turned to the use of the wartime Pit Production Committees. These committees were to:

deal with cases of absenteeism first, by persuasion and, where that failed, by reporting the offender to the National Service Officer, who was then left to take action if he chose under the Essential Work Order. Events subsequently

[12] Arthur Horner to the Conference at Blackpool, July 1951; *Conference Report,* p. 147.

proved that the existing procedure for dealing with absenteeism was too cumbrous. National Service Officers had not taken a sufficiently firm line in their treatment of offenders, and through long delayed and often ineffective action they had tended to destroy the influence and authority of the Pit Production Committees.[13]

This arrangement, however, did little to help attendance and much to discredit the Pit Production Committees. Workmen's representatives were not comfortable in their roles as disciplinary tribunals sitting in judgment over their workmates, and the committees became so preoccupied with absenteeism cases that they neglected their main responsibility, which was production problems.

A somewhat more workable arrangement was developed shortly after the establishment of the Ministry of Fuel and Power in June 1942. Investigation officers, frequently chosen from local officers of the miners' unions, were appointed in each of the Ministry's regional offices. Pit managers were instructed to report cases of persistent absenteeism to the Ministry's regional office, which would hold periodic meetings at individual pits. The investigation officer sat as chairman of a committee which also included the pit manager and a workman's representative, called an assessor, who was paid 22/6 per sitting by the Ministry. Neither the manager nor the assessor shared any responsibility for the disposition of cases, an arrangement which was supposed to deflect the natural resentment of offenders from these individuals to the independent outsider sent in by the government. The investigation officers were empowered either to warn offenders or to fine them up to £1. If the defendant refused to accept the fine, he could be taken to court by the employer for damages.

The Essential Work Order as applied to industry generally in 1941 made absenteeism and persistent lateness offenses, that is, crimes punishable by fine or imprisonment. The EWO as applied initially to coalmining, however, did not make absenteeism and lateness offenses in that industry — presumably because other measures were then in force to deal with the problems. When the investigation officers were appointed in the summer of 1942, the EWO applicable to coalmining was altered so that absenteeism and persistent lateness were treated as punishable offenses just as in any other war industry. This was the legal sanction on which the investigation officers relied to get the bad cases into court.[14] If a man who had been fined subsequently attended faithfully for a specified period, his fine was remitted. Forfeited fines were usually given to some local charity (usually a miners' charity).

The Agreed New Procedure (as the scheme using investigation officers was called) worked reasonably well during the three years it was

[13] From the Report of the Select Committee on National Expenditure, Feb. 19, 1942, reprinted at pp. 224–231 of the *NUM Conference Report*, 1942.
[14] Cf. Harold Wilson, *New Deal for Coal*, pp. 63–64.

in operation (summer 1942 to summer 1945), but it did not prevent absenteeism from rising by one-quarter or one-third during this period. Sam Watson, the astute General Secretary of the Durham Miners' Association, has commented as follows:

> As the Agreed New Procedure developed, it was given increased authority to deal with cases of indiscipline, breaches of the Coal Mines Acts, and in general, all cases where it was deemed essential to take action to preserve discipline and order in the industry.
> The method worked smoothly, and we had no complaint from either the Lodges or the members who were dealt with under the New Procedure.
> Absenteeism had tended to rise gradually in 1943, 1944 and 1945, but nothing like it had in those coalfields where no such scheme was in operation.[15]

When the Labor Government was returned to office in July 1945, Emanuel Shinwell (then a miners' M.P. from County Durham) was made Minister of Fuel and Power. Within two months, Mr. Shinwell announced that in future no miner was to be subject to any fine for absenteeism. The men were to be put on their honor — even the men without honor. In place of fines, responsibility for policing absenteeism was placed on the lodge secretary — a scheme promulgated by the Minister without consulting the union. The Shinwell approach was characterized as follows by the leader of the Durham Miners, a man who became chairman of the Labor Party for 1950:

> A new arrangement was devised whereby the Manager would report persistent Absentees to the Pit Production Committee. The Committee would send a limited number of the worst cases to the Lodge Secretary. The Lodge Secretary would send for these members to attend the Lodge Committee or the Lodge Meeting. If they came (and a great number did not) all the Lodge Officers could do was to impress upon them the necessity for regular attendance at work . . .
> If the Absentee did not attend work more regularly following his attendance at the Lodge Committee or Lodge Meeting and was again reported for Absenteeism, he could be classified as an "incorrigible" and dismissed from the industry; a step we would not agree to under the old procedure, as it was exactly what some of the Absentees wanted, particularly the Optants, Volunteers and Bevin Boys.
> The Lodge Secretary was allowed a maximum of £5 per year to cover the cost of postage and stationery. He got nothing for the hours of patient labour put in under the new arrangement.
> Many Lodges refused to operate the scheme . . . the present farcical method will probably remain until the Essential Work Order is lifted from the Industry.[16]

[15] *Information Pamphlet No. 16*, NUM, Durham Area, May 1946, p. 5. Mr. Watson's final sentence suggests that the scheme applied only to certain coalfields. However, it was certainly the most common procedure, though certain districts may have secured the Ministry's permission to operate modifications of the scheme. Indeed, some districts may have refused to operate it.

[16] Sam Watson, *Information Pamphlet No. 16*, pp. 5–6.

This comment was written by a distinguished miners' leader after only eight months of the new Shinwell approach to absenteeism.

Few of the old coalowners could have leveled a more forthright indictment of the Shinwell policy, the effect of which was to do away with any real effort to control absenteeism from the summer of 1945 until the introduction of the five-day week in May 1947.

During the period when Mr. Shinwell's honor system was in effect, absenteeism rates climbed higher than they have ever been. However, this period began with the ending of the war in the summer of 1945 and it seems inevitable that much of the rise in absenteeism must have come from the combination of psychological release attributable to the end of the war and of managerial laxity due to the time-lag between the election of the Labor Government and Vesting Day in 1947. But at least Mr. Shinwell's policy did nothing to moderate these influences.

It was after the shortcomings of the Shinwell policy had been revealed that the Coal Board and the NUM negotiated the automatic attendance bonus embodied in the Five-Day Week Agreement. We have already reviewed experience under this Agreement and it will be recalled that after a substantial but short-lived reduction in absences, the rate had again risen to disappointing levels by the fall of 1947. This determined the Board to add a system of administrative controls to the continuing pressure automatically exercised by the bonus.

By late fall of 1947 the National Coal Board had determined to urge the Divisional Boards to adopt new measures to control attendance. The following letter from the Divisional Labor Director for the Northeastern Division was sent to Area Labor Officers in December:

> The Board have had under consideration for some time the question as to the procedure to be adopted with regard to dealing with habitual absentees and individuals who are not playing the game. It is considered that Colliery Managers should make the fullest use of Pit Consultative Committees by seeking their advice as to what is best to do with these individuals.
>
> It is suggested that the Pit Consultative Committees should give the individual concerned *a very strong warning* and they should be informed that unless their attendance and/or conduct improved within a given time *drastic action will be taken.* If this warning is disregarded the Pit Consultative Committees should then *advise the Colliery Manager that they agree to the dismissal of the individual concerned. Appropriate action is then left to be taken by the Colliery Manager.*
>
> The question then arises if an individual dismissed at one Colliery should be placed at another Pit in the Division. With regard to this matter the Labour Relations Department at Headquarters have intimated that such individuals should not be placed elsewhere, but the Ministry of Labour and National Service should be advised that they be released from coal-mining employment as being of no further value to the industry.
>
> Please advise Colliery Managements to act accordingly [italics mine].[17]

[17] Letter from Tom Smith, Divisional Labour Director, Northeastern Division, to Area Labor Officers (*ALO No. 22*), Dec. 6, 1947.

This procedure, probably designed to protect the colliery manager by arming him with recommendations of the consultative committee, unavoidably placed the control of absenteeism in the arena of joint consultation. Consequently it could not trade on the advantages of consultation without falling afoul of its shortcomings. These shortcomings are partly political, partly administrative — and the latter seem even more important than the former.

The political difficulties involved in tying absenteeism-control to the consultative machinery consist of (1) the manager's relation to the local branch officers and (2) the relation of these officers to their own membership. If the manager's relations with the lodge secretary are taut and hostile, the manager is not likely to make much headway in persuading the consultative committee (on which the lodge secretary sits *ex officio*) to agree to unpopular disciplinary measures. True, a forceful manager would be within his rights in bypassing the consultative committee and in imposing his own discipline on persistent absentees; but such instances are rare. If, on the other hand, the manager's relations with the branch are only "normally" cool or even good, the way is at least open to do something. The limit to what can be done will then depend primarily on the political strength or courage of the branch officials, mainly the secretary.

The political position of local branch secretaries is the weak point in the system, for there has not been any significant refusal to face-up to the whole problem on the part of district leaders, in most districts anyway. They know the problem is serious, they publicly acknowledge it, and they encourage pit committees to fulfill their responsibilities. But they do not feel the heat which scorches the local officers who must enter into the face-to-face relationship involved not only in administering a policy but in living among the people who must be disciplined. Though it is unsafe to generalize broadly, one does not get the impression that local political pressures had boiled up so strongly that the consultative committees in most districts were formally refusing to deal with the question; but this may largely have been because no undue strain had been put on the system.

There were some districts, however, where the fear of political repercussions had frustrated the work of the CCC's on this subject. Below is reprinted a scheme explicitly designed to get around this political difficulty, together with a report of the fate of this proposal.

The cases the committees are normally asked to consider are so definitely clearcut that no great political risk is involved in assenting to the inevitable. This is not always true, but it often is. Thus the manager may not ask the committee to take action except in the cases of men who have worked only a half, a third, or a quarter of the shifts over the past month or six weeks. How can a union official, concerned to "make nationalisation work," oppose taking action in such cases? What excuses can the offenders themselves put forward?

The answer is not always as negative as these rhetorical questions suggest. In many cases, the absentee subcommittees (for most consultative committees appoint subcommittees from among their members to administer whatever scale of discipline they approve) discover that even men with such apparently indefensible records can often give convincing explanations for their conduct. To a certain extent, therefore, the subcommittee interviews of persistent absentees do teach the management and union officials something about valid personnel problems confronting men in their pit. This enhances general understanding of the problem even if it fails to bring down the absenteeism rate.

This is the scheme for handling absenteeism in a South Yorkshire Area which was devised by an Area Labour Officer specifically to protect branch representatives from suffering political consequences as a result of honest efforts to coöperate with managers in reducing absenteeism.

Step 1: Preliminary Investigation. The reporting of absentees to the manager, who selects the worst cases and arranges for these individuals to attend the next meeting of the pit consultative committee.

Step 2: Pit Consultative Committee. Appraisal of individual cases by consultative committee on basis of their interview. If a majority believe the man has no acceptable excuse, he is to be placed on probation for a number of weeks and told of the subsequent step.

Step 3: Joint Committee. "This Committee will consist of two independent persons, one representative of the Management and one representative of the workmen, chosen from different units not in the immediate vicinity of the unit concerned." It is their function to impose specified disciplinary measures, which were to be according to an agreed scale of fines and, finally, downgrading or dismissal.

This proposal was sent out to the various pits with the following explanation: "The idea of the Joint Committee being chosen from individuals not connected with the Unit concerned is to avoid any concerted action by a particular section of the workmen against a member of the Joint Committee at the ensuing branch elections."

The procedure was agreed to by the Divisional Consultative Committee, but only after the then President of the Yorkshire Miners had stipulated that the final fine should be £5. This amount proved so severe that the local branches refused to accept the scheme and it was never put into operation — a result some felt was intended. The result is that the procedure actually followed is very unsatisfactory in this Area, where the problem is left in the hands of the full consultative committees: "They definitely duck the problem. They call a man in and tell him to 'try to improve' and he says 'yes' and he doesn't and that's usually the end of it."

There are few penalties that might be imposed on absentees which have not been thought of by people in the industry. While additional ones are undoubtedly conceivable, the following ten penalties are those which have received the most attention; the first seven have been used extensively, but the last three do not appear to have been used anywhere. We shall list each penalty separately and comment briefly on the problems involved in using it.

1. *Down-grading:* This alternative is possible in the case of facemen or contract workers, but not in the case of daywagemen who man the haulages or work on the surface, since their wages are commonly at the weekly minimum which is guaranteed them. The acute shortage of facemen has made this penalty almost as costly to management as it is for the men concerned.

2. *Consigning men to night work:* Such work is often disliked by many men (especially the younger ones, who are the worst offenders) and if they become sufficiently irked by their penalty they are free to quit. Many managers have undoubtedly hesitated to invoke a penalty that runs the risk of losing manpower, and consequently this device has not been widely used.

3. *Warning men:* This very common device consists either of summoning offenders before the Colliery Consultative Committee or an absenteeism subcommittee and, if no convincing excuses are offered, warning the men to improve their attendance lest "drastic action" be applied. The warning process often involves sending men remonstrative letters which hold out the possibility of further action by the CCC or by the pit manager.

4. *Fining or bonding men:* The difference between these two methods is that when a man is fined, he forfeits the money permanently while if he is "bonded" his "fine" is returned to him upon subsequent good attendance. The level of fines runs from about 5s. to £1, often varying with the man's record. Fines were commonly used during the war by the Ministry's Investigation Officers.

5. *Denying men week-end work if they have missed midweek shifts:* It is not easy to say how much absenteeism arises from men taking off days during the week in the knowledge that they can make up the loss by volunteering for week-end work — an opportunity presented at the many pits where a voluntary Saturday shift has been worked. Such week-end work is paid at time and one-half. My impression is that relatively few pits have used this penalty, the reason being that either the CCC members felt that it would unduly penalize men who had "unavoidably" lost a "voluntary" midweek shift (and such "voluntary" shifts are, indeed, getting the week-end services of key men, for example, cuttermen, pan-turners, if they had been out during the week).

6. *Posting the amount of wages not earned:* Here is a device used by certain pits in Yorkshire, as described in the minutes of its Consultative Committee: "The total number of bonuses lost in June was 164 representing a cash value of £202. Wages lost in addition totalled £367. The figures for May were, respectively 124, £152 and £322. The Committee agreed that the statement be posted on the notice board." Very occasionally, the names of individual absentees have been publicly posted — but the practice has been anything but common and there have been cases where the repercussions put an end to the practice.

7. *Dismissals:* These have not been as rare as one might infer from their low place in this list. This ultimate penalty, however, had to be used only as a "last resort" until manpower controls were lifted on January 1, 1950, as prior to that dismissal for voluntary absenteeism was just what some men hoped for. But even in the previous year, the Board dismissed 8000 men "for

bad attendance, undisciplined behavior and so on" (the number for 1948 was 4900).[18] In using this penalty, the Board has once again been faced by the manpower shortage: if it dismissed men without very bad attendance records, it ran the risk of accentuating the shortage. Hence dismissal has been used chiefly only in cases where the industry had "given up" on the man and no longer felt him any use to the industry; this means that it has not been possible to use the dismissal weapon liberally to make other men toe the mark. We shall have something more to say about dismissals when we come to discuss the relationship of the CCC and the colliery manager in the administration of discipline.

8. *Court prosecutions:* Although prosecution before local magistrates was used during the war and had apparently been used occasionally by some of the private companies before the war (the charge then being breach of contract of service), it has not been used since nationalization — as it has been, occasionally, in the case of unofficial strikes. The Board's attitude, like the union's, has been that they could impose privately any penalties that a local court might impose.

9. *Disqualifying men from vacation rights:* Certain labor officers of the Board say that it might be a good idea to tie vacation payments to good attendance during the previous year. There are two difficulties with such a proposal: (1) local people who might be persuaded to work out such an arrangement are discouraged from trying to do so because it would require a renegotiation of the national vacation agreement — a reform local officials often feel powerless to initiate; (2) any such arrangement would require the announcement of standards of eligibility and this might lead a great many men to feel that they were entitled to a certain amount of absence before falling from grace.

10. *Tying "home coals" privileges to attendance:* This has not been widely discussed, possibly because concessionary coal is such a sacred subject that no interference with existing customs is possible. However, these agreements are all local and would not require any national action — though they would require the announcement of eligibility standards (a difficulty that may be more theoretical than real). But the tenacity with which men in the industry cling to their home-coals privileges suggests that making these privileges conditional on good attendance would exercise an effective incentive.

The difficulty of devising adequate penalties under conditions of full employment, when men can "ask for their cards" if too harshly pressed in the knowledge they can easily sign on at a nearby pit, is obvious. Absenteeism is simply another area of conduct where full employment has removed the disciplinary techniques and pressures that most managers were accustomed to before the war and which were largely automatic. The task of working out a new system of incentives and disincentives (which will require the cultivation of a general disapproval of absentees as well as specific penalties) has not been easy. Witness the vigor with which this view has been put by Arthur Horner:

"A new consciousness must be created, a new morality, a new understanding of our true interests which will present these irresponsible minorities with such a wall of opposition on working-class grounds as

[18] NCB, *1949 Report,* p. 74.

will discourage and make impossible success in their endeavours to interfere with production" (from a speech to the NUM's 1947 Conference).

At least one coalfield (North Staffordshire) has experimented with the use of special workmen's representatives — "group disciples" picked from outside the consultative machinery — whose job it was to speak to absentees "on the job," admonishing them to better attendance and seeking out the reasons for their absences. This device, which grew out of a suggestion made at an Area-wide union-management meeting called by the AGM in August of 1949, was an attempt to improve on what was regarded as the indifferent effectiveness of CCC's in handling absenteeism. The great difficulty was in finding good men who were willing to assume this unpleasant task and after a short trial the method fell into disuse and main reliance was once again put upon the consultative committees.

When at the end of 1947 the Coal Board decided to supplement the attendance bonus with a return to some kind of administrative control, it will be recalled that the Board suggested to Divisions that Pit Consultative Committees warn offenders and that if these warnings were subsequently disregarded the committees were then to advise the pit manager that they agreed to the men's dismissal. The manager was then free to dismiss or not, as he believed best. The plain implication was that the manager was not free to impose his own discipline: cases first had to go before the pit committee. Precise procedures may have varied among and within Divisions, but the clear conception was to give the pit committees a central role in policing the problem. There is little available information about the role of the pit committees in this area during the first half of 1948; there was, however, a rather significant turn of events during the latter half of that year.

In the spring of 1948 the Board and the Union, ever anxious to increase output, had set up a Joint Production Committee at national level (it was a subcommittee of the National Consultative Council); this group concerned itself with manpower, the rate of output, overtime working, the implementation of the Five-Day Week Agreement, and attendance. Among other things, the committee was confronted by the fact "that there were workmen who systematically work only four days a week." To attack this problem it framed a set of proposals in which the union representatives on the committee acknowledged management's right to deal with men who failed to honor their contract of service but "we decided that it was in the best interests of the workmen themselves that the Union should participate in any procedure which was designed to bring home to irregular attenders their responsibilities to their fellow workmen, the industry and the country rather than that the management should have a free hand." The NUM Executive Committee then reports as follows:

It was with this end in view that we accepted the proposals of the Committee to adopt a scheme for the setting up of Joint Attendance Committees. These Committees were to interview workmen who had been reported by the management for unsatisfactory attendance and in cases where the management's complaint was found to be justified the workman concerned was to be warned of the consequences of a failure to attend work regularly. In the case of a workman who on a second occasion was interviewed by the Joint Committee, provision was made for the Committee to recommend the imposition of a fine or some other disciplinary action. In the event of a fine being imposed on a workman, provision was made in the scheme for such fine to be remitted to him if he worked all the shifts available to him for the next 12 weeks.[19]

Technically, the Joint Committee was not to impose the discipline, which lay with the manager; but when the manager acted on a case finally turned over to him by the Joint Committee, the union pledged itself not to support the men concerned.[20]

In October 1948, the foregoing proposals on the control of absenteeism, together with several others all designed to boost output, were submitted to the NUM's districts for acceptance or rejection before taking effect. Somewhat to the embarrassment of the national officers of the NUM, the Area Unions overwhelmingly rejected the suggested procedure. Areas accepting the proposal were Durham, Northumberland, and North Wales; twelve other Areas explicitly rejected it, with the remaining Areas unreported.

Some districts replied that they felt they had superior procedures either in operation or contemplated; but the more fundamental explanation seems to be that the Area Unions were reluctant to commit themselves officially to a system that involved union representatives in the disciplinary process. The upshot of the rejection was to remit the whole matter to the Joint Production Committee in London which decided to proceed with its original plan in those few coalfields which had accepted it and, in the others, to authorize the Coal Board to tell local managements to impose their own discipline — following a procedure very similar to that which it had been intended the Joint Attendance Committees would follow.[21] In practice, this has meant that in the overwhelming majority of pits, managers have not acted unilaterally but have moved cautiously or even timidly, mindful of how the Colliery

---

[19] Report of the NUM Executive Committee to the 1949 Conference, *Conference Report*, 1949, p. 223.

[20] From a summary of the Joint Production Committee's Report carried in the NUM's *Information Bulletin*, Oct. 1948, p. 110.

[21] The instructions sent out to Divisional Boards from London are reprinted as Appendix IV. These are useful not only in throwing light on the handling of absenteeism but also in bringing out the varying nature of the "contracts of service" under which men are employed.

Consultative Committee viewed the whole absenteeism problem. This brings us back to the previously noted political difficulties that have sometimes made the committees unwilling to tackle the problem — and to the somewhat limited range of penalties which the committees and the managers have invoked even when they were willing to move in on the problem.

Have we any evidence as to the number of absenteeism cases actually handled, either by the CCC or by the colliery manager acting on his own? A summary of dismissals for various reasons in the Yorkshire coalfield from January 1949 through June 1950 revealed the following points: Of the more than 2100 disciplinary dismissals during the eighteen-month period, over two-thirds were for "persistent absenteeism." Dis-

TABLE 16

Formal Role of Colliery Consultative Committees and Pit Managers in Making Dismissals "For Cause" in the Yorkshire Coalfield, January 1949 through June 1950 [a]

Dismissals Recommended by the CCC			Dismissals Other than Those Recommended by the CCC		
	A[b]	C	A	B	C
Jan. 1949	44	5	14	19	28
Feb.	56	14	20	16	36
March	65	19	19	3	18
April	43	4	7	11	17
May	48	1	64	13	31
June	31	0	32	13	22
July	64	0	57	20	23
August	39	0	39	17	28
Sept.	42	4	34	13	21
Oct.	48	0	79	10	25
Nov.	80	4	40	11	25
Dec.	72	0	33	14	44
Jan. 1950	28	1	19	6	10
Feb.	25	0	14	7	15
Mar.	52	0	14	16	44
April	25	1	23	10	11
May	33	0	26	10	38
June	23	1	24	10	17
Totals	818	54	558	219	453

Source: "Divisional Summary of Dismissals — Releases — Engagements" NCB, Northeastern Division, monthly, mimeo.

[a] The above figures are available on an Area basis but there seems little significance in such detail unless one has made a close study of the problems and procedures in each Area.

[b]

Cause of Dismissal, Regardless of Method (18 months)

		% of total
A = Persistent Absenteeism	1,401	66
B = Misconduct	221	10
C = Other Causes	507	24
Total	2,129	100

missal for habitual absenteeism is much the most frequent cause of dismissal from the industry. Of the 1400 dismissals for absenteeism, over 800 were made on the recommendation of the CCC; all we know about the remaining 40 per cent is that they were dismissals "other than those recommended by the CCC" — this might include some cases where the manager and the committee disagreed and the manager went ahead with dismissal on his own. It might refer mainly to pits where the CCC had agreed to leave the whole matter in management's hands; or it might refer to the fact that at many pits managers have dismissed men with the worst records on their own without prior reference to the CCC; finally, it might mean that at some pits the CCC accepted the necessity for discipline on these cases but declined to name a specific penalty — and the managers then went ahead and "sacked" the men.

The CCC's have taken a far more active role in disciplining men for poor attendance than for other reasons: while they made recommendations in 60 per cent of the absenteeism cases, they recommended dismissal in less than 10 per cent of the nonabsenteeism cases, which were handled mainly by management alone. This probably reflects a lesser concern for nonattendance cases and not simply a greater reluctance to dismiss for these other causes.

It may give us a somewhat better understanding of the way in which the CCC's approach their handling of absenteeism if we quote a few excerpts from minutes of CCC meetings at particular collieries, selected at random from the summaries prepared at Divisional headquarters in Doncaster, Yorkshire.

Kiveton Park (July 1950): The manager said he intended taking action against a number of men — "several of them appeared to work just sufficient to ensure them receiving Home Coal and Holiday Pay and the position was going to be altered." Interviews and letters had failed to improve their attendance. "The men in question were not aged, the majority of them had another job which took precedence over their work at the colliery." (Here was a situation where the manager appeared to be letting the CCC know that he planned to take action; if the workmen's representatives agreed to this action, the manager, presumably, could expect them not to object when the discipline was imposed.)

Nunnery (July 1950): ". . . due to absenteeism over the week-end it was not possible to work a proper cycle until about Wednesday . . . he [the manager] wished the men would awaken to their responsibilities."

Grimethorpe (July 1950): The manager pointed out that figures showed "that fillers were the worst absentees. He suggested that figures be got out for one unit showing the individual absenteeism record of every man on that unit, from fillers to wastemen, etc., together with production figures . . . The Committee agreed that this was a sound idea. . . . Mr. Hawkins [a contract worker] mentioned the fact that workmen had a habit of playing when their working places were in bad condition."

Shuttle Eye (July 1950): "Bonus and approximate amounts of wages and bonus lost for five-weekly period ended 1st July were discussed. Little improvement has been shown for the publishing of these figures."

Bullcroft (July 1950): The percentage among haulage men has been serious lately. The men's representatives said that the "recently adopted system of posting the number of bonuses lost through absenteeism under the Five Day Week Agreement would discourage this practice, and whilst they agreed that the situation was most serious, they considered it was a national and not a local problem." (Here is a pit that pins its hope on a remedy which the preceding pit had found ineffective; more significant perhaps was the refusal to face up to the fact that national problems are the sum of local problems.)

Silverwood (June 1950): The management was "very concerned at the high rate . . . among cuttermen and erectors (conveyor-men). This was causing serious dislocation of the pit." The men's representatives wanted the manager to let these men out on Sunday morning in order to get the faces properly prepared for Monday, but the manager pointed out that this would negate the spirit of the Five-Day Week Agreement. The committee agreed to send letters to the worst offenders "informing them that if no improvement of attendances were made, steps would be taken to replace them, and they would be down-graded . . . It was also agreed that full support would be given to any official who down-graded a man who was a persistent absentee." (Here is some subtle fencing between the men's representatives and the manager over the failure to observe the Five-Day Week Agreement; also the Committee — and the local branch? — is put on record to support downgrading.)

Wath Main (June 1950): After subcommittee interviews with men who had "extremely bad attendances after notification . . . it was evidently time some drastic action was taken . . . It was agreed the system was evidently so familiar to the men and this was leading to some slackness and complacency . . . it was eventually agreed to discontinue the sub-committee interviews and call selected cases in front of the full Committee for prompt action." (One senses the futility of "warnings" given to habitual absentees; one also wonders what kinds of "prompt action" the full committee will invoke — as well as about the amount of time the Committee will spend in collectively interviewing individual cases.)

Wharncliffe Woodmoor (May 1950): "The Secretary of the Committee reported that the subcommittee met on Friday, May 5th, and considered at length the record of 51 workmen whose attendance was unsatisfactory, particularly with respect to Monday absence which causes considerable disorganization. The names of 11 workmen were selected for interview and 5 others previously repeatedly interviewed and warned were selected for address by letter intimating that if there were no improvement, the subcommittee would have no alternative but to recommend the management to take action." (The subcommittee is performing some of the routine screening which might well be left up to a full-time colliery clerk; one senses, too, the number of "second chances" many of the committees give to the men they interview before they finally bring themselves to turn the case over to the manager.)

If one realizes that several hundred pit committees throughout the industry concern themselves with the problem of absenteeism every fortnight, one begins to sense the huge amount of time and thought that has been devoted to the problem at the grass roots, as well as in London. Without this effort by the committees, it seems certain that absenteeism would have been higher; but one cannot down the feeling that the committees are not well adapted to the one thing that the problem of absenteeism most requires: namely, getting down to individual cases. The committees realize the need for this, but their ability to do so is inherently clumsy, involves delays, consumes more time of more men than seems necessary, and frequently results in "no action" because of divided responsibility (in practice if not in theory) between the committee and the management. Yet the easy answer — that responsibility for disciplining absentees ought to be fixed unambiguously on management — has more appeal in logic than in practice: few pit managers would be much better able (by training or by time) to "get down to cases" than are the CCC's; furthermore, they would be confronted with the same difficult problem of what sanctions to invoke that has so often eluded the committees. New penalties are needed fully as much as new procedures — and we have suggested above that the two most eligible ones seem to be (1) relating the employee's rights to "miners' coal" to good attendance, and (2) relating eligibility for vacation pay to good attendance.

Not until pit managers and their underofficials are considerably better trained in handling personnel matters could much improvement be expected from giving them a freer hand in the matter. At such a time, it might be hoped that the CCC's could divest themselves of the frustrating, thankless task of sitting as absenteeism courts and could turn over this task to management on the one hand (who would initiate discipline) and the trade union on the other (which could challenge managerial action through the grievance procedure). Such a change would require a thoroughgoing reappraisal of the functions of joint consultation and of collective bargaining (two functions now often closely fused); but it is hard to believe that any such reappraisal is likely in the foreseeable future. In the meantime the main attack on attendance will continue through the committees and this effort is likely to show all the variation in effectiveness with which we already know the committees operate.

CHAPTER NINE

THE TRANSFER OF MANPOWER
IN SCOTLAND

One of the most strongly claimed advantages of nationalization was that it would permit "rationalization," or the closing of inefficient pits and the concentration of production at more efficient ones. In doing this, nationalization promised to give a new urgency to an old process; it also promised to effect this transfer of resources (mainly labor) with a greater regard for "human values" than had been characteristic of pit closures in the past.

The Board had closed about 150 pits during the first seven years of its life. This represents one out of every seven pits taken over in 1947, and compares with the target of about 400 closures hoped for by 1965. The number of closures by divisions (through 1952 only) was as follows:[1]

East Midlands	11
Northeastern	5
Southeastern	0
Durham	8
Northern (N & C)	11
West Midlands	15
Northwestern	8
Southwestern	36
Scottish	44
Total	138

Many of these closures have been very small pits, and the majority have occurred in two of the Board's nine Divisions — South Wales and Scotland. The present chapter is a report on the progress of the closure and labor-transfer program of the Scottish Division; for it is there that the largest effort has been made to transfer manpower from a declining

[1] Compiled from the Board's 1951 and 1952 *Reports;* the 1953 *Report* does not specify closures by Divisions. The number of men affected and output directly lost are not given. All but six of the "closures" were outright closures; in six reported cases, pits lost their separate identities through merger with adjoining pits. The figures exclude the small, privately operated, licensed mines employing fewer than 30 men.

to expanding areas. Published material on this experience is scanty, and most of the information here reported was gathered in a four-week visit to that coalfield in the fall of 1950.

The Scottish Division has been fairly well pleased, though scarcely elated, by the progress of its rationalization program. On the positive side is the record of a fairly substantial number of permanently transferred miners and their families; on the negative side must be entered a failure to recoup in the expanding areas the amount of output lost at the closed pits in the declining area around Glasgow. Perhaps all that can be claimed is that, given the closures, the loss of output would undoubtedly have been somewhat larger if there had been no carefully planned transfer program.

In saying this, we make the assumption that the productivity of the transferred men is higher in their new pits than it would have been if they had taken jobs at other Lanarkshire pits when they were laid off.

The Changing Location of Output

The map facing page 1 shows the location of Scotland's coal deposits in the lowlands that lie generally between Glasgow and Edinburgh. It was the conjunction of iron ore and good coking coal near Glasgow in Lanarkshire that allowed that city, Britain's second largest, to become such an important center of the engineering trades. The chief problem confronting the Scottish coal industry today is the physical and economic exhaustion of the Lanarkshire coalfield, that is, the area within a radius of fifteen to twenty miles of Glasgow (see Figure 4, page 251).

A coalfield or coalmine is scarcely ever abandoned because it is physically exhausted — the point of abandonment is almost always determined by the rising costs of production incurred as the high-rent seams are extracted, the price of the product being taken as constant or as rising less rapidly than costs. This fact is important, for it means that the point at which a pit becomes "exhausted" is dictated by economics and not by nature; people can argue about economics where they could not about nature. Total Scottish output has been declining for more than a generation: today, at 23 million tons, it is just over half the 41 million tons produced in 1910. Most of these losses have been experienced during the two World Wars, with little or no recovery after either. Not all four Scottish fields have suffered equally, but the field which has declined most has been the largest of the four, the so-called Central coalfield.

The county of Lanarkshire includes both the large Central coalfield and the much smaller, rural, and isolated Douglas Valley field. When speaking of the Lanarkshire coalfield, we will mean this Central field unless Douglas Valley is explicitly named. Douglas Valley is not in the declining part of the Central field; indeed, it is an expanding area, though a small one.

Between 1910 and 1939, output in the Central coalfield declined 45 per cent. In this Central field, those pits which fall within the county of Lanarkshire (pits which accounted for three-quarters of the Central field's output in 1910) dropped from 18 million tons in 1910 to 8.5 million in 1939, a decline of 53 per cent. Note, however, that despite the heavy loss of output over the past generation the Central field still produced more coal and employed more men in 1939 than any two other Scottish coalfields put together.

Within the Central coalfield itself, there are some ten districts. While production is not expected to fall off in every district, it will in a majority of them. The area which will suffer most heavily during the next decade or two is the Clyde Valley, stretching the twenty miles from Lanark to Glasgow and including such well known mining towns as Wishaw, Motherwell, Larkhall, Hamilton, Blantyre, Bothwell, and Cambuslang (see Figure 2). Of the thirty or so Clyde Valley pits operating in 1947, only two or three are expected to remain open until 1965; even during the second World War about one out of every three pits in the valley was closed. Taking all past and prospective closures in the Central field between 1947 and 1965, roughly a fifth of all mining jobs in Scotland will have been discontinued unless expansion can be achieved in other areas.

The prospective pit closures in Lanarkshire are not being accepted without a fight by some interest groups (local businessmen, Lanarkshire M.P.'s, and the adversely affected branches of the NUM). Certain technical conditions afford these groups a basis for argument, making closure decisions anything but smooth sailing. For example, there are two areas in West Lanarkshire that were once mined under water conditions so difficult that several pits had to be abandoned before all the usable coal had been extracted. These areas have now been flooded underground for nearly twenty years, but the knowledge that good coal remains has led to agitation for the dewatering of these areas to protect employment opportunities for Lanarkshire miners. The two areas lie between Glasgow and Coatsbridge and Wishaw and Cleland, respectively. One is thought large enough to support a colliery of 1000 men for about twenty years, the other a colliery of about 400 men for nearly thirty years. These 1400 prospective jobs represent about 20 per cent of the 7000 Lanarkshire coalminers in 1950. Engineers who have twice surveyed these areas report that it would be technically possible to dewater them but that the added pumping costs would raise the already high production costs by 5 to 7 per cent. A state subsidy to cover this additional cost has been suggested, but the suggestion has not gone far. In the face of this situation, it is hardly surprising that the Scottish Area of the NUM has

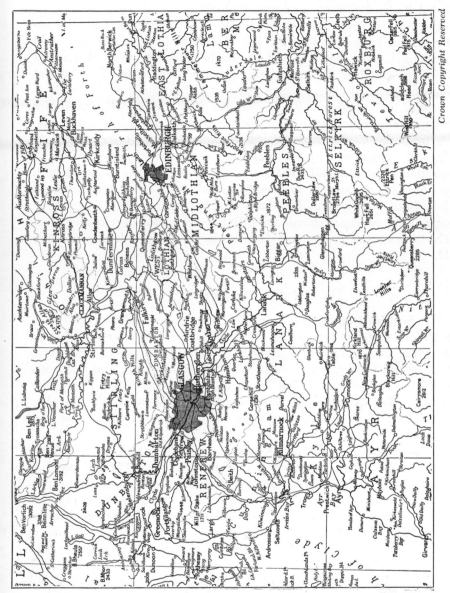

Fig. 4. Towns and Counties of the Scottish Lowlands

Scale: 1 in. = 1 mi.

agitated for the dewatering scheme on the ground of its technical feasibility, while the Coal Board (like the private owners before it) has rejected its feasibility on economic grounds.

The NUM has also sought to mobilize the support of technical experts behind its feeling that the life of several Lanarkshire pits could be extended if the Board would decide to carbonize some of Lanarkshire's high-cost coal. By carbonizing the coal, output could be sold at a higher price and at least some pits would be transformed from "uneconomic" to "economic" pits. In October 1950, the NUM organized a conference of groups interested in such a prospect, including representatives from the Coal Board. Wakefield Adam, an engineer invited to address the conference by the NUM, urged the establishment of a carbonization plant to handle 500 tons daily — a small beginning at best. The Board has not seen fit to accept this judgment, but it has continued to bore in the region around Glasgow; one hopeful report in 1952 proved disappointing on closer examination. But no fortuitous developments are likely to erase the prospect of large-scale closures in the Lanarkshire field.

Future Prospects

The Scottish transfer program is an attempt to facilitate the movement of mining families out of the Central coalfield near Glasgow to three other areas: (1) southward, into Ayr and Dumfries; (2) eastward, into Fife and Clackmannan north of the Firth of Forth; and (3) eastward into the Lothians, which lie on the south side of the Forth (see Figure 2). The extent of the shift in output hoped for by the Board is shown in Table 17, which details the changes in both output and employment.

These figures are sufficient to indicate the direction of the relocation which the Board will try to effect over the next ten to fifteen years, a movement which has already begun. Naturally, not all parts or pits within the expanding "areas" will share in this expansion; some closures are bound to occur even in the expanding districts; and there will be some internal shifts in output and employment within these same areas. For example, the expansion in Fife and Clackmannan will occur mainly at the eastern end of the Fife coalfield, along the coastal strip between Kirkaldy and Leven, where the workable coal ends. A large part of the expansion will occur at just three large pits (the Michael, the Wellesley, and the Frances), all undersea pits which will experience major reconstructions. In Ayr and Dumfries, where the coal is "spotty" and mining conditions bad enough to make large capitalization inadvisable, expansion will occur at five sites: the Douglas Valley, Cumnock, New Cumnock, Sanquahar, and Dalmellington.

The added output of the extra manpower planned for the three expanding coalfields differs quite considerably: in Fife and Clackmannan 1500 more men are expected to add 3.7 million tons (about 250 tons per man per year); in the Lothians, an extra 1500 men will add only 21 million tons (or 140 tons per man-year); in Ayr and Dumfries, 2400 more men are expected to add only 2.6 million tons (or about 110 tons per man-year). The marginal productivity of this extra manpower will be well below the average productivity of everyone employed in these districts, which by 1965 is expected to be around 400 tons per

TABLE 17

Annual Output and Employment in Scottish Areas of the NCB,
1949–1950 Actual and 1961–1965 Planned [a]

1. Production (mn. tons)	1949	1961–65	*% increase or decrease*
Total, Scottish Div.	23.8	30.6	+29
Fife and Clackmannan	7.3	11.0	+51
Lothians	3.7	5.8	+57
Central West	4.6	3.8	−17
Central East	3.7	2.9	−23
Ayr and Dumfries	4.5	7.1	+59
2. Employment	mid-1950	1961–65	*% increase or decrease*
Total, Scottish Div.	81,900	75,800	− 7
Fife and Clackmannan	23,300	24,800	+ 6
Lothians	12,500	14,000	+12
Central West	17,300	10,900	−37
Central East	13,200	8,100	−39
Ayr and Dumfries	15,600	18,000	+16

[a] *Plan for Coal: The National Coal Board's Proposals*, Oct. 1950, pp. 10 and 12. (The number of Areas in the Division have been increased from five to eight since these projections were made.)

man-year, or about the same as the expected average for the entire industry. The relationship between annual production and average employed manpower per year is scheduled to change very markedly over the Board's planning period. In 1949 and 1950, each man produced about 290 tons of coal per year; in 1961–1965, each man is expected to produce about 400 tons per year — an increase of about one-third. This increase in productivity (which applies to Scotland) is on the same order as that assumed for the industry as a whole. The OMS implied by such an increase in productivity is about 1.6 (against 1.21 early in 1952); probably the Board's expectations (which are never explicitly stated in *Plan for Coal*) are unduly optimistic: the industry will be lucky to have an OMS of 1.35–1.4 by 1965.

The scale of the Scottish closure program is not at present its most significant aspect. In the three years, 1942, 1944, 1945, about 2400 men were made redundant by closures in Scotland; in the first three years following nationalization, 1947–1949, about 3200 men were made redundant. The postnationalization closure program, however, has not been one of gathering momentum: 1949 has been the peak year for closures since nationalization; in 1950, fewer men were affected than in either 1948 or 1949.[2]

The main interest of the closure program lies rather in the way in which closures have been handled and in the success which the Board has had in moving men to new locations.[3]

The Closure Process before Nationalization

Before nationalization, the customary method of closing a colliery was to post a pithead notice announcing that on a certain date the pit would cease operations, the men being duly paid off and then left to seek alternative work wherever and however they could. Undoubtedly the more conscientious employers took steps to explain the necessity for closure to branch officials and their men and used their trade connections to try to find jobs for their men at other pits. Firms with more than one pit were better able to mitigate the hardships involved, but during the 1930's there was almost nothing that managements could do.

Men who lost work before finding a new job received only their normal unemployment benefit; the only job agency to help them was the public employment exchange, and if they could find work only in another district they did not receive any transportation assistance or settling-in grant to tide them over until receipt of their first pay at the new job. A system of transfer allowances was developed, it is true, by the Ministry of Labor during World War II, but housing was so difficult to find during that period that the help given by this "mobility money" was largely offset by the discouragement of finding accommodations; this difficulty was naturally greatest among men with families.

The Closure Process under Nationalization

The necessity for a geographical shift in the location of output in Scotland had been so well established before nationalization that on

---

[2] See the speech, "Transfer of Men," by John Colthart, Deputy Labor Director, Scottish Division, delivered to the Area Labor Officers on Sept. 17, 1950, 3 pp., mimeo.

[3] It is worth while calling attention at this point to a study entitled *Scottish Mining Communities*, New Series, No. 61, mimeo., prepared by The Social Survey (a branch of the Government's Central Statistical Office) in the autumn of 1946, just before nationalization.

Vesting Day all interested groups (managements, the union, M.P.'s, local councils, housing authorities, and so on) were well aware that closures were coming.[4] The Divisional Coal Board, however, was under obligation to carry out closures in much closer collaboration with the union and with other groups, such as local councils and housing authorities, than had been customary in the past.

In an early statement entitled "What I Think of Migration," Area President Abe Moffat noted that mass transfers of communities was something new to the industry but that it was in the men's interest "where absolutely essential." He expressed the belief that the men who elect transfer "will be treated with the greatest kindness by fellow workers" — a reference to the problems involved in the allocation of job rights in the developing pits. The main task of the union, he said, "is to protect the men's interest. Where concentration is inevitable, we must ensure that all available men shall be provided with jobs, and all married men transferred shall be guaranteed houses." He also took note of some complaints from his membership with respect to the fact that wages in the declining Central field were slightly higher than in the developing fields; he said that "if the Coal Board want to make a success of the scheme, they must give some consideration to this aspect. In some coalfields — particularly Fife — production is high, and there is no reason why the economics of such coalfields should not provide more favourable scales." Quite apart from how much of an actual discouragement to movement this "uphill" wage differential constituted, the transfer program obviously strengthened the NUM's hand in eliminating wage differentials among the Board's five Areas.[5]

It did not prove difficult to secure a general endorsement of the transfer program from top officials of the Scottish Area of the NUM. However, the union reserved the right to treat each closure "on its merits"; in practice, this has meant that the NUM has periodically been given lists of pits in the Central coalfield which the Board proposes to close during the coming one to three years. The union has thus known approximately where and when the "lightning would strike." The sharing of this planning information with top union officials has not created embarrassment for the closure program; but when it comes time to close a particular pit, the normal experience is the development of considerable local resistance.

The Board's procedure is to make a "full explanation" of the particular circumstances requiring closure to the Colliery Consultative Committee, giving local union representatives, and others, every opportunity

---

[4] A first-rate technical discussion of the problems faced in the Scottish coalfields will be found in the *Report of the Scottish Coalfields Committee* (Scottish Home Department, Edinburgh, 1944, Cmd. 6575, 184 pp.).

[5] See *Coal*, Feb. 1949, p. 18, for Mr. Moffat's statement.

to satisfy themselves of the justice of the Board's decision. The local people affected can almost always think of a great many reasons why the pit should be kept going: other pits ought to be closed before this one; the high costs reflect poor management more than anything else; some new coal-cutters or some larger pumps or cheaper tea in the canteen would lower costs; or a fairer price for the pit's output would eliminate losses or, at very least, the pit ought to be given a year's grace to "see what can be done." [6] It is not uncommon for meetings on a particular closure proposal to drag on for half a year or so before the decision is made definite and a particular death-date fixed. In the meantime, many local pressures are brought to bear on the Board (which sits in Edinburgh) and on the Area management, for during these deliberations, the chief responsibility for carrying through the closures in Lanarkshire rests on the Area General Managers and their Production Directors in the Central West and Central East Areas.

The key question, of course, is what happens when local pressures threaten to balk the Board's decision to close a particular pit. It should be understood that in theory, at least, the Board has not acknowledged (nor has the union claimed) any union right to veto closure decisions. The interesting problems, therefore, concern the unofficial procedures and tactics invoked by the Board and, especially, by the NUM. Does the Board, for example, invite the top union officials in Scotland (e.g., either Moffat or Pearson or the Miners' Agent who resides in Hamilton, Lanarkshire, James McKendrick, C.P.) to try to persuade their local members to accept the inevitable? If so, do these relatively disinterested union leaders side with the Board in its planning or with their own membership, the victims of planning?

Apparently, the Board has had to accept local opposition, including the opposition of James McKendrick, on almost every one of the Lanarkshire closures, and it has not been able to invoke the active help of the union's Edinburgh headquarters in securing local consent to the Board's proposals. However much this situation may differ from similar American situations (in which coalowners often find the International Office of the United Mineworkers helpful in downing local opposition to moves calculated to increase the industry's efficiency), I did not gather an impression that the NUM's Scottish headquarters had deliberately set out to sabotage the rationalization program: by maintaining an ambiguous neutrality, they were simply leaving the "dirty work" up to management.

The relative neutrality of the Area headquarters of the NUM in Scotland may have been helped by the fortuitous transfer of the union's headquarters from Glasgow to Edinburgh early in 1948; though the new

---

[6] See Appendix V for an account of a closure case that involved one of the small West Tyne pits, under the Northumberland Divisional Board.

location across the lowlands is only about an hour and a half bus ride from the mining villages around Glasgow, the transfer removed Moffat and Pearson from intense local pressures in the affected region. Both Moffat and Pearson reside in expanding areas and this may make the Board's job slightly easier.

The test of the union's sincerity would be whether or not it supported any strikes that might develop in Lanarkshire over disputed closures; but no such instances had arisen during the first four years of nationalization. Nevertheless, Board representatives in Edinburgh did report feeling that local union opposition to closures in Lanarkshire had increased in strength every year, so that closures were becoming more difficult to carry through. But so far no proposed closures appear to have been abandoned as a result of local resistance.

MATCHING UP JOBS AND MEN

If the obligation to "consult" has made the closure process somewhat more difficult for management, the transfer techniques worked out by the Board have made moving decidedly easier for the men.

The Coal Board has not relied primarily on the public Labor Exchanges to effect its transfers; instead it has organized its own labor market outside and parallel to the Labor Exchanges, on an industry basis. This procedure has operated as follows: When a pit is closed, the men are given their "cards" (that is, their insurance books). No formal attempt is made at this point to inquire into a man's availability for transfer or to describe to him the Areas in which he may expect to find another job in the industry, though general reassurances would have been given during the consultations preceding closure. In order to secure unemployment benefit, redundant men must go to their local Labor Exchange and register for work. The Board has assigned an assistant Area Labor Officer to the local Labor Exchanges (not continuously, but at those particular Exchanges where and when redundancies have occurred), and these men ask unemployed miners about their willingness to accept transfer to another area. Up to the middle of 1949, arrangements for transfers were made during these interviews between redundants and the Board's labor officers at the regular employment exchanges.

This procedure did not prove satisfactory, partly because there were too many men for the labor officers to interview satisfactorily, but mainly because there were too many intervening steps between the pits that were actually offering specific jobs and the individual applicants. Hence, a system of pooled interviews, called "panels" was developed.

Under this system, the labor officers continued to interview redundant men at the Labor Exchange shortly after their layoff, but they limited themselves to inquiring whether or not the men were interested in transferring. The names of those so interested were forwarded to the Board's Divisional headquarters in Edinburgh and, when 100 to 150 names had been accumulated in the file, the Deputy Divisional Labor Officer arranged for Agents or managers from pits in expanding areas (the line officials hiring labor) to be present at a mass interview at a central point in the redundant Area. Men who had signified an interest in moving were notified to be present on that date. The Board paid such men their travel expenses to and from the interview and paid for any shift lost in order to attend. This latter provision was made because not only men who had already become unemployed through pit closures but any men working in the redundant Area were eligible for the transfer privileges. This important provision allows men who know that their own pit, though still working, is likely to close within the next few years, to move earlier than they actually have to. The result is to create vacancies in pits still working in the redundant Areas which are thus able to absorb more local redundant men than they otherwise could.

At the panel interviews, usually held in a big room in a public hall, there will normally be two or three Agents or managers from each of the three main development areas (Fife, the Lothians, and Ayr and Dumfries) plus a representative of the Labor Department familiar with the pits represented by management. No trade-union representatives are customarily present. Although the NUM originally sought to post representatives at these panel interviews, the Board declined to permit it on the ground that its own labor officers (mostly ex-NUM officials) constituted a "neutral" observer competent to prevent line officials from misrepresenting details of the jobs being offered, housing prospects, transportation arrangements, and so on.

Applicants go to the table representing the Area of their choice and discuss available jobs with the management representatives. If the applicant has a skill for which the manager has a job-opening, a firm offer of employment is almost routinely made. If the applicant is satisfied with the job, its pay, his prospects for housing, and so on, he will accept; if not, he is free to proceed to a second or third interview with representatives from other Areas. If the applicant accepts an offer of employment, the manager gives him a card confirming the details of the job and its rate of pay, plus any "conditions" that may be attached (this is to prevent misunderstandings arising when the man shows up for work at the new job). When the man does show up at the new job, the manager notifies the Divisional Labor Department in Edinburgh so that their records can be completed.

This procedure has worked well and is considered a marked improvement over the initial exclusive reliance on the public employment exchanges. The advantage of having the Board organize its own labor market (in the sense of a place where vacant jobs and job applicants physically meet) is that the Board, unlike the public employment office, has a direct economic interest in the result and can bring to the market place that degree of detailed knowledge and intimacy of appraisal which is necessary to sales of labor in any labor market. Furthermore, the Board brings to its panel meetings a complete transfer program, including moving grants and housing arrangements; these items are usually critical in persuading men to accept geographical transfers and are not normally within the competence of the public employment offices.

MOBILITY BENEFITS

Under the transfer program, the Board, not the miner, assumes the costs of moving. This important system of grants was not "invented" by the Board itself (a system of grants to encourage labor mobility had been introduced by the Ministry of Labor during World War II); indeed, the Board only assumed this added cost in November 1948, after the Ministry declined to continue paying for a program that was directly benefiting the Board.

The particular "mobility benefits" payable include travel pay to the new job, a lodging allowance for married men until they are able to set up housekeeping, and a settling-in grant for single men. Men normally spend a few weeks living in a miners' hostel before getting a house. The Board also pays moving expenses for household goods (or "removal" expenses, as they are called in Britain); this includes the costs of travel expenses for a man's family. The travel and moving payments are made at cost, as determined by the Board (which, in the case of moving, makes the arrangements). The lodging allowance in 1950 was payable at the rate of 4s. per day to married men separated from their families and was payable until the transferred man's family joined him at the new location; the settling-in allowance is payable only to single men and consisted of a single payment of 24s. 6d.

The total cost of these transfer payments in 1949 (when 1800 men participated — the highest number of any year) was £14,603; this represented about one three-thousandth of the Division's total production costs for that year. Sixty per cent of the total transfer costs were represented by the lodging allowance for married men, 25 per cent for moving expenses, and the remaining 15 per cent for travel, settling-in grants, and lost wages.[7]

Redundant men who do not transfer are entitled both to the general unemployment benefit and the Board's own redundancy benefit, both of which are payable for a maximum of twenty-six weeks.

[7] William Willox, Divisional Coal Board Office, Edinburgh, Sept. 13, 1952. An account of the original introduction of these benefits in 1940 by the Ministry of Labor will be found in the *Ministry of Labour Gazette*, June 1940, and Jan. 1941.

THE SCALE OF PIT CLOSURES

We need to get some idea of the scale of pit closures in the declining areas of Scotland, of the number of men affected by these closures, and, especially, of the number of men who have been successfully transferred to the expanding Areas. The quantitative information in this section is derived from a study of unpublished Divisional figures; the data studied covered the period 1947–1950, but were most complete for the two years 1949–50.

From 1947 to 1951 the Board closed over 40 of the 187 pits it took over in Scotland (a few new ones, mostly drifts, have been opened). Not all these 40-odd pits were in the worked-out Central coalfield, but the great majority of them surely were. By the end of 1950, 26 pits in the Central field had been shut, and undoubtedly a few more there were closed in 1951. It is mainly what happened to the men laid off at the 26 pits of the Central field that interests us in what follows.

About 4000 men were made redundant by the 26 closures just referred to. Approximately three-quarters of these men were successfully placed in other jobs in the coal industry, but — and this is the point of most interest — most of these successful transfers were made within the same Area where the layoff occurred; only a small proportion (on the order of one-quarter) were transferred out of the declining field and into an expanding Area, a transfer which requires a change of residence. The proportion of men laid off at a closed pit who transfer to an expanding Area seems to vary greatly from one closure to another. An examination of records for seven closures in 1950 showed from 6 to 41 per cent of the redundant men placed, as being placed in an expanding Area requiring a change of residence.

It thus appears that, on the average, three out of every sixteen men (a quarter of three-quarters) laid off as a result of closing a pit migrate to an expanding Area. Of the remaining thirteen men, three or four leave the industry because they want to or cannot be absorbed by other nearby pits, and the remaining nine or ten find work at other pits in the area surrounding their closed pit. Some may feel that the attempt to "transfer resources" by closing high-cost pits and transferring men to low-cost pits in other Areas is not very successful if only three out of sixteen redundant men actually transfer. The subsequent paragraph partly corrects this impression, but one may fairly ask: what proportion of the men laid off at any pit ought to be considered as potential migrants? When one makes allowance for the older men not in a position to start life anew, for boys too young to go out on their own, for injured men whom other pits will not hire, for men who decide to retire, for men tied to a particular region by family responsibilities, then the proportion who do in fact agree to migrate does not seem surprisingly low.

The process of consciously relocating labor through the transfer program, however, does not end with the migration of men who have been laid off as a result of closures. The benefits of the migration program are open to anyone in the declining Areas who wishes to use them, and the record indicates that more employed people have migrated than unemployed. From one-half to maybe two-thirds of the migrating men have moved from pits that were not undergoing immediate closure. Many such men are, of course, migrating "ahead of unemployment," for the dim future of the Central coalfield's pits is widely appreciated among the mining population. When employed miners migrate, it makes easier the task of distributing men during each subsequent closure.

By the end of 1951, the Board reported that some 2750 men had availed themselves of the mobility benefits under the transfer program begun by the Division in 1949. Although this number represents only 3 to 4 per cent of the total numbers now at work in the expanding districts, it would represent a significantly higher percentage of the new hires in those districts. These men, relatively young and already trained, represent a class of labor which the Board very much needs to retain in the industry, particularly if they can be shifted to the lower-cost districts.

The Critical Role of Housing

Housing, not higher wages, has been chiefly responsible for tempting men to uproot themselves from their native Lanarkshire and move to another part of the Scottish coalfield. Although wages were not uniform throughout all the Scottish coalfields in 1950, the 4 to 5 per cent differentials which could be found in certain occupations were not considered by either Coal Board or union officials to have any discernible effect on the transfer program. These differentials had been much greater (on the order of 20 per cent for faceworkers between, say, high-wage Ayr and Fife), but had been progressively narrowed in the interest of standardization and not because the differentials threatened the transfer program. In late 1948 and early 1949, wages for some Lanarkshire occupations were higher than wages in some of the expanding Areas, and this condition led some militant critics of the Board's transfer program to charge that the Board was "out to break up the high-wage districts"; but the differentials were so small that not even Mr. Moffat or Mr. Pearson took the charge seriously. There have been no suggestions that high wages ought to be used to encourage transfers — they would be useless without houses. This fact largely defines the class of men who have participated in the transfer program.

For generations, employers in the coal industry have had to build houses in order to persuade men to come and work for them. But while housing has long been a necessary consideration in attracting labor to this often isolated industry, it has frequently not been a sufficient condition; and employers in expanding areas have often had to offer higher wages than the established districts (such wage inducements would, of course, be necessary only in periods of rather full employment). In postwar Britain, the attraction of housing has had a much more exclusive pulling power than ever before, as housing has probably been the commodity in shortest supply all over the country. Generally speaking, however, the Labor Government was exceedingly wary of using housing as an economic incentive, even though the hope of a house was one of the strongest possible incentives.

The Government did give preference to mining communities in the distribution of several thousand aluminum prefabs immediately after the war and many mining Areas are today distinguished by clusters of these temporary dwellings. But this allocation was a special "one-shot" operation; the British have an abiding preference for the "traditional" type of brick house (usually duplex), and the overwhelming proportion of all postwar housing has been of this type. It is generally true that the government has allocated new houses in a rough proportion to the existing distribution of population, though a few exceptions have been made, often rather late in the day. Thus, only in the winter of 1950–51, when 10 to 15 per cent of the population were faced with possible unemployment due to the extreme shortage of manpower in the pits, did the government initiate some cautious steps to give mining communities preference in the allocation of new, traditional-type houses.

The Scottish transfer program is an outstanding exception to the foregoing remarks about housing policy, for the Divisional Coal Board, in conjunction with Scottish housing authorities, has used housing as an incentive as was not done nationally before 1952 (see page 191). One cannot say whether this Scottish policy has arisen from a more realistic appraisal of economic needs during the postwar period or from being confronted with a major closure program which had long been foreseen and which has been more threatening both to output and to prestige than any prospective closures in England and Wales. In any event, the housing program in Scotland has operated as a sufficient condition of mobility, in the sense that the Board has not had to raise wages in the developing Areas above those in the declining Areas in order to induce movement.

In Scotland, as in all other Divisions of the Coal Board, the Board did not enter the housebuilding field directly. Instead it put detailed production and manpower forecasts before the normal housebuilding agencies which serve the general population. About 80 per cent of all houses built in Britain under the Labor Government were built by public bodies under the general supervision of the Ministry of Health (in Scotland, the agency governing allocation and design of housing is the Department of Health for Scotland). The remaining 20 per cent represent houses built by private individuals who have been able to secure permission from their local authority.

The bodies normally responsible for doing the building and allocating the completed houses were the local authorities (that is, town governments); these may (1) enter into contracts with private builders, who build in accordance with the authorities' specifications (the most common procedure), or (2) employ building labor directly. In addition, there is in Scotland an organization known as the Scottish Special Housing Association, a private nonprofit organization which was set up before World War II in order to build low-cost houses with government aid. The SSHA was set up before local councils engaged in house building to quite the extent they came to do after the war; it is in no sense a special organization set up to meet postwar needs of the transfer problem in the Scottish coal industry. From the point of view of the kinds of houses it builds, where it builds them, their design and facilities, and their rentals, SSHA may best be regarded as little different from the local authorities.

THE SCALE OF THE BOARD'S HOUSING PROGRAM

A special housing program for Scottish mineworkers has been in existence since after the second World War. Of the 13,800 miners' houses planned from the beginning of the program through 1953, some 5000 were included in the "pre-1948" program, a period with which we shall not be concerned, since this pre-dates the Coal Board's assumption of responsibility for the Scottish transfer program. Our interest lies in the location and allocation of the 8800 houses scheduled for the 1948–1953 period; the location of these houses, by districts, for each year from 1948 through 1953 is shown in Table 18.

The entire mining labor force in Scotland has hovered around 80,000 ever since nationalization, and the building of 8800 new houses in the six-year period under review suggests that more than one miner in every eight will have been rehoused under this program. Actually, the proportion would be considerably higher than 10 per cent, since many households contribute more than one member to the industry. In practice, the Miners' Special Housing Program was somewhat behind schedule in 1951, but not so seriously that we cannot assess the effectiveness of the program in facilitating the transfer program.

It will be noticed from Table 19 that not all the new houses have been allocated to the developing districts, though more than 90 per cent have been. Table 19 compares the distribution of new houses, 1948–1953, with the distribution of the labor force as of 1949. Areas 1, 2, and 5 are the expanding districts; Areas 3 and 4 the declining dis-

TABLE 18

Scottish Miners' Housing Program, 1948/49–1953

Area	1948/49 Program	1950 Program	1951 Program	1952 Program	1953 Program	Total
East Fife	412	300	498	300	000	1,510
West Fife	292	260	000	000	200	752
Alloa	288	364	110	200	66	1,028
Lothians	602	570	394	510	488	2,564
Central West	210	102	100	000	000	412
Central East	246	36	100	000	000	382
West Ayr	126	122	30	198	228	704
East Ayr	370	364	342	302	68	1,446
Totals	2,546	2,118	1,574	1,510	1,050	8,798

Source: NCB, Scottish Division.

tricts (especially Area 4).[8] The two declining Areas together accounted for 41 per cent of the Division's manpower in 1949, but were scheduled to receive only 9 per cent of the housing allocations for the whole Division. Furthermore, all the Central West and Central East houses were scheduled for completion by the end of 1951; no new houses for these Areas were planned for either 1952 or 1953.

TABLE 19

Proportional Allocation of New Houses (1948–1953) in Relation to the Existing Distribution of the Labor Force in 1949 [a]

Area	Per Cent of Labor Force	Per Cent of New Houses
1. Fife and Clackmannan	27	37
2. Lothians	14	30
3. Central West ⎱ Declining	23 ⎱ 41	4.7 ⎱ 9
4. Central East ⎰ Coalfields	17 ⎰	4.3 ⎰
5. Ayr and Dumfries	19	24
	100	100

[a] Derived from the "Summary Schedule" of the *Miners' Housing Program*, Scottish Division, NCB, dated June 22, 1950, and NCB manpower figures at Division offices in Edinburgh.

Not all the new houses allocated to the mining community are intended for transferred men, although this was the Board's original intent. However, this intention proved so unacceptable to local interests that the Board altered its policy by agreeing to allocate 50 per cent of its special allocation to transferred miners and 50 per cent to local miners. Thus, only about 4400 of the 8800 houses were intended for transferred men. The political importance of this "50–50" rule cannot be overestimated. Only this rule for allocating extra housing between local and imported manpower has allowed the Coal Board to use housing as an incentive to labor mobility in Scotland. The same device has, apparently, not been suggested, or for some reason has not been adopted, in certain English coalfields where its use would seem, on the face of things, just as feasible as in Scotland. The following quotation from the *Manchester Guardian* of December 20, 1950 refers to the "political" problems involved in expanding housing allocations in the Yorkshire coalfield:

A regional official of the Ministry of Health said that another tour of local authorities would be made. He hoped that homeless miners would be housed within five years. Admitting that some councils were not prepared to accept miners from other areas until this was done, the official said that newcomers would have to take their place on the lists.

[8] The number of Areas in the Scottish Division has been increased from five to eight since 1951. This chapter uses the 1951 designations except in Table 18.

This has been one of the difficulties of increasing allocations. So far as information given to me showed, the Ministry could order special building if the new houses were needed for imported labour, but not if they were needed for an existing labour force. Local councillors would soon lose the confidence of the electorate if they allowed newcomers to "by-pass" the housing lists, and thus the situation arose in which some authorities have refused increased allocations.

INTEGRATION OF THE CLOSURE AND TRANSFER PROGRAMS

The coördination of pit closures in Lanarkshire with manpower transfers to the expanding districts does not have to be closely regulated, since over half the men who transfer do so without any pit being closed. However, there must be close coördination of the volume of manpower transfers wtih the progress of housebuilding and the development of pit-room in the receiving districts. The lack of any necessity for close timing between closures and transfers is fortunate in view of the almost inevitable battle, real or ceremonial, between production officials and the trade union (often backed by local political interests) whenever the Board proposes to close a particular pit. After allowing for the fortuitous slack in the closure-transfer relationship, an attempt is made to keep a rough complementarity between housing and closures; if this were not done, the Board would run a much greater risk of losing men from the industry when closures do occur, and, if housebuilding fell behind closures, the political problem of carrying through shutdowns might become even more difficult. It has certainly been true that the Board has not imposed a harsh closure program in Lanarkshire during the first six years of nationalization, closing pits rapidly without regard for housing prospects in the developing areas. So long as housing has been in short supply, the house-building program has tended to regulate the pace of the closure program.

NEW HOUSING AND NEW TOWNS

The relocation of labor in Scotland will not give rise to more than one wholly new town within the foreseeable future. Immediately after the war, it was first thought that four new towns would be needed in Scotland to serve the changing geography of the mining industry — two in Fife, one in East Lothians, and a fourth in Ayrshire. These were not to be purely mining towns but "mixed" towns in line with the universally held belief that it is socially and economically healthier for towns not to be based exclusively on a single industry. The Social Survey found that about two-thirds of their sample would prefer to live in a "mixed" community rather than a one-industry mining town. A substantial minority, however, preferred the "mining village." Significantly, the reason for the preference for "mixed" towns was put down not to the social broadening attendant upon coming into contact with families in other industries, but to the hope that only by living in "mixed" towns could the miner or his sons avoid mining.[9] By September 1950, however, all but one of these "new towns" had been abandoned; the one remaining is Glen Rothes in East Fife, on the northern edge of that coalfield about five miles in from the coast.

The Board's special housing program contemplates building at 83 separate sites in its five Areas — 26 in Fife and Clackmannan, 13 in Central West, 8 in Central East, and 18 each in the Lothians and Ayr and Dumfries. Of these sites 37 will provide 100 or more houses, the largest being an 800-house site at Kirkaldy, Fifeshire; 34 sites will involve 50 or fewer houses.[10]

Most of the house-building will consist of housing "estates" (a "development" is called an "estate" in Britain) added on to existing villages and towns. In some cases the new estates will be big enough to absorb the villages in which they will be built; in others they will not greatly change the character of the present towns. But wherever 250 to 300 houses go up, there arise the full-scale problems of any new community's transportation, shopping and recreation facilities, school space, representation on local government bodies, relations between old residents and the "newcomers," and the adjustment of people to new houses, unfamiliar surroundings, and new social relationships.

Anyone familiar with the postwar housing shortage in Britain will understand why the average expectation regarding housing is so much different from what it is in the United States, where a certain freedom of choice, even among wage-earners, is common. People who have been on housing lists for five to ten years are not particular about the details of any house they may be lucky enough to get (a five-year wait for a new house is not unusual in Britain: a couple marrying even in 1950 had to expect to wait several years for a new house — and old ones are simply not available for renting). Hence the freedom of choice problem is not only not acute, it is largely irrelevant. When the Divisional Housing Officer in Edinburgh notifies a transferred man that he is to have such-and-such a house at such-and-such an address and at a specified rent, the man has to be very dissatisfied before refusing to take the house offered. However, the Board's rules do allow a transferred man three refusals before he disqualifies himself from further consideration.

Most houses within a particular development carry a common rent and have equal facilities, though this is not always the case. Where rent differentials exist (for example, at Kennoway, where different groups of houses were built to different designs by different builders), this naturally produces some dissatisfaction among neighbors who are in the

[9] *Scottish Mining Communities*, pp. 84–89.
[10] Scottish Division, NCB, *Miners' Housing Program* — schedule dated June 22, 1950, 6 pp. mimeo.

same approximate income group; but so far these differentials have not constituted a major problem.

Far more important than local rent differentials within the new estates is the much greater differential between average rents in new postwar houses and the rents customary in the prewar substandard housing to which so many Lanarkshire miners have been accustomed. In 1950, Lanarkshire rents varied between 2s. 6d. and about 7s. a week in miners' houses. New Council houses in the development areas are rarely under 19s. and sometimes get to 23s. Even at these relatively high rentals, a high subsidy is involved. It has been estimated that it costs about £80 a year to carry all charges for an average postwar house in Scotland. Of this amount, the tenant's rent covers only about £28, the local authority contributes about £23 out of local taxes, and the national exchequer contributes about £29.[11]

It is probably true to say that on an average transferred men face a rent increase of roughly 300 per cent — for example, from 5s. a week to 20s. There is no great resentment, apparently, at this state of affairs, for certainly there is no question in anyone's mind that the new houses are far better than the old. But this does not eliminate a straightforward economic problem of some magnitude — families having to get used to very much higher rents. It must also be realized that a pieceworker moving from Lanarkshire, where he may have been accustomed to a weekly income of 9 to 12 pounds, must serve a probationary period in his new pit in the development area at a considerably lower wage — they are assured only the national minimum (£5 15s. in September 1950, subsequently raised to £7 1s. in December 1951). Thus the jump in rents comes at a time when incomes, in some cases, are seriously reduced. However, the problem had not become so serious as to embarrass the transfer problem; and the increase in the national minimum must have substantially relieved the "squeeze." Again, people are usually so glad to get a house that the resulting shock of a substantially increased rent is not unmanageable, and of course the prospect of higher rents is always vaguely realized by anyone who is moving into a new postwar house.

The new public authority houses into which transferees are moved are not "tied," though the older prewar houses which the Board inherited are. New tenants entering these old houses owned by the Board do have to sign a "missive" committing them to move out if they leave the coal industry. However, this obligation is never enforced unless men fall in arrears on their rents: the Board's Housing Factor (manager) in Fife reported that the Board could count on the sheriff never evicting a man if his rent was paid up, regardless of "tied house" commitments on paper. In Fife, there have been about ten cases since nationalization

[11] The Economist, May 20, 1950, p. 1113.

where the Board has threatened to evict a man unless he returned to mining, and in each case the man did return.

There has been a great deal of pressure to eliminate completely the "tied" house, whereby a man employed by a coal company and occupying a coal company house was legally obliged to surrender his house if he left his employment. In recent years this issue has been more important in principle than in practice, as many coal firms did not enforce their rights to the letter, the general custom being not to disturb disabled miners forced to leave mining employment, miners' widows, and similar distress cases. This is the present NCB attitude, not only in Scotland but in the English coalfields as well. The significant point for our interest is that the absence of the "tied-house" principle in the new public houses for transferess does not seem to have led to difficulties. Transferred men are not getting houses on the pretext of taking up mining jobs in expanding areas, working in a pit a few weeks, and then "getting out of the industry." Thus the housing program is not contributing to wastage from the industry. Undoubtedly the problem would be somewhat more serious if there were more nonmining jobs in the development areas, but there are relatively few such jobs; indeed, this is one of the major problems of the whole transfer program and deserves separate comment.

Coal Board and trade union officials in Scotland are almost unanimous in agreeing that the provision of more subsidiary employment in the developing mining areas would not only diminish the dependence of those areas on mining but would persuade more families to transfer than have yet done so. None of the three expanding areas has any large number of nonmining jobs and at least one, the Douglas Valley, has practically none. This is a serious deterrent to Lanarkshire mining families who often have family members in nonmining employment.

So far practically nothing has been done to encourage the building of light industry in the development areas. Indeed, the government's policy, raised to a requirement of law, has been the other way — to allow new industries to build only in areas where there is unemployment, which in Scotland means chiefly in Lanarkshire. The NUM officials in the expanding areas and the NCB have both sought to bring pressure on the Scottish Home Office to authorize new industries for Fife, the Douglas Valley, and the expanding parts of Ayrshire and Dumfries; but government authorities say they are bound by the law, which limits new industries to areas with unemployment. Local government and Labor M.P.'s in Lanarkshire naturally press for local interests and are not sympathetic to any attempts to assign new firms to other districts in the interest of the miners' transfer problem. Any quantitative estimate of the effect of this reluctance to adjust the distribution of general industry in the interest of facilitating the relocation of the mining industry is impossible; but Coal Board and NUM officials feel it is a significant factor.

With labor shortages in mining characteristic of the developing areas, it might be supposed that men willing to transfer would have no difficulty in getting the kind of job they want. One might even believe that the hope of upgrading would be an effective motive in persuading some men to move. Such a hope, however, does not seem to be a factor of much significance; on the contrary, local agreements between managements and branches in the expanding pits customarily provide that preference for upgrading shall go to local men over any men transferred in from outside. In short, plant — not industry-wide — seniority governs job allocations, though the rule is more likely to be customary than contractual. If a pit in a developing area has a significant number of potential facemen who have been retained on haulage jobs simply because there were no others to man these jobs, the introduction of transferees serves to give the local men an opportunity for upgrading and to consign the transferred men to haulage jobs, at least in the short run. However, these problems have not produced any cascade of grievances — after all, men transfer with a written agreement that they will be employed for a particular grade of work, so the number of misunderstandings is kept to a minimum.

The NUM has not taken any special steps, or set up any special machinery, to "police" the transfer program. Thus, when a transferred man starts at a new pit in a development area he may not be told who the branch secretary is and the latter may not be informed that the man has started. Only if the man gets a sufficient grievance to move him to seek out the branch secretary is the problem likely to come to the notice of the union.

It is generally true that mining conditions in the development areas are easier than in Lanarkshire, but of course it is not always true (a man may leave a good working place in Lanarkshire and get a steep, watery place in Fife). What probably counts most is not how good or bad conditions are but what a man is used to. Thus, it probably exercises little effect on the volume of transfers to tell men in Lanarkshire that if they move to Fife they will be working in 4-foot seams instead of 20-inch. Nor is the effect all positive when men are told that Fife pits are "non-gassy" and that they are in fact "naked-light" pits: some men are uneasy working in naked-light pits, no matter how non-gassy they are reported to be. On balance, it seems doubtful that generalizations concerning mining conditions have any significant effect on persuading men to move or not. The certainty that women will move into "better conditions" by transferring is probably greater than in the case of their husbands: all the kitchens in Fife have running water and new "cookers." Eighty per cent of all Scottish miners work at pits with baths, and

nearly all the developing pits have them. This is not an important factor
in the general movement, though in the few cases where men have been
offered transfers from a pit with a bath to a pit without, the lack of
baths at the latter has almost invariably dissuaded men from accepting
employment at those pits.

With housing the undisputed key to the transfer program, it follows
that those whose circumstances do not make housing a main need have
much less incentive to move. Unmarried young men who move must live
in lodgings or a miners' hostel, and most of them quite naturally prefer
to remain with their families if the latter do not move. Men past the
age of forty to forty-five, whose children may be self-supporting, often
do not feel the pressure for housing that the younger married men feel.
In fact, the program has largely drawn on married men in their twenties
and thirties.

As we have seen, probably three-quarters of the Lanarkshire men
placed following redundancy do not migrate; we have also seen that
the proportion of total redundants placed has so far been high, even
though most of the placements have been within the remaining Lanarkshire pits. This means that the problem of derelict Lanarkshire mining
communities has not yet risen to acute proportions and certainly the Coal
Board has not yet had to face up to the problem of creating an economic
vacuum in Lanarkshire. It is too early to tell what will become of the
hundreds of small shopkeepers, publicans, doctors, dentists, and so on
whose incomes have been based on spending power generated by mining employment. Though such repercussions may already have appeared
in some localities, they are still largely problems for the future. The
closures and transfers to date have not yet been on a sufficient scale to
leave Lanarkshire a wasteland.

Even though realism advises that the proportion of redundant men
from a closed pit who may be thought eligible for transfer is relatively
small, by no means all of this eligible group have exercised their options.
Undoubtedly one reason why some have not is that, even though they
may not find alternative mining employment in Lanarkshire, their incomes do not entirely dry up. In addition to the basic unemployment
benefit of 26s. a week (supplemented by 6s. for a wife, 7s. 6d. for the
first child, and 5s. for each additional child), unemployed surfacemen receive £1 13s. 8d. (£2 3s. 8d. for underground men) a week for
twenty-six weeks as redundancy benefit from the Coal Board; in addition,
family men get the family allowance of 5s. a week for each child after
the first. All such income is tax free. Thus it was possible for some men
to receive, even though unemployed, up to nearly £5 per week — which
represented then a decrease of only 15 or 20 per cent from their previous
earnings, if on day rates (as about 60 per cent of miners were). When
unemployment benefit stops, together with the redundancy benefit from

the Board, families can often qualify for "relief" from the National Assistance Board — whose aid includes shoes and coats for children in addition to minimum subsistence allowance for food and rent.

But there is no evidence that any significant proportion of redundant men were refusing to move simply because they preferred to "live off the state" for six months or more. The problem is at bottom one of judging motives; for men who refuse transfer and accept state benefits during a period of unemployment are just as likely to be "waiting for something to turn up" in Lanarkshire as they are to be "malingering."

Labor officers of the Board appeared to feel that many more potential migrants who, instead, sought jobs at local pits with a limited future, did so for "family reasons" rather than because they knew they could live on welfare benefits for half a year. Many men turned down transfers because "their wives don't want to move" or because "they have aged parents to support" or because "they have two daughters working in Glasgow" — reasons which sometimes appeared to NCB transfer officers as a form of "buck passing." But obviously this involves personal situations about which general judgments are very difficult to substantiate.

Almost all the redundancies in Lanarkshire have occurred at pits within fifteen miles of Glasgow. This is a built-up, cosmopolitan region; and even though many Lanarkshire miners live in "pit villages," they feel themselves, in part anyway, citizens of Britain's second largest city. The expanding areas in East Fife and southern Ayrshire and Dumfries are nearly a day's journey away from Glasgow by bus. The Lothians coalfield is within ten to fifteen miles of Edinburgh, but Edinburgh is not Glasgow, especially to a Glaswegian. Furthermore, the development areas are largely rural and the character of life is that of the isolated village, not the city. These facts mean that the transfer program is forced to run against the normal direction of migration in most industrialized countries, namely, from countryside to city. The psychological difficulty is probably the biggest the transfer program has to cope with among the redundant men who might be considered eligible for transfer. Dalmellington, East Fife, and Edinburgh may have good football teams, but the "Celtics" and the "Rangers" will always be in Glasgow.

The number of Lanarkshire men who have moved to a development area and have felt dissatisfied enough either to return to Lanarkshire or to seek to move to a second development area has been very low, probably not over about 5 per cent of the total transfers. Officially, the Board's transfer scheme does not allow a man, say, to move to Fife and then, if dissatisfied, to move from Fife to the Lothians or to Ayrshire. This is because men are eligible for the financial grants attendant upon moving only if they have "last worked in a redundant area," or have been unemployed in any area over three months. However, the Board has not enforced this rule rigidly, though there has been some skirmishing between the local labor officers who do the actual interviewing of intending movers and the Divisional administrative officials whose job it is to watch the costs of the transfer program. The former tend to wink at cases of men who turn up at Lanarkshire interviews who had formerly moved to a development area but who have returned and now signify a desire to move a second time. Technically, these men would have to work in a Lanarkshire pit for three months before reëstablishing their eligibility to move, or would have to wait out three months of unemployment in a development area, but the rule is too technical to stand in the way of men who can present valid reasons for desiring a second chance. The upshot of the rule and its administration seems to be that the presence of the rule has prevented "job shopping" by trans-ferees, while it has not stood in the way of a "second chance" for men who can persuade the interviewing officers that they deserve it.

The second group of men who have moved to a development area and wish to leave it are those who do not wish to try out a second development area but who simply want to return to Lanarkshire. Though "homesickness" is a common malaise in the developing areas, it has not been so strong as to lead more than perhaps 1 per cent of the men to give up the idea of staying and to return "home." "But," as one burly stripper confessed, "if they'd build any houses in Lanarkshire . . ." This problem will undoubtedly diminish as more people "from home" move into the expanding areas and as the latter come to feel more "like home" to those who have moved out to them.

CHAPTER TEN

LABOR AND TECHNOLOGICAL CHANGE

Between 1950 and 1965, Britain plans to invest two and one-half times as much capital in her coal industry as was invested in it on January 1, 1947. Some of this gross investment will take the form of revolutionary new face machinery; some of it will permit radical reorganizations of underground transport and of coal-winding; some of it will occur on the surface, chiefly in the form of new coal cleaning and preparation plants. But all of it will have as one of its effects a reduction in the number of men required to perform specific operations. By substituting capital for labor, the Board hopes that 1965 output can be increased by about 20 per cent over 1947–48 levels, while the labor force is allowed to fall by about 11 to 12 per cent (output would rise from around 200 million tons to 240–250 million; manpower would fall from about 700,000 to about 620,000). To achieve these results, output per manshift will have to rise by about 25 per cent, that is, from 1.2 tons to 1.5 or 1.6.

Labor displacement is by no means the only labor problem which this ambitious program will raise: new machinery and equipment will change the "job mix" of the industry, the proportions of the various skills required; at the face, particularly, new machinery is already changing the time-honored method of wage payment for facemen, with day rates being the rule on all the new machines now being tried out; again, the organization of work-teams among facemen is being profoundly affected by the new methods of work that go with the new equipment — a change of great potential importance to morale and productivity. Among underground haulagemen (who account for one-quarter of all underground miners) and among surfacemen the effect of new transport and coal preparation arrangements is confined much more narrowly to displacement alone: their method of wage payment is not likely to be affected, and the sociology of group organization among these grades has never played nearly as important a role in the industry's productivity as is true among facemen.

One development the modernization of the British industry will not entail is any large-scale introduction of American mining methods and coal-getting equipment. During the last war, when British output had to be raised as rapidly as possible, a team of American mining engineers was invited to Britain to see if they could make a contribution on the basis of American methods. A considerable effort was made to introduce American equipment (mainly Joy-loaders, "duckbills" and shuttle cars) and to persuade British engineers to adopt room-and-pillar systems on a much wider basis.

The technical controversy surrounding the attempt to "Americanize" the industry during the war is now settled: British engineers are today solidly agreed that their technical future does not lie in adapting their methods of working to standard American equipment. This should not be taken to mean that the American industry has nothing to teach the British; but important as such contributions may be, they will not be central to the technical reconstruction of the industry. The British have already begun to adapt to their own conditions certain promising items of face equipment used on the Continent; but the main contribution will be indigenous if the present promising experiments are any key to the future.

It should be obvious that, if the NUM and its membership do not accept the changes which the engineers recommend, the outlook for modernization is dim indeed. The attitude of the mineworkers is certainly not the only factor in setting the pace of technological change: but a coöperative attitude toward change is a necessary if not a sufficient condition of a fruitful investment program. On the record to date, students of the industry might agree that the rate of introducing new methods and equipment and of disseminating them throughout the industry suffers at least as much from the cautiousness and conservatism of managerial personnel as from labor opposition.

Management's cautious attitude, in turn, is not a reflection of the industry's organization but rather of the psychological attitude towards "change" which is part of the British national character. It was the difference in attitudes towards technological change between Britain and the United States that most impressed the Britons who visited the United States in 1950 as members of the Anglo-American Productivity Team on coalmining.[1] This is not to say that there has not been a significant change in the pace of technological change since nationalization: union leaders, mining engineers, professors of mining engineering, teachers in local technical colleges, and machinery manufacturers are all agreed that since nationalization the process of capital investment has gone

[1] *Coal: Productivity Team Report,* 1951, p. 42. Suggestive insights into Anglo-American differences in attitudes toward change will be found in Geoffrey Gorer's study, *The Americans,* p. 117.

forward with much more energy than ever before. The only reasonable doubt is whether change is going ahead with all the energy that might reasonably be expected; the answer to this is by no means clear.

Despite the accelerated pace of capital investment, it has gone forward not nearly as fast as the Board had originally planned. As the Board's 1952 *Report* states, most of the big schemes have "fallen behind schedule, some seriously so." For 1950–1952, actual expenditures on colliery capital account were less than two-thirds the rate outlined in *Plan for Coal* (£89 million spent *vs.* £138 million planned). Germany, with a coal industry about half the size of Britain's, is reported to have invested 50 per cent more than Great Britain during this period; and France, whose coal industry is only half the size of Germany's, has invested four or five times as much again as Germany (derived from the statement of Dr. Heinrich Kost, director general of the German Coal Management Board, as reported in the *Financial Times,* March 5, 1953). The oft-made explanation for the lag in NCB investment is the severe shortage of technical personnel sufficiently experienced to carry out the basic planning that must precede capital expenditures.

The Main Directions of Technological Change

The character of the changes that will affect haulage and surface workers is not difficult for the layman to imagine. For example, the Coal Board anticipates that the greatest savings of manpower can be made on the haulages, by substituting locomotives or main trunk conveyors for the present antiquated system of rope haulage which is now so general in most districts. The introduction of skip winding will likewise greatly reduce the number of attendants needed at the pit bottom and in handling coal as it comes off the cage at the top. Likewise, the construction of modern coal cleaning plants will mean a fairly extensive substitution of automatic cleaning for the hand-cleaning methods now so common.

The direction in which changes at the coalface are moving requires somewhat more explanation. Broadly speaking, there are two great changes toward which the industry is working. The first of these is the elimination or reduction of hand loading; the second is the extension of continuous mining. These changes are usually related, but not always so: at present, the most commonly used machines which eliminate the need for hand loading still tie the method of work to a definite cycle of operations, with all the inflexibility that cyclical working imposes. There are, however, a number of promising face machines and conveyor developments which permit the all-important coal-getting operation to be performed on all three shifts. Not only does three-shift working make more economical use of all the pit's capital equipment, but it greatly re-

TABLE 20

Amounts of Selected Forms of Investment in Coal-Getting and Coal-Handling Equipment during the First Six Years of Nationalization [a]

(Items in Use at End of Year)

	1947	1948	1949	1950	1951	1952
I. Coal-Getting Machines						
Longwall loading machines						
Huwood Loaders	n.a.[b]	n.a.	30	35	n.a.	n.a.
Tons loaded	n.a.	n.a.	0.9 mn	1.1 mn.	n.a.	n.a.
"Cutter-Loaders"						
Meco-Moore	34	42	51	70	81	112
In East Midlands Div.	n.a.	36	41	51	(?)	60
Millions of tons loaded	3.2	3.0	3.6	4.4	5.3	6.3
Joy "Continuous Miner"	0	0	1	1	3	3
Joy "Closter-Getter"	0	0	0	4(?)	n.a.	n.a.
Wedge-type Machines						
Coal "Ploughs" & "Scraper boxes"	0	1	2	3	5(?)	14
"Samson Stripper"	n.a.	2	3	3(?)	6	7
II. Transport						
Locomotives underground (80 in 1946)						
Diesel	n.a.	170	279	n.a.	390	n.a.
Electric battery	n.a.	15	52	n.a.	60	n.a.
Total	n.a.	185	331	n.a.	450	over 500
Feet of conveyors	9 mn.	13 mn.	n.a.	n.a.	n.a.	n.a.
Skip Winding Schemes	(a handful)	same	9	11 (7 under constr.)	n.a.	n.a.
Mine cars (1½ to 6 ton capacity)	n.a.	n.a.	n.a.	**n.a.**	3800	5000
III. Coal Preparation Plant (to be completed the next year)						
Washeries and extensions	5	19	10	16	17	17
Froth Flotation Plants	2	5	3	11	16	12

Source: NCB, *Annual Reports*, 1946–1952.

[a] There are numerous additional items, particularly in the coal-getting category, that are not included in the above table. Such machines include the Uskside Mechanical Miner, the Grassmoor Goblin, the Bolton "Claw," the Logan Slabber, the Trepanner, the Multi-Disc machine, the Lambton Flight "Waffler," and an American Augur. Most of these are still in the experimental stage and it is not believed that more than a very few of any one type have yet been tried.

[b] n.a. = not available.

duces the risk of those interruptions to production which are the bane of cyclical work, where no shift can begin its work unless the work of the preceding shift has been completed. Furthermore, faces that can be worked continuously allow a pit to wind coal on more than one shift from fewer faces than where each face is on a cycle (much less expense is involved in working five faces for two or three shifts a day than in working five faces on one shift and five additional faces on a second shift). One great and final advantage of the new machines which eliminate hand loading (and which may permit continuous mining) is that they allow coal-getting to be performed by relatively small work-teams. This is an advantage of great psychological importance, as we saw in discussing faceworkers' price lists (Chapter Six).

An approximate notion of the amounts of gross investment in various types of new plant and equipment is afforded by Table 20. At best, the table can establish only a few quantitative measurements, which must be used with great caution. With the present fluidity of the industry's technology and the number of experimental projects being tried, the qualitative aspects of new investment may have far more significance with respect to the industry's creativeness, future efficiency, and safety than any quantitative report can convey. Nevertheless, Table 20 does make clear, for example, the commanding lead which the Meco-Moore now enjoys in the critical field of cutter-loading machines: developed before World War II, its use has expanded nearly fourfold since 1947. Also reflected is the decreasing geographical concentration of Meco-Moores in the East Midlands Division. Other face items that show evidence of increasing though still very limited use are the coal plough, scraper boxes, and the Samson Stripper. But clearly these new machines, not to mention the others listed which have not spread rapidly, have a long way to go before they achieve the significance of the Meco-Moore.[2]

With respect to coal-handling or transport equipment, the use of locomotives has been extended quite rapidly and mine cars and conveyors are likewise being used on an expanded scale. These are the forms in which capital is substituted for labor in the reorganization of underground haulages. Skip-winding is spreading, but not very rapidly. The data on coal preparation plants give us some notion of the number of reorganized pits at which redundancies among surfacemen may have occurred.

[2] There is no systematic way of following in detail the number of new items of equipment brought into use year by year; nevertheless, until 1953, the narrative accounts of technical developments in the industry contained in the Coal Board's *Annual Reports* included enough figures so that a reasonably accurate estimate of the rate of change can be established. Furthermore, the main technical developments within each division were separately described in the Board's *Reports* (though not uniformly each year), enabling students to get some idea of the geographical distribution of key pieces of equipment. Unfortunately the Board's *1953 Report* does not contain comparable information.

Labor Attitudes

The rapid mechanization of American coal mines during the past quarter century has been strongly encouraged by the aggressive wage policy of the union in the industry, the United Mineworkers of America. Ever since the first World War, that union has tried to force high wage rates on American coal owners. Under John L. Lewis' firm leadership, the union has understood and accepted the natural consequences of its policy: substantial unemployment among the mining labor force. This policy has not always been embraced with enthusiasm at the local level, but the international union, dominated by Mr. Lewis, has been willing to intervene in local situations in a way which has protected and not frustrated the employers' attempts to mechanize. John L. Lewis is not blind to the difficulties of unemployment in his industry, but he has realized that causes other than mechanization have been primarily responsible for unemployment and he has sought relief through means other than the discouragement of technological change.

Mr. Lewis, who takes pride in his union's liberal policy towards mechanization, has criticized his British counterparts for what he calls their long-standing opposition to technological change.[3] His indictment, however, is not accurate, for the old Miners' Federation did not have the clear and unified policy against mechanization which Mr. Lewis assumes. What is true is that while the British miners' unions have not prevented mechanization by a policy of open opposition, neither have they adopted the aggressive wage policies favoring mechanization which have characterized the United Mineworkers of America. One appraisal stated that British coalowners called the miners and their unions "extremely reasonable" in their attitudes towards mechanization in the 1930's.[4] A somewhat contrary impression is given by the following quotation from the 1945 *Reid Report:*

[After 1926] mechanisation, though seldom encountering active resistance, was not generally received by the men with enthusiasm, no doubt in the fear that more machines meant fewer men at a time when unemployment was widespread; and, where machinery was installed, its potential savings seem largely to have been dissipated by a quiet but effective determination that the number of men discharged should be kept as low as possible . . . In addition to this, they have steadfastly required the observance of old customs and traditions which are inappropriate to the conditions of mechanised mining, and thus have put a brake upon the modernisation of the Industry (p. 36).

The *Report* goes on to note the much more favorable attitudes characteristic of certain Midlands districts, attitudes still characteristic today.

[3] Interview in *U.S. News and World Report*, Nov. 19, 1948.
[4] John Hilton (ed.), *Are Trade Unions Obstructive?* (1935), p. 51. Unfortunately there is no indication of the type or extent of the investigation which produced this judgment.

Since World War II, the NUM has not only endorsed but it has "demanded" extensive capital investment; this affirmative policy received expression as the first of the twelve demands embodied in the "Miners' Charter" adopted early in 1946:

(1) The modernisation of existing pits and the sinking of new ones as rapidly as possible whilst strictly observing as a minimum the standards laid down in the Reid Committee Report; the provision of adequate compensation for those who become redundant; and at the same time aiming at the general application of the day-wage system.[5]

The real problem is not the general attitude of top union officials toward new investment, which they unquestionably welcome, but the attitude of specific groups of men who are threatened by specific changes and the provisions which are made to allocate the jobs which remain after the changes have been introduced. The first question involves the wages at which working miners will agree to work the new machines, and their views as to how much work they can reasonably be expected to perform with these machines. The second question — what rules govern the allocation of jobs in the event of technological change and what provision is made for laid-off employees — is equally important, though it can be answered more simply.

There are no formal seniority clauses in agreements between the Board and the union, though strict respect for this factor is customary in many districts. The only formal requirement embodied in any national agreement is that foreign workers shall be dismissed before any British miner. This means that managements would normally select employees for layoff alone or in coöperation with the local branch, possibly after consultation with the Colliery Consultative Committee. The point is that each situation is handled administratively, not according to contractual "legislation." In practice, men displaced from jobs by technological changes are offered vacant jobs at their same pit (and if the displaced employees are relatively few in number, there would almost always have been such vacancies during the past decade or so); they do not enjoy job rights which allow them to "bump" other employees off similar jobs if they hold longer service. Indeed, at one rather extensive reorganization in Durham where the manager proposed that seniority be the basis of deciding which men should have to transfer to other pits, the miners' lodge decided that "all men ought to be treated alike and the names be drawn from a hat"!

In the statement on modernization in the "Miners' Charter," the NUM laid it down as a condition of support that men made redundant should be given "adequate compensation." Such compensation was agreed to by the Board in an industry-wide contract signed in December 1948, and made to run for five years. The agreement was renewed, with slight

[5] NUM Conference Report, 1946, p. 289.

modifications, early in 1954 for a further five years. Clause 1 of the 1948 agreement explicitly noted that far-reaching plans of reorganization and reconstruction would be necessary and that these might give rise to "redundancy of an exceptional nature." The union then pledged itself to "coöperate and collaborate with the Board to bring about the prompt and effective implementation of all projects of reorganization and reconstruction undertaken by the Board."

In return for this pledge, the union got the Board (and the Minister of Fuel and Power) to agree that anyone employed for three years or more who was made unemployed by major reorganization projects undertaken by a Divisional Coal Board would be entitled to receive 33s. 8d. per week if a surfaceman (43s. 8d. for underground men), this benefit to remain payable for a period of twenty-six weeks or until the employee accepts other employment. This payment was in addition to the state's regular unemployment benefit. Not only does such an agreement soften the mobility problems of individuals faced with technological unemployment: it puts union officials in a much stronger position to persuade members to accept specific changes than they would otherwise be in.

The union has not been altogether happy with the agreement (for example, the latter applies only to redundancies arising out of reorganizations and not to those arising when a pit closes down independent of a reorganization project) and naturally sought improvements at renewal time early in 1954. The benefit was converted from a flat payment to a formula tied to the national minimum weekly wage and the Divisional Boards' obligation to consult the union when undertaking reorganizations was spelled out more explicitly. But the conditions of eligibility and the 26-week benefit period were left unchanged.

In contrast to stay-down strikes and unofficial action of other kinds against Coal Board and union attempts to carry out the reassessment of work tasks or, more commonly, against decisions to close down pits, there has been a notable lack of overt opposition to reorganization schemes involving new methods of coal transport or of face working that threaten unemployment. Indeed, the pressure sometimes runs in the other direction, especially among facemen: they sometimes want to install face machinery before management or higher union officials think it advisable.

Facemen want "mechanization" because they believe it will lighten their work, not because they long to become more productive. It is not always true that mechanization of faces increases productivity; for example, there are several faces in Yorkshire where the Barnsley seam is worked by hand (indeed, this is the most common method of working this particular seam) and many experts think this method more efficient than the conventional mechanization, where cutters are used and the coal is fired down. For example, Maltby Colliery in South Yorkshire works the Barnsley by hand-got methods and in 1950 enjoyed an OMS of 40 cwts.; Rossington, only a few miles away, works the same seam by conventional mechanization and has an OMS 20 per cent lower. But there has been popular pressure at some pits (usually near others that are mechanized) for management to "cut this coal and fire it down for us."

But to generalize on how specific projects have been received at individual pits without firsthand knowledge of such pits is risky: we must leave this aspect of the subject with the cautious report that there is certainly no significant degree of antimechanization feeling among British mineworkers today. But at individual pits where the men and the management do not get on well, it seems likely that modernization proposals have no chance of being discussed "on their merits," with an understanding that the chips shall be allowed to fall where they may.

The favorable report so far given of the general attitude toward technological change must be balanced by a somewhat less favorable report of the terms on which facemen have been willing to operate new face machines (disputes over work standards do not arise among the day-wage haulagemen and surface workers). Before the problem of work standards at the face can be understood, however, we need to know something about the way in which face mechanization is changing the method of wage payment.

Technological Change and the Wage Structure

Wherever the new type of power-loading or cutting-loading machinery listed in Table 20 has been introduced, the men who operate it have been put on day rates. This is true of Meco-Moores, Gloster-Getters, Samson Strippers, the Coal Plough, the Joy "Continuous Miner," and other similar machines. Wherever these machines are introduced, they are necessarily put to work in pits where there are other faces being worked either by hand-got methods or, more commonly, by the conventional mechanization of drilling, cutting, shot-firing, and hand-filling. More important, the volunteers who man these machines are always ex-pieceworkers who have been accustomed to normal pieceworkers' earnings under the pit's normal method of work.

It might be thought that since new face machines are installed only where they increase labor productivity and (at the old wage rates) reduce labor costs, some arrangement would be negotiated for sharing the estimated financial gain of the particular machine under consideration. However, this is not the case: what determines the rate to be paid on a Meco-Moore is not the potential saving which the machine is expected to produce but the average level of earnings for faceworkers in the pit or in the particular NCB area or division.

There are Divisional agreements in Yorkshire, Durham, and Scotland; and there is probably also one in the East Midlands, where the Meco-Moore has been pioneered and where labor relations have been uncommonly good. The Durham and Yorkshire agreements specify a uniform wage rate for Meco-Moore teams through their respective divisions (the rate is not the same in both divisions). In Scotland, a slightly different arrangement prevails, as noted below. The West Midlands Division, using six Meco-Moores in 1950, had not yet negotiated a Divisional agreement covering these machines. The NUM has not attempted to negotiate with the Coal Board any industry-wide agreement covering rates of pay on power-loaders in general or on any specific machines. The union did try to secure such an agreement from the private owners before nationalization, but the attempt (which went to arbitration) failed.[6]

The first comprehensive agreement on Meco-Moore power-loaders was signed in July 1947, after nationalization. At that time there was only one Meco in all of Yorkshire's 115 pits; by the end of 1948 there were four; a year later there were six; and by the end of 1950 there were thirteen — still only a quarter as many as were then in use in the East Midlands Division just to the south. There is a reason for the slow extension of Meco-Moores in Yorkshire which we shall discuss later; here we are interested in the method of wage payment.

The original Yorkshire agreement provided a straight shift rate of 35s. 2d. for one class of men in the 12–15 man team and 34s. 2d. for a second class; a bonus of 4s. for both classes was payable "when a full strip is filled off." No provision was made for changing these rates and no indication was given in the agreement as to what principle had governed their determination. In the spring of 1950 a subcommittee of the Divisional Disputes Committee made an investigation of the working of this agreement on the basis of experience gained during its first three years of operation. Their findings included the following comments:

During the time the Meco-Moore Power Loading Agreement was under discussion . . . the average earnings of pieceworkers employed at the coal face in the Division was 33s. 6d. and consideration was given to this fact when 35s. 2d. and 34s. 2d., respectively, were agreed upon as being the shift rates to be paid . . . From information at our disposal it is estimated that the average earnings of pieceworkers during the three months ending 31st December, 1949, was 38s. 2d. a shift, 4s. 8d. a shift more than the average earnings of pieceworkers at the time the agreement was made.

The men employed on Meco-Moore machines are those who were previously employed at the coal face on contract rates and they would have been

[6] Thirteenth Porter Award, June 13, 1945, *NCB Memorandum of Agreements*, pp. 129–131.

in receipt of the higher average earnings had they not been transferred to the Meco-Moore machines. In view of this we are of the opinion the shift rates of men employed on Meco-Moore machines should be related to the average earnings of pieceworkers employed at the coalface, any adjustment, increase or decrease, being made at 12-month intervals.[7]

It is clear from the above that the operators' earnings were meant to be related not only to strippers' earnings but to all faceworkers employed on contract jobs, and that the basis of this relationship was to be the Divisional average, not merely the average for the pit in question. Furthermore, there were now to be regular reviews, annually. It also appears that at the time of signing the original 1947 agreement, Meco-Moore operators were given a rate a few shillings above the average earnings of pieceworkers; this premium was abandoned in the 1950 revision.

The occasion that had led the NUM to seek a national agreement for power-loader operators in 1944 was the offer of "scandalously low" rates of pay on the American machinery (presumably Joy loaders) which it was then proposed to introduce. It was then that top union leaders first began to think seriously of how men should be paid on the new machines. They decided to press for (1) a day wage and (2) a "composite" method of payment, whereby everyone in the team was paid on the same basis, though not necessarily the same amount.

The Scottish Division-wide Agreement was signed in April 1948, a year later than the Yorkshire Agreement. The agreement explicitly covers both Meco-Moore machines and another type known as the "Logan Slabber." The basis of payment in Scotland differs from that used in Yorkshire: "All members of the Team will be paid on a Day Wage Basis at the rate of the Average Earnings per Shift of Pieceworkers in the Colliery, or in Scotland, whichever may be the higher . . ."

In addition, grade 1 operators receive an additional 3s. per shift and grade 2, or assistant, operators 2s. This contains two important advantages for operators working on these machines: (1) men working at high-wage pits run no risk of having to accept a machine on which the rate might be lower than their current earnings (which could well happen under the Yorkshire Agreement tying their wages to Divisional average earnings alone regardless of the pit's earnings) and (2) men working at low-wage pits get a very substantial advantage, jumping up not only to the Divisional average but getting a further differential as well. In such circumstances, there might be a quite large differential between the earnings of Meco-Moore operators and other facemen in the same pit. Thus the Scottish agreement would seem to hold out the best advantages for the men and might be assumed therefore to make easier the introduction of the machines — though it must be remem-

[7] Yorkshire Disputes Sub-Committee Report, May 16, 1950.

bered that the large number of low and highly inclined seams in the Division limits the technical possibility of using these machines much more severely than would be true of Yorkshire. Indeed, though the Meco-Moore is made in Scotland, there were probably not more than two or three in use there in 1951.

The Durham coalfield has mechanized some of its room-and-pillar work with Joy loaders or duckbills and shuttle cars (introduced during the second World War), but it has relatively few conditions where the Meco-Moore longwall machine is suitable. The American method of working room-and-pillar layouts (that is, with coal cutters, firing of the coal and Joy loaders or duckbills filling into shuttle cars which carry the coal to the main haulage) did not appeal to Sam Watson, the Durham miners' leader, any more than to most British mining engineers. American equipment is fine for driving headings, but for use in rooms it requires that too much roof be left exposed. A county agreement covering power-loading in bord-and-pillar work was signed between the private owners and the union in September 1945. The scale of pay, which was to be uniform over the whole coalfields, established day rates for four job categories which were all from 2s. to 4s. per shift higher than the average earnings of various grades of facemen at this time:[8] it thus appears that the owners gave a premium of roughly 10 per cent for men working on the new equipment.

In 1946, the Durham coalowners offered the union an identical agreement for Meco-Moore work on longwall faces, but the union refused to accept it. The basis of the refusal was that pieceworkers' earnings had risen quite markedly between the fall of 1945 and the fall of 1946, so that men on power-loading schemes no longer enjoyed any advantage: the union undoubtedly wanted to raise the price before signing a new agreement. As a result, no county-wide Meco-Moore Agreement was signed until June 1948 — eighteen months after nationalization. This June agreement provided day rates which were higher than the customary wages by about 5–8 per cent. But they were also higher than for men working under the 1945 power-loading agreement for bord-and-pillar work just quoted above. This situation created an "inequity" which was corrected two months later, in August, when a "master agreement" covering both types of power-loading work was signed: in effect the bord-and-pillar men were simply brought up to the scale of pay granted the Meco-Moore men in the latter's June agreement.

In all the power-loading agreements quoted from Durham, there has been a "local option" clause which allowed any pit to put power-loading on a "composite" basis — that is, all men drawing equal pay on a day-

---

[8] The average earnings of pieceworkers (grouped into 15 to 20 classes) in the coalfield are included in the *Annual Reports* of the General Secretary of the Durham Area of the NUM.

wage basis. Note that in Durham, as in Scotland, there is no bonus arrangement to act as incentive for the men to complete their full cut.

At the end of 1949 there were three Meco-Moores operating in the West Midlands Division and three more were installed during 1950. The introduction of these machines, however, has not always proceeded smoothly and the main difficulty has been one of negotiating wage rates for machine operators.

The first Meco-Moores used in the West Midlands went into pits on Cannock Chase — which, it will be recalled, pays considerably lower wages than the highly productive Warwickshire coalfield twenty miles away. The kind of contract worked out at the Cannock pits called for a day rate of 35s. per shift for a cut of stipulated length, plus a bonus for extra yardage. Such a rate yielded operators a good 10 per cent more than facemen working on conventionally mechanized faces.

But what rate for Meco-Moore operators ought to apply in the higher-wage Warwickshire district twenty miles away and in the same Division? The Divisional Board felt that there ought to be a single rate for this kind of work throughout the Division — and it felt that this price ought to be 35s. per shift, the one they had got accepted in the Cannock Area. Not surprisingly, this price proved unacceptable to the high-wage War-wickshire miners, who were accustomed to earn 38–40s. Indeed, the first machine scheduled for Warwickshire was actually delivered to a pit, but the men wanted something more than the Divisional Board was willing to bid, with the result that the machine was taken away to an-other Area in the Division. The labor officer of the Warwickshire Area, a former branch secretary, urged the Board to raise its bid by the rela-tively small amount he felt necessary to secure union approval, but the Board refused to compromise its hope for a uniform rate for the whole Division. Six months later, a proposal to install a Meco-Moore at another colliery was held up because by now, with piecerates having risen, the men were demanding 45s. a shift plus 3s. a shift for the chargeman. One difficulty with the negotiations at this second pit (Birch Coppice) was that the Board wanted to install the machine in a seam of coal that had a bad dirt band running through it — which meant that the coal would not fetch a good price in the product market. This condition kept the Board's bid lower than it might otherwise have been.

The following report appeared in the *Birmingham Post* on December 18, 1950:

The installation of a £12,000 power loading plant at Birch Coppice Col-liery has led to a wage dispute between the Board and the Warwickshire NUM. The loader, which requires six to eight operatives and can do the work of 25 to 30 men, was due to start operating this morning . . . A meeting of NUM representatives yesterday discussed the wages aspect of the new plant . . . it is understood that the meeting expressed grave concern at the pro-posed decrease in shift rates, and passed a resolution calling on miners in the coalfield not to operate the new machinery until the Board and the NUM had negotiated a new wages agreement.

Thus a failure to agree on a price had delayed the introduction of the machine by at least a year in the Warwickshire coalfield. Quite obviously one major difficulty was the fact that if the Board gave Warwickshire miners a Meco-Moore price in line with the earnings of other facemen there, the Cannock Chase Meco-Moore operators would try to get the same rate — which it might be hard to deny them. And if Cannock men got the same rate as Warwickshire men, the Cannock Meco-Moore rates would be all out of line with the average earnings of 95 per cent of the regular facemen on the Chase. In this Division, then, the widespread introduction of Meco-Moores seems to require a greater equalization of wages among all the facemen of districts with historically different wage levels; until this happens, the machines are likely to be accepted only by the men in the low-wage districts.

There is a widespread feeling, found in many Divisions, that productivity has not risen as much as it should have done in view of the large amount of capital investment that has gone on since nationalization. Even some NUM officials and extrade union officials now working for the Board make this lament. Nobody has a convincing, simple explanation for this disappointing result, which applies not only to the new face machines described above but to new investment generally. It is possible, however, to tie down this judgment with respect to the particular item of equipment to which we have paid most attention, the installation of Meco-Moore cutter-loaders: in several districts, but particularly in Yorkshire, there is a widespread feeling among mining engineers that the men have not agreed to work the machines in a manner which makes their installation worth while.

These doubts do not apply to the Nottinghamshire coalfield, where the Meco-Moore was first introduced (in the late 1930's) and where its use has been extended most widely. The Notts miners were led by George Spencer when the Meco-Moore was introduced and Mr. Spencer, who had led the Notts miners out of the MFGB after 1926, believed in coöperating with the owners. Spencer was undoubtedly able to secure high wages as a result of the atmosphere which his coöperative attitude helped to produce. In return for high wages, Notts miners have apparently been willing to accept work-loads which have made new investment worth while.

Given a day-wage method of payment, the factor which determines whether or not it "pays" to introduce a Meco-Moore is the amount of coal it produces per shift. Within the types of Yorkshire seams for which the machine is suitable, differences in the price at which different coals are sold appear to be relatively unimportant.

Since the thickness of the "cut" it takes is fixed, the crucial variable is the length of cut which the men will accept as a "reasonable day's work." Because of the great variety of seam conditions, this question can only be settled pit-by-pit through the bargaining process, though the day rate for whatever becomes the "normal" cut at any pit is the same for the entire coalfield, as we have seen.

The men thus know what they will earn before the machine is installed, but management does not know how much work it will get for this payment. True, a manager and his prospective operators will agree on the length of cut they will hope to consider "normal" after the machine gets down the pit; but a new installation will require a few months of trial operation before the men become accustomed to the new method. During this period, they cannot be expected to complete a cut as long as that which it has been agreed will ultimately become "normal" — and if they get it into their heads that the original bargain was unreasonable, the manager may find that he has to redefine the "normal" stint the machine will work.

The length of the face must usually be tailored to the length of cut taken by the machine, since the machine is expected to travel the whole length of the face during one shift. This is a difficulty from which a continuous-mining machine, not tied to a cycle, would not suffer: the next shift could simply carry on from where the preceding shift leaves off.

It is a common but not universal complaint in Yorkshire that Meco operators will not take as long a cut as the machine is capable of accomplishing with reasonable effort on the part of the men during a shift: the men would rather take a short cut and work easily or finish their day's work early. So far, the Board has not had much success in overcoming this attitude, an attitude which robs the investment of most of its attraction. It is this fundamental problem which chiefly explains why Meco-Moores have not been introduced more rapidly into the Yorkshire coalfield, a major coalfield where technical conditions would permit its adoption on a substantial scale.

The psychological problems which have tended to frustrate face mechanization in Yorkshire have not, apparently, been present in the adjoining East Midlands Division. In this most progressive Division, Divisional leaders for the Board have repeatedly paid public tribute to the leaders of the district unions and to the men themselves for their coöperative attitudes which have permitted management to innovate and experiment. Some of these experiments have been dramatic in their results and the Division, never blind to the value of internal morale and external public relations, has encouraged considerable publicity. The most widely reported experiment has been the new system of continuous mining developed at Bolsover Colliery, where the men and their union were persuaded to accept a bold experiment that required laying off and transferring 800 of the pit's 1600 employees.[9] The new system of work (which involves more variety in work tasks, smaller work groups, and a change over from piecework to day wages at the face)has succeeded to the point where it had been extended to some dozen surrounding pits by late 1952.

Another widely heralded innovation occurred at Donisthorpe Colliery, South Derbyshire, where a Joy "Continuous Miner" was installed late in 1949. Though this machine does not appear to have wide potential use in Britain (four were in use by early 1953 and four more were on order), its installation at Donisthorpe illustrates some of the labor problems that are likely to become of increasing importance as face mechanization gains momentum. The manufacturer's representative attached to the pit (an American who had once been a union official in the Illinois coalfields) reported that when Arthur Horner visited the pit to inspect the machine, the operators complained that the machine made them work pretty hard. Horner's reply was that he "couldn't see it," and that was the end of the latent grievance. The chief labor difficulty that impressed the Joy representative was the wage structure, which he felt was out of joint with the requirements of efficient operation: deputies earned less than the men operating "the Miner," and the operators earned 50 per cent more than the mechanics and electricians on whom the machine's maintenance depended. The result was that "everyone wanted to be an operator and nobody wanted to be a deputy." Furthermore, the deputy's status had not impressed this man favorably: the deputy's lack of status had hindered effective supervision, on which mechanized mining with day wages greatly depends. The problem present at Donisthorpe has been general throughout the industry: supervision does not have the extra pay and status and training that it ought to have, and skilled mechanics and electricians are undervalued in terms of the job-mix which modernization implies. The NCB has recognized these problems, and in 1951 and 1952 took steps to raise supervisors' pay and status.

An unpublished paper by John A. Mack of the University of Glasgow reviews some of the work that has been done in the field of technological change up to 1947. While Mack himself warns against undue romanticism about the old hand-got methods of work, he does believe that "one potent and neglected cause of trouble in the pits has been the substitution for a psychologically satisfying underground craftsmanship of an

[9] See *The Bolsover Story*, published by Area No. 1 headquarters East Midlands Division (1950), 71 pp. By the end of 1952 there were still critical mechanical problems to be solved in connection with the "Gloster Getters" on which the Bolsover system partly depends. The effect of such rapid extraction on roof control, a vital factor in safety, has also to be appraised over a longer time-span. The *Daily Worker*, for example, has used this issue to snipe at the Bolsover experiment.

unsatisfying semimechanization; and that the evil effects of this substitution can be redeemed only by the completion of mechanization in such a way as to restore the interest and variety of mining, calling forth the personal initiative of each man in the collective operations of small teams of omnicompetent and interchangeable technicians."

Among older miners it is not hard to detect nostalgia for the methods of work that were prevalent in the 1920's; this arises most commonly in discussions of modern training methods for juveniles: the old-timers don't like the "vestibule" system (special training pits or part-production faces in regular pits) which are almost inevitable under orthodox longwall work if there is to be any formal training at all — and the law requires it.

One of the great promises of the kind of technological changes we have been discussing is that they may allow an escape from the frustrations — for managements and men alike — which inhere in the present method of work. The promise is one of small groups; it is one of quite new types of face machinery, mainly of British design, though some of it is adapted from continental and American designs; it is one of day rates instead of piecework; it is one of varied individual tasks rather than rigid specialization; and it is one of operating flexibility. The significance of these advantages has just begun to be appreciated generally in Britain within the past seven or eight years, and it may be expected that the next half generation will see an increasingly conscious effort to attain these conditions and to assess their results in terms of mining costs, individual earnings, group morale, the "tone of the pit," the quality of first-line management, the state of labor relations, and even in terms of nervous disorders among facemen (there is some evidence that present methods lead to a fairly high incidence of such maladjustments). If changes in these directions can be mounted on a large scale in many coalfields over the next ten to twenty years, Britain will have duplicated many of the features of American mining without having transferred American technology to British conditions — a development which, as we have seen, is now regarded as technically impossible.

An unpublished paper by John A. Mack of the University of Glasgow reviews some of the work that has been done in the field of technological change up to 1947. While Mack himself warns against undue romanticism about the old hand-got methods of work, he does believe that "one potent and neglected cause of trouble in the pits has been the substitution for a psychologically satisfying underground craftsmanship of an

° See The Bolsover Story, published by Area No. 1 headquarters, East Midlands Division (1950), 71 pp. By the end of 1952 there were still critical mechanical problems to be solved in connection with the "Closter Getters," on which the Bolsover system partly depends. The effect of such rapid extraction on roof control, a vital factor in safety, has also to be appraised over a longer time-span. The Daily Worker, for example, has used this issue to snipe at the Bolsover experiment.

THE TECHNICAL REQUIREMENTS OF MODERNIZATION: A SUMMARY OF THE REID REPORT [1]

In September 1944, the Ministry of Fuel and Power appointed a seven-man Technical Advisory Committee "To examine the present technique of coal production from coal face to wagon, and to advise what technical changes are necessary in order to bring the Industry to a state of full technical efficiency." Since the recommendations of the Reid Report have been accepted in the main by the National Coal Board as the basis for its initial fifteen-year plan for modernizing the industry, one must know what is in this Report if one would understand what types of specific projects are involved in the general increase of "capital investment" which nationalization has made possible. What follows is necessarily a very general summary of the Committee Report.

I. Real Investment During the First Half of the Twentieth Century: The Forms and the Results

There had been two main types of investment in the industry from 1900 to the outbreak of war in 1939. The first was the development and spread of *coal cutting* by machinery, the second was the adaptation and spread of *conveyors* to underground coal transport. As a result of these changes, a large-scale change-over occurred in the system of mining coal — i.e., from "room and pillar" to "longwall advancing." There was also a significant extension in the use of electricity as the form of power and a limited amount of progress was made in the construction of surface plant for coal cleaning and sizing. The timing of these developments fell into three phases.

1. 1900 — WORLD WAR I

The number of pits fell between 1900 and 1914 from 3,089 to 2,734, but average annual tonnage per mine rose from 73,000 to 97,000 tons. The period was one of expanding total demand, total output, exports, and profits. Technical progress, however, was slow, due mainly to the small size of the average

[1] So called after the Committee's Chairman, Charles C. Reid, former Managing Director of the Fife Coal Co. and an original member of the NCB. The full title of the document is *Coal Mining: Report of the Technical Advisory Committee*, Cmd. 6610 (March 1945), 149 pp.

firm, the geographical isolation of the separate coalfields, the relatively limited amount of international travel by British engineers and the comparisons to which that leads, and a frequent shortage of financial capital. The combination of strong demand, capital shortage and sharp competition among separate small firms imposed an insistent pressure for output upon mine managers which was often in conflict with sensible long-term planning. Though the Reid Committee was tolerant of the reasons for this short-sighted outlook, they could not fail to observe that it "left the mining engineers with a legacy of mines not easy to reconstruct to fit the requirements of today . . ."

The principal technical advance in the decade preceding World War I was the development of coal-cutting machines. Between 1908 and 1913, the percentage of total output which had been cut rose from 5 to 8, while the number of *coal-cutters* in use rose from 1,700 in 1909 to 2,900 four years later. The number of coalface *conveyors,* on the other hand, was only 180 in 1909 and rose only to 360 by 1913. The coördination of conveyors and cutting machines (and the use of longwall mining implied thereby) did not take place until the period between 1920 and 1925. Face haulage relied on hand- or pony-putting, with only slow replacement by rope haulages moved by auxiliary engines. A gradual replacement of the traditional room and pillar system of mining by longwall advancing took place — especially at the newer and more highly capitalized pits (where the pressure for early returns was greater).[2]

2. WORLD WAR I — 1925

During the war, the number of cutting machines in use almost doubled (rising from 2,900 to 5,100 between 1913 and 1920); by 1925 their number increased a further 22 per cent, to 6,650. The number of conveyors rose from 360 in 1913 to 800 in 1920 and then nearly doubled again, to 1,500 in 1925. This immediate postwar period was one of widespread experimentation with *intensive* mining, meaning the concentration of production into smaller areas of the seams being worked through the use of face conveyors on machine-cut longwall faces. This development fell short of expectations, largely because it "dealt with but part of the problem of winning coal." The chief shortcoming of this postwar reliance on intensive mining was that it did not lead to a reëxamination of traditional haulage systems, which remained largely the same as they had been in the nineteenth century. Nor was there any significant attention paid to the training of either men or officials in the proper use of the newly installed machinery (beyond routine demonstrations by the equipment makers' representatives). By 1925, OMS was only 89 per cent of the 1913 figure (though it is true that the hours of work per shift had been reduced from eight to seven), prices were high and profits "negligible." With labor constituting about 71 per cent of total production costs, the owners looked to lower wages to get costs down.

3. 1926 TO 1939

After the 1926 strike, the industry reverted to longer hours and lower wages. The rise of German and Polish competition in export markets did something to spur on the extension of *face* mechanization, but "In the light

[2] Longwall advancing permits almost complete extraction as soon as coal is reached, whereas room and pillar mining requires that substantial proportions of the coal be left untouched while the pit is developed, the pillars being extracted several years later, usually. And "untouched coal" represents "unrealized income."

of subsequent events, it is regrettable that no adequate steps were taken, either by the Mines Department, or the Mining Association, to bring home to the Industry the serious competitive position which was developing against it." [3] During the 1927–1939 period, OMS overall rose by a modest 11 per cent, with face OMS rising by 20 per cent. This improvement was achieved not by the small increase of 9 per cent in the number of cutting-machines installed, but mainly through the wider use of conveyors[4] (both at the face, where their use had begun before 1926, and in the gates), for getting the coal away from the faces to the main transportation arteries. Thus the number of face conveyors in use in 1927 was 2,100; by 1939 this figure was 5,900, but the tonnage conveyed on the faces rose from 28 million to only 34 million (1928–1939). The proportion of total output machine-cut also rose over this period from just over 20 per cent to just under 60. But despite an undeniable correlation between the proportion of output cut-and-conveyed and face OMS, the Reid Committee noted that "it is evident that the output has not been increased as much as might have been expected from the percentage increases in the proportion of the total output cut and conveyed." Furthermore, the correlation varied greatly from district to district — and in such crucial exporting districts as Durham, South Wales, Northumberland, and Scotland, there was little or no improvement in face OMS for the twelve-year period under review.

The state of haulage arrangements: In 1944, one out of every four men employed underground was engaged on haulage and tub loading — a very high proportion when compared with other countries. This figure, like almost all figures for the industry, varied considerably among the different districts — from 20 per cent for South Wales and Fife to a high of 34 per cent for Lancashire and Cheshire. More significant is the amount of coal handled per man engaged on haulage and tub-loading operations. The over-all figure for Britain was 4.9 tons per shift per worker in 1944, varying from under 2.5 tons in Cumberland, Bristol, and Somerset to over 8.2 tons in South Derbyshire and Leicestershire. Not only was this 1944 figure low, but it had not risen significantly since the late 1920's despite three developments which might have been expected to raise OMS "outbye," namely:

1. The increased use of belt conveyors for gate haulage.
2. The replacement of hand tramming and pony putting by both conveyors and rope haulage.
3. The concentration of working on fewer faces, which "should have resulted in each haulage road dealing with larger tonnages." [5]

[3] *Reid Report*, p. 7.

[4] During this same period, the two countries whose natural conditions are closest to Britain's (the Ruhr and Holland) adopted the *pneumatic pick* as the characteristic method of winning coal. It is almost unknown in the United States and rarely used in Poland, Upper Silesia, or other coalfields. Limited use was made of this development in Britain and by 1943, only 10 per cent of her output was won by such picks — 85 per cent of this figure being accounted for by Durham, South Wales, and West Yorkshire. Despite the many advantages of this method of coal-getting, the Reid Committee did "not consider that pneumatic picks will come to rank as one of the principal methods of winning coal" in Great Britain. Instead they felt that the main line of future development must lie in the *mechanical loading* of coal won by other means — a development "which appears to be impracticable where these picks are used."

[5] There were practically no gate conveyors in use in 1927; by 1939 some 2,400 had been installed. During this same period, the number of horses "on colliery books" declined from about 55,000 to about 30,000; horses would find their most intensive use in gate haulage. (*Reid Report*, p. 13.)

This poor result suggested "that haulage arrangements between the gate conveyors and the shaft-bottom were extravagant in manpower, and were getting worse" (since the per cent of coal gate-conveyed rose from 40 to 59 between 1929 and 1940 while the tonnage handled per outbye worker actually fell).

Underground electrification: The installed horsepower underground rose from 878,000 to 1,224,000 between 1927 and 1939 — a rise of 40 per cent. There was a much greater proportionate increase in the horsepower of conveyors and loaders during this period (amounting to a 400 per cent rise), than in the amount used for coal cutters, which rose only 78 per cent. The amount of horsepower per man employed underground rose from 1.6 in 1927 to 2.02 in 1939.

As for winding, the number of pits using electric winding (as opposed to steam) rose from 252 in 1927 to 350 in 1939, so that about one-third of all pits taken over by the NCB in 1947 had electric winding. Skip-winding had been installed only "at a few collieries."

On the surface, a considerable rise in the consumption of electricity occurred as a result of the progress made over the period in the installation of cleaning and washing plant (the proportion of total output washed or dry-cleaned during the period rose from 20 to 47 per cent and the proportion of total shifts worked on the surface rose from 21.4 to 23.6 per cent, reflecting the growing attention being given to coal preparation — albeit much of it by hand). But the rise in connected electrical load on the surface over the period was only 34 per cent, though the increase was almost double that figure for screening and washing equipment. Horsepower per man employed on the surface rose from 4.2 to 6.6, 1927–1939.

The Reid Report summarizes the technical changes of the 1927–1939 period in the following words:

It was clearly the opinion of mining engineers, at that time, that in the mechanisation of the coalface, and in the improvement of roadways, was to be found the key to the higher OMS which was being sought. Equally clearly, this policy has failed to yield the results anticipated, for OMS, overall, in 1927, from which year the installation of conveyors was intensified, was 20.61 cwt., and, in 1939, was only 11 per cent more at 22.88 cwt.[6]

II. Specific Proposals for Raising Productivity in the Future

The following section will summarize only the principal recommendations submitted by the Reid Committee.

1. *Systems of mining:* In 1944, three-quarters of all pits in Britain used the longwall advancing system of extraction, with a substantial number (about 15 per cent) using room and pillar and only a small number using longwall retreating. After reviewing the advantages and disadvantages of these three systems, the Committee urged that room and pillar mining be used wherever nature permitted and that where longwall mining was advisable, the retreating system should be preferred to longwall advancing. Since room and pillar mining seems to be possible in only a fairly small proportion of British pits, the major future change should be one from longwall advancing to longwall retreating.

2. *Methods of "getting" the coal:* The small proportion of *hand-got* coal still found in Britain is naturally expected to die out as soon as the seams where it remains advantageous are exhausted. No great increase in the use of

[6] *Reid Report,* p. 15.

pneumatic picks is anticipated, largely because it is hoped that much greater benefits can be secured through further developments in power-loading, particularly in combination with the cutting operation. Although the Committee foresaw that mechanical cutting would continue to be widely used with hand filling, it stated its conviction "that development in mining technique must be directed, first and foremost, to a reduction in, and ultimately to the practical elimination of hand loading." This may be done by the further extension of mechanical loading machines and by the development of combination cutting-and-loading machines. As the Committee observed, "The Industry is passing through a crucial stage in regard to face machinery development, and we can imagine no object upon which money could more wisely be spent than this."

3. *Underground transport:* The Committee believed that "no single operation associated with coal production in Britain offers more scope for improved efficiency than that of underground transport." [7] The necessity for "revolutionary changes" in haulage stems not only from the promise of increased over-all efficiency but from the fear that forthcoming developments in technique at the coalface may be so satisfactory that the traditional haulage systems will be unable to cope with the increased output made possible by the new methods of facework.

To modernize the British haulage system, the Committee recommended that *locomotives* be used wherever possible.[8] It went on to note that "In the majority of British mines a great deal of work will be required on the construction of new roads in stone before locomotive haulage can be installed." Furthermore, the layout of a pit depends very largely on the type of haulage system decided upon — and, in the words of the Committee, "the adoption of locomotive haulage would require fundamental changes in the layout of the majority of British mines." Significantly, the Committee observed that "In the past this work has been considered as of minor importance compared with the production of coal, but this *attitude* needs to be radically changed." It is evident, therefore, that a great deal of road-building will be required to permit the widespread introduction of locomotive haulage. In many pits, *trunk belt conveyors* will prove more satisfactory than locos and their use is expected to spread quite considerably. The main disadvantages of "trackless mining" is that conveyors cannot be used as easily for man-riding or the transport of materials inbye as is possible with mine cars, two subjects which the Committee felt deserved a great deal more attention in British mining.[9]

4. *Size and design of mine-cars:* The normal size of "mine cars" in use in British pits in 1945 varied between 8 and 23 cwts. — a heritage of hand-got mining when the tubs (the Committee defines "tubs" as anything below 25 cwts. and "cars" as anything above this capacity) had to be taken right to the face for direct hand loading. The adoption of cars of 2–4 tons, of standardized design, would carry obvious advantages in underground transport. The chief

[7] Some measure of the potential increase in OMS that lies hidden in haulage reform is indicated by the fact that if the standard of underground transport in British pits were brought to that achieved in the similar Dutch conditions, OMS — underground — would have risen by one-third. *Reid Report,* p. 66.

[8] Electric trolley locos are standard in most American pits whereas Continental pits — normally laid out for locomotive haulage — use diesels. Diesels promise to be much more common in British mines, though trolley locos may be used in certain conditions and storage battery locos in others.

[9] A 1943 survey showed that of 615 pits employing over 250 men, 407 of them worked faces more than 2000 yards from the shaft bottom without any provision for man-riding — a great waste of productive time and energy.

difficulty is that larger cars will require much re-tracking of roadways and in many cases an alteration of shaft-bottom and winding arrangements to accommodate larger equipment.

5. *Underground power supply:* A considerable extension of electricity, which must replace compressed air in many schemes, is anticipated. In the past, safety regulations have hampered the spread of electricity — within the existing standard of mine ventilation. However the Committee emphasized that there would be very few pits which, *if properly ventilated,* could not be made safe for electricity and it laid down the principle that "The aim must be to make the mines safe for electricity rather than to install machinery which can be safely worked in badly ventilated mines." The chief bar to first-rate ventilation is poor planning and insufficient attention to detail; it is not a gap in the "state of the art." It should be pointed out, too, that the spread of electricity is desirable not only because of its efficiency as a source of power but in order to raise the standard of underground lighting.

6. *Winding:* The standard of winding in Britain is generally high, being both efficient and safe. But the Committee favored (1) experimentation with the Continental Koepe winder, (2) the electrification of more winding engines, (3) the introduction of many more skip-winding schemes, and (4) a general review of shaft-bottom and pit-top layouts with a view to the installation of more automatic handling equipment.

7. *Surface plant and layout:* After noting that the efficiency of surface manpower, as measured by tons handled per manshift, remained substantially unaltered from 1927 to 1939 at the low level of 5 tons per shift, the Committee stated that "if the economies in manpower which are necessary are to be achieved a general reconstruction of the surface plant will be required at the majority of mines." The great cost of such investment must, in the Committee's view, lead to the adoption of double-shift coal winding wherever possible.

The foregoing constitute the principal technical recommendations of the Reid Committee. The Committee also made important observations concerning the training of workmen and officials and concerning the improvement of labor relationships in the industry.

AGREEMENT RESPECTING DUAL MEMBERSHIP BETWEEN THE AMALGAMATED ENGINEERING UNION AND THE NATIONAL UNION OF MINEWORKERS [1]

With the object of preventing disputes between the two unions, and to promote complete organization in the coalfields, it is agreed:

(1) That members of the Amalgamated Engineering Union employed in and around collieries shall also become members of the appropriate Area of the National Union of Mineworkers. The National Union of Mineworkers undertakes to provide similar protection and benefits to such members as it gives to others paying similar contributions.

(2) The National Union of Mineworkers raises no objection to its engineering craftsmen becoming members of the Amalgamated Engineering Union. Such membership would entitle them to benefit according to their section and full recognition and assistance should they terminate their employment in the coalfields.

(3) That when negotiations on the wages of colliery craftsmen are contemplated, consultations shall take place between the two organizations.

(4) Any question with respect to membership or other matters affecting the two organizations shall be dealt with by a meeting of three representatives from the Executive Council of each body.

[1] Quoted from NUM *Information Bulletin,* March 1946, p. 39.

THE MINERS' CHARTER: 1946

(Approved by the National Executive Committee of the NUM and presented
to the Minister of Fuel and Power)

The Minister of Fuel and Power having raised with the National Executive Committee of the National Union of Mineworkers the question of the recruitment of manpower to the coalmining industry, and the Committee having undertaken to give it the most serious consideration, has now arrived at the following conclusions.

In the view of the National Executive Committee, the extent to which the following changes are introduced will determine the rate of entry of new recruits to the industry:

1. The modernisation of existing pits and the sinking of new ones as rapidly as possible whilst strictly observing as a minimum the standards laid down in the Reid Committee Report; the provision of adequate compensation for those who become redundant; and at the same time aiming at the general application of the day-wage system.

2. The adequate and careful training of youth in the various phases of mining operations, and the establishment of a clearly defined scheme of promotion; the provision of further training and tuition required in cases where workers desire to enter for a colliery technician's career.

3. The introduction of new safety laws to meet the conditions of modern mining and especially to suppress the development of industrial diseases.

4. The payment of compensation rates to meet incapacity due to industrial injury or disease which shall guarantee the injured person from financial loss and the provision of an adequate income for the dependants of those killed as a result of injury or who die from an industrial disease. (This paragraph will require to conform to the policy on Workmen's Compensation, as is accepted by the National Executive Committee at its meeting to be held this week.)

5. The average wage standards shall not be permitted to fall below those of any other British industry.

6. The restoration of the 7-hour day for underground workers; the introduction of the 40-hour week for surface workers; and the establishment of the 5-day week without loss of pay.

7. The continuation of the principle of the guaranteed weekly wage when the Essential Work Order is withdrawn.

8. Payment to be made for two consecutive weeks' holiday and six statutory holiday days in each year.

9. The provision of pensions for mineworkers who cease to be able to follow their employment after 55 years of age and the payment of a subsidiary pension from the industry in addition to pensions provided from other legal enactments.

10. The building of new towns and villages of a high standard and situate at places calculated to enable miners to have increased opportunities for social facilities and to break down the segregation of mineworkers and their families from the rest of the community, accompanied by the provision of adequate transport services at reasonable rates.

11. The complete reorganisation of health and welfare services so as to put a brake upon the wastage of manpower due to ill-health.

12. Compulsory medical examination with training arrangements at full wages pending employment as a skilled workman in another industry if withdrawn from the coal mining industry on medical grounds.

The National Union of Mineworkers having in mind the manpower crisis which exists and recognising the complete dependence of our country's economy upon coal production, calls upon the Government through the Minister of Fuel and Power to give guarantees that effect will be given to the foregoing measures in accordance with a time-table and a progressive plan.

It is realised that whilst the manpower requirements of the industry will tend to fall as the industry is modernised, during the immediately ensuing years it will be necessary to depend upon the technical equipment now in existence. The only permanent source from which new manpower can be drawn and adequately trained is youths under 18 years of age. In the past the maintenance of the industry's manpower by the continuous supply of youths to compensate for wastage has come mainly from the ranks of mineworkers' sons. The mining community, however, is not willing to accept a special responsibility for the supply of new mining labour. Quite rightly, miners and their wives have come to regard their sons as citizens entitled to seek a livelihood in more congenial, less dangerous and better paid employment in the same way as do the sons of other people. Young persons will only be attracted to the coal mining industry in sufficient numbers when it offers to them conditions of employment which compare favourably with those offered in other industries, and a higher standard of living than has been the lot of those who have toiled in the industry in the past.

PROCEDURE FOR CONTROLLING ABSENTEE-ISM SUBMITTED TO DIVISIONS BY THE NATIONAL COAL BOARD AT THE END OF 1949 [1]

General

1. The recommendations of the Joint Committee on Production for the setting up of Joint Attendance Committees have not been accepted by the constituent organisations of the National Union of Mineworkers. The Union have, however, stated that there is (i) an overwhelming recognition of the need for increased production; (ii) a general recognition that steps must be taken to deal with workmen who, without good reason, absent themselves from work on the normal shifts of the week.

2. After further discussions by the Joint Committee, it has been arranged that their recommendations as to the setting up of Joint Attendance Committees [2] shall remain open for acceptance by individual pits or by the area organisations of the Union, but, failing acceptance, the Board as indicated in Clause V of the recommendations will themselves introduce a similar procedure not involving the Union in Joint responsibility for the assessment or infliction of penalties.

3. This procedure of which the Union has been informed, and which is set out [elsewhere], ought clearly to be administered on similar lines in the various areas, so far as the differing conditions at different mines allow. These notes are intended as guidance to this end.

Procedure Where Joint Attendance Committees Are Not Set Up

4. The Management should deal first with the worst cases of absenteeism. As standards of attendance improve, it will be possible to give attention to the less serious cases.

5. When a workman is first interviewed as a presistent absentee, the im-

[1] Quoted from the *1949 Annual Report* of the General Secretary of the Durham Area, National Union of Mineworkers, pp. 114–116.

[2] Joint Consultative Committees may perform the functions of the Joint Attendance Committees where the Divisional Board and the Area organization of the NUM agree.

portance of regular attendance to efficient production and the effect of irregular attendance on the standard of living of the other workmen should be emphasised. He should be warned that, next time, a penalty will be inflicted, which may be dismissal or a fine, suspension, downgrading, transfer to another colliery or proceedings for breach of contract. The Lodge Secretary (or other appropriate Union official) should be informed that the warning has been given, since the Union have agreed not to contest disciplinary action in such cases.

6. Where the workman's attendance does not improve substantially as a result of the warning and he offers no satisfactory explanation, the management should take disciplinary action by dismissal or otherwise according to circumstances.

7. The alternative forms of disciplinary action available will depend largely on the workman's contract, the terms of which may be express or implied. General rules cannot therefore be given to apply in every case. The following paragraphs may help but in doubtful cases Divisional Legal Advisers should be consulted.

8. *Summary Dismissal.* During the currency of the Essential Work (Coalmining Industry) Order, a man could only be discharged without notice for serious misconduct, and even then had a right of appeal to the local Appeal Board. Now, however, the Board's right to dismiss summarily depends on the workman's contract of service and the common law. If, for instance, the contract of service provides that, for a violation of the rules or bye-laws in force at the colliery or for infringing the Coal Mines Act Regulations, or for disobeying an order of an official of the mine, the workman may be dismissed without notice, the manager can act accordingly. Even without such a provision summary dismissal may be justified for misconduct so serious as to amount to a failure by the workman to perform an essential part of his duty under the contract. Wilful disobedience will, except in trivial matters, amount to such misconduct, as will neglect or gross mishandling of machinery. Absenteeism may justify summary dismissal, if it is so frequent and persistent as to show a virtual disregard of the contract of service.

9. *Dismissal with Notice.* Dismissal with notice in accordance with the contract is always open to the Management; but under the Agreement between the Board and the NUM, dated 19th August, 1948, the Union reserve the right to take action if they consider that dismissal with or without notice is not reasonable under the circumstances. The Union did, however, agree under Clause VII of the Scheme of the Joint Committee for securing more regular attendance, that they would not contest disciplinary action taken by the management or countenance any concerted action by other workmen in support of the workman concerned.

10. *Imposition of a Fine.* A fine can only be imposed if — (i) It is provided for in the contract; or (ii) the workman agrees to submit to it in lieu of another penalty.

If the fine is to be deducted from wages, care should be taken to comply with the Truck Acts. To secure this, there must be a written agreement to submit to a fine signed by the workman and specifying the amount and the act or omission for which it is imposed. The fine must, of course, be fair and reasonable and must not exceed the actual or estimated loss occasioned to the Board.

11. *Suspension.* Before the war, power to suspend a workman was either expressly or by custom included in many contracts at particular pits. In particular, a man's lamp was often stopped for one or two days on reporting

after being absent without satisfactory excuse. This power was in abeyance during the war, but, *where the contract permits*, stopping the lamp after a day's absence may be found useful. This should, however, only be done during the week in which a man has absented himself, because a workman so suspended might otherwise claim that he was capable of and available for work and therefore entitled to a guaranteed wage.

12. *Downgrading.* In many contracts of employment, the management have the right to transfer a man from one class of work to another and to pay wages appropriate to the new work. Where this is so, no difficulty arises; but in other cases the workman will have to be given notice to terminate his contract and offered a new contract in the lower grade, or engagement at another pit.

13. *Proceedings for Breach of Contract.* There remains the remedy of proceedings for breach of contract. This in suitable circumstances and particularly in some Divisions, may be an appropriate alternative to other penalties short of dismissal.

Conclusion

14. If an improvement in the standards of attendance is to be achieved, firm measures must be taken to deal with bad attenders in accordance with this procedure. In taking them, Divisional Boards and lower managements will have the full support of the National Board.

CLOSING UNECONOMIC PITS: AN EXAMPLE OF THE "COMMUNICATIONS" PROBLEM TAKEN FROM NORTHUMBERLAND [1]

N.C.B. THREAT TO WEST TYNE PITS?

Acomb and W. Wylam May Be Victims of Long-Term Policy

The existence of a number of small collieries west of Newcastle, including the Acomb and West Wylam pits, may be threatened as a result of a reported National Coal Board Plan to close Ventners Hall Colliery, Bardon Mill, at the beginning of October, as part of the Board's long-term policy.

Sixty Acomb men have already been transferred to the Mickley area — a move which may foreshadow the closing of the Acomb pit, despite the fact that efforts are even now being made to open new seams. The Ventners Hall men are to resist the threatened closure of their pit by every means at their disposal.

Ventners Hall Colliery is believed to be the first pit marked out for closing under the Coal Board's policy and at a meeting held behind closed doors at Haltwhistle on Sunday the eighty men employed there decided to make theirs the test case for all the West Tyne small pits, and to resist the N.C.B. plan with every means to their hand.

The men complained at their meeting that the administrators were bungling the whole nationalisation scheme and were in consequence jeopardising the future of the local mining industry and threatening large-scale unemployment and a mass diversion of labour.

While they refused to disclose their plan of campaign, which will take fuller shape at another meeting in the next fortnight, to resist any attempt by the Board to transfer them to other pits and eventually to close Ventners Hall, the miners said that they intended to fight to the last ditch. They said such interference was dangerous at an hour when the nation was in desperate need of every ton of coal it could obtain and when the miners were putting all they knew into the job and topping their 900 ton target easily every week with an average of 34 hundredweights per man shift — an output far in excess of that at thousands of other collieries in the country.

They complained, too, that the National Coal Board had kept neither themselves nor the management adequately informed of the progress of their plans for the closure of the local collieries.

[1] From the *Hexham Courant,* May 11, 1951. Hexham, Northumberland.

The men were told that the National Coal Board were contemplating the closure of the colliery and had fixed October 1st as a provisional date for production to cease because the coal being produced was of an inferior quality and it had been difficult to dispose of it at an economic price.

The Board got its answer on Sunday when the men drew attention to the fact that last week lorries which arrived from the West of England for supplies had to be sent away empty because the demand had far exceeded the supply and there was not sufficient coal at the pithead to fill them.

The men acknowledge the presence of large quantities of sulphur in the unscreened coal which is all that the colliery produces, but they are doubtful about the accuracy of the analyst's report which indicates that 14 per cent of the coal is sulphur.

If that report is correct, however, the men put forward two reasons why they should still continue to work the colliery.

They think that with the nation in dire need of sulphur some scheme could be devised for the extraction of this mineral and they also point to the fact that a non-sulphur coal seam in the pit is standing unworked.

The unscreened coal from Ventners Hall is sold in the main for £2 5s. a ton compared with £3 5s. for normal good house coal and the National Coal Board claim that they are suffering a heavy loss on every ton that is produced in spite of the high output of the colliery, which averages almost 200 tons a day and 35 hundredweights per manshift including all non-production staff.

The men are hoping that the National Coal Board will also have second thoughts about the plan and that the Miners' Union and other organisations will not have to be called into the fight.

They see in the introduction this week of a £1,500 electric cutter and the promise of two more similar machines a glimmer of hope for the colliery, for they cannot understand why the Coal Board should be introducing modern equipment and at the same time talk of closing the pit.

Men from Birkshaw colliery, Bardon Mill, are in sympathy with their miner neighbors because they feel that their future, too, may be affected by the closing of the Ventners Hall pit.

They fear that if Ventners Hall is closed the labour force will be transferred to their coal faces and they have the impression that this will seriously reduce the life of their colliery.

Miners from both pits also fear that the merger of labour forces might mean loss of pay, for many of the skilled piece workers have the impression that they may be forced to accept datal work while trainee pitmen work at the coal face and thus earn the higher money.

A Haltwhistle man who was concerned in the distribution of coal from Ventners Hall said that the National Coal Board had only themselves to blame for the fact that their income from Ventners Hall colliery had dropped.

He said that when the Coal Board assumed control of the pit they had immediately cried "stinking fish" by reducing the price of the coal.

For years the people of Carlisle and elsewhere had accepted the Ventners Hall coal as satisfactory. Occasionally there had been complaints but in the main the consumers took what supplies were offered them.

When the Coal Board reduced the price and offered only unscreened supplies the domestic and industrial consumers were naturally suspicious and demanded supplies from other collieries.

He said that the coal from Ventners Hall was still extremely valuable and but for the hundreds and thousands of tons that had gone from the colliery during the past winter there would have been many empty grates on Tyneside.

It is not only in their own colliery at Ventners Hall that the miners allege bungling on the part of the N.C.B.

Many of the older miners this week told a "Courant" reporter that there was hundreds of pounds worth of equipment lying idle at Lambley colliery which would probably never be used.

They call the colliery, into which the N.C.B. has sunk thousands of pounds "Coal Board Folly," because they contend that the enterprise in search of new coal seams will be a complete failure.

They point, too, to Acomb colliery, from which 60 of the 290 workers are being transferred to the Mickley area. They see in this the first move towards closing the colliery where money is being spent in an attempt to find new coal faces.

They allege that the men sent from Acomb are not needed in the Mickley district and will simply have to stand about instead of getting on with the job of coal production.

They condemn the Coal Board officials for administering the scheme from remote places and not making thorough inspections of the pits before deciding on their actions.

The Haltwhistle miners say they will not take closure of their colliery lying down. They will fight, not only because it is their livelihood that is at stake, but because they do not want to leave their native district for other work.

They contend that if there is to be a mass exodus of pitmen the district should be informed and local councils should not be fooled into sinking thousands of pounds of ratepayers' money into housing schemes that might after all never be needed.

A well-known local miner said: "We shall fight to the last ditch. It may mean taking this a long way, but however far it does go we shall follow it, and if we don't win then there may be something doing."

VENTNERS COLLIERY AND THE N.C.B.

Statement by Board and the Miners' Union[2]

The "Hexham Courant" report last week of the projected closing of the Ventners Hall colliery under a long-term plan of the National Coal Board, has drawn a joint statement from the Northumberland and Cumberland Division of the Board and the Northumberland Area of the National Union of Mineworkers.

In a "prepared statement," which is given below, the Board deals item by item with the "Courant" report, and admits finally and definitely that the Ventners pit is to be closed.

The article published in the "Hexham Courant" of May 11th under the heading "N.C.B. Threat to West Tyne Pits?" is so misleading that we feel that it must be answered jointly by the National Coal Board and the National Union of Mineworkers. Many of the statements made are inaccurate and seriously misrepresent the true position whilst the inference likely to be drawn from the article is one of complete lack of co-operation between the N.U.M. and the Board. Nothing could be further from the truth and we propose therefore to answer the points made in your paper in the order in which they occur.

[2] From the *Hexham Courant*, May 18, 1951. Hexham, Northumberland.

1. There is no threat to the West Tyne Pits as a whole. Some of these pits do suffer from certain disadvantages of long haulage, thin seams, inferior quality of coal, or exhausting reserves. But there is every prospect of a very considerable life at all except Throckley, Ventners Hall and possibly Acomb.

2. You state in your article that "sixty Acomb men have already been transferred to the Mickley area — a move which may foreshadow the closing of the Acomb pit." The transfer of these men has not yet taken place, it is in fact scheduled for the end of May and has been made necessary as all concerned are aware by the deterioration in the quality of coal in one of the main developing districts which has had to be abandoned.

The men concerned in the move are those men who now live in the Mickley area and who have to travel daily to Acomb to work. The question has been the subject of the fullest and closest consultation between the Board and the N.U.M. and has only been agreed after approval at all levels.

Contrary to the suggestion in your article, the transfer will help prolong the life of the pit and give added security to the Acomb families who work there.

3. It is true that the Board has decided to close Ventners Hall. The decision has been made because of the difficulty of disposing of the coal. The Thirlwell Seam at Ventners is of marginal quality. The normal outlet for the smalls was the Carlisle Electricity Works. These works have found themselves unable to accept it owing to the damage caused by the sulphur to their plant. The only alternative was to make an unscreened coal, for which there is not a sufficient market at prices fixed by the Ministry of Fuel and Power. The "zoning" of these prices makes it uneconomic to send it far afield, and the local demand is not sufficient to keep the pit going without heavy stocking during the summer.

4. According to the article, the men at Ventners Hall are prepared to resist the N.C.B. plan with every means to their hands whereas on the contrary the Northumberland N.U.M. and the Ventners Hall Lodge have shown full co-operation in the plan. Nobody likes the closure of a pit, but all responsible on both sides of the Industry have agreed that this has to be faced, and that the plan which has been made involves no loss in output.

5. It is true that Ventners Hall has a high output per manshift when geological conditions underground allow it. The workmen at the colliery are as good as any in the country and, given a trouble-free week, can produce the 900 tons and almost 34 cwts. per manshift which you twice quote as a regular output. But the fact remains that faulting and ramble (following stone) have continually baulked production. The figure of 900 tons per week has only been topped once this year (909 tons for the week ending 3rd March). The average weekly output for the year to 5th May, excluding all weeks affected by holidays and including six working Saturdays, is 775 tons. The O.M.S. for the first Quarter is 24.8 cwts.

6. The complaint alleged to have been made by the men that the Board has kept neither the management nor men informed is best answered by reference to the meetings which have taken place. The future of Ventners Hall has been the subject of discussion between the N.U.M. and the Board.

7. In your article you state that last week lorries which arrived from the West of England for supplies had to be sent away because the demand had far exceeded the supply. The truth is that on one occasion a lorry from Maryport arrived, as arranged, at the pit and, unfortunately, found that production had temporarily ceased owing to a breakdown underground. After waiting for four hours and having been informed that prospects of a resumption of pro-

duction that day were not good, it went away. Actually production was resumed later in the day, and other lorries took away their loads.

8. The men would have good reason to doubt any analyst's report which gives the sulphur content of the seam as 14 per cent. A recent spot sample (April 20th) gave 14.2 per cent as the sulphur content of the top 6 inches of the seam. (Usually this top stratum averages 1 per cent of the sulphur). The seam as a whole averages 5.2 per cent. The Board has been trying for some time to fix an economic method of extracting the sulphur, which here is mainly inherent in the coal, but have so far been unsuccessful. The sample showing 14.2 per cent was the latest which has been sent away for investigation.

9. The men it is said point to the fact that there is a non-sulphur coal seam in the pit standing unworked. There is, in fact, no other seam in the pit. The only other seam known to exist at Ventners Hall is at a considerable depth and is only 1 foot 3 inches thick, including a stone band.

10. The new type of coal cutter which has recently been introduced is one of two which was expected to be delivered in December of last year. It was hoped at the time of ordering that by cutting immediately below the top stratum of high sulphur coal it would be possible to reduce the average of the seam to a reasonably low level. In view of the increase in sulphur content of the seam as a whole, it is not expected that the improvement will be sufficiently marked. The object in sending the machine even at this late hour to Ventners Hall is to gain experience particularly in keeping up the ramble with the help of the top coal.

11. It is stated that the men at Birkshaw were at first doubtful about the ability of their pit to take on extra men. At a meeting at Burt Hall the men went over the pit plans with Board representatives and they agreed that it was a practicable move. Indeed, one of the main development places has been standing for some months as sufficient pit room has already been won out. In addition there are very large reserves of Little Limestone coal at Bardon Mill.

12. When transfers of men take place there is always a fear of loss of wages. The Board and the N.U.M. are agreed that the plan as it has been designed will reduce this hardship to the minimum and that, although some face workers may have to change their class of work, e.g. from filling to stonework, they will all be engaged on face work.

13. The article quotes a Haltwhistle man as saying that the Board have only themselves to blame for the fact that their income from Ventners Hall has dropped. It is agreed that the Carlisle people had accepted the coal for years. But the coal they accepted was Best coal, and the production of Best coal depends on a market for the Small. Once Carlisle had refused to accept the Small it was impracticable to produce Best. There is a stock of 1,566 tons of Small on the ground still, which has lain throughout the winter without a market. As for the price, that is fixed for the domestic market by the Fuel Overseer.

14. It is agreed that the unrationed coal from Ventners Hall has filled many fireplaces on Tyneside. But, from a national point of view, it is surely preferable to produce a similar tonnage of coal which can be used by industry generally rather than continue to produce a coal which virtually lacks a market.

15. Lambley is reported to be called "Coal Board Folly," into which the N.C.B. has sunk thousands of pounds. It is probably not appreciated that when the Board took over on Vesting Day there was barely two years' life in the seams above the line of the water that lies against the Stublick Dyke. The

continued livelihood of 400 families depended on quick action by the Board in making arrangements to lower the water level and exploit the lower part of the seams which have been worked to the water level during the last 250 years.

One approach, at Whitescut, to the seams at the west end has been a failure owing to heavy faulting. This accounts for some machinery being temporarily out of use. The other two at Midgeholme and Lambley still have their difficulties, but will overcome them in time. In the meanwhile, those 400 families are working, and coal is being produced. If the Board had failed to act to meet possible social consequences they would undoubtedly have been charged with disregard or neglect, but when the Board act as they have done to ensure the continued employment of this community they must still be the target.

Any stick will do to beat Nationalisation.

16. It is alleged that the men transferred from Acomb to Mickley will simply have to stand about instead of getting on with the job of coal production. We protest most emphatically at the injustice and untruth of this allegation. Faces are ready for them from which they will be able to produce high quality coal instead of the inter-banded coal which they have been working at Acomb, 32 per cent of which has to be discarded in the process of cleaning at the surface. All the faceworkers transferred will be immediately placed to face work.

17. The Coal Board officials are condemned for administering the scheme from remote places and not making thorough inspections of the pits before deciding on their actions. This is a completely erroneous impression. Such schemes as these are not, in fact, hatched around the Board Room table. They are put up to the Board for approval by Area and Pit officials who know the pits intimately and have weighed up the situation carefully before any proposal is made.

18. The Haltwhistle miners are reported as saying they will fight not only because it is their livelihood which is at stake, but because they do not want to leave their native district for other work. There is no question of any miner being asked to leave Haltwhistle. Instead of being taken by bus to Ventners Hall he will be taken to Birkshaw, or wherever it may be. The local councils have not, therefore, been fooled into sinking thousands of pounds of ratepayers' money into housing schemes that might, after all, never be needed. We regard it as mischievous that any such suggestion should have been made.

We regret the necessity of this long reply to your article, but think it important in the interests of the Nation that the good relations existing between the N.C.B. and the N.U.M. should not be endangered by allowing ill-informed criticism and misrepresentation to go unanswered.

The statement is signed by: J. Bowman, Chairman, Northern (N. and C.) Division; and R. Main, Secretary, Northumberland Areas, N.U.M.

SELECTED BIBLIOGRAPHY

A NOTE ON THE BIBLIOGRAPHY

The unannotated bibliography is concerned primarily with items that bear on the coal industry as such, though a few introductory references to works on socialism and nationalization are also included. No entries have been made from the unpublished documents and pamphlet material, often primarily of local interest and sometimes confidential, which any investigator will find in the coalfields. The printed but usually confidential *Minutes* of the NUM's National Executive Committee and of its Area Executive Committees are particularly valuable examples of this class of materials.

An important document not separately listed is the *Report* of the NUM's National Executive Committee, presented annually to the union's Conference. This is published independently in advance of the Conference and is reprinted with the verbatim report of the *Conference Proceedings*. The latter has not generally been considered a public document, but is usually available to responsible students of the industry. It is invaluable, and not only for an understanding of the industry's labor problems. Verbatim *Reports* are also published following Special Conferences. The NUM publishes a bimonthly *Information Bulletin* devoted mainly to reporting selected industry statistics and Parliamentary news of importance to the mining industry.

The following official publications, not listed in the bibliography, are important: the NCB's *Daily Press Summary,* which draws from a large sample of British newspapers; *Coal,* the NCB's monthly newspaper intended mainly for employees of the industry; detailed statistics are available in the Ministry of Fuel and Power's *Weekly Statistical Statement* (mimeographed). At irregular intervals (seven times since 1944), the Ministry has published a booklet of summary tables under the title *Statistical Digest.* Financial figures, arranged by some twenty major districts, will be found in the Board's *Quarterly Statistical Statement of the Costs of Production, Proceeds, and Profits of Collieries.* All national labor agreements between the Board and the several unions are published in the *Memorandum of Agreements, etc.,* a cumulative set of annual volumes. Detailed maps showing the location of every colliery and information concerning the names of managerial personnel at all levels of the industry will be found in the handbook, *Guide to the Coalfields,* of which a new edition is published annually by the Colliery Guardian Co., Ltd.

With some arbitrariness, I have put asterisks on those items which I feel, taken together, give the most useful introduction to the industry.

SELECTED BIBLIOGRAPHY

1. Books and Pamphlets

Acton Society Trust, *Studies in Nationalized Industry.* (Nine pamphlets.) G. R. Taylor, ed. London, The Acton Society, 1950–51.

Allen, G. C., *British Industries and Their Organisation.* London, Longmans, Green, 1945 ed. 338 pp.

Anglo-American Council on Productivity, *Coal.* London, n.p., 1951. 107 pp.

Arnot, R. Page, *The Miners: A History of the Miners' Federation of Great Britain, 1889–1910.* London, Allen and Unwin, 1949. 409 pp.

Beer, Max, *A History of British Socialism.* London, Allen and Unwin, 1 vol. ed., 1948. 451 pp.

*Benney, Mark., *Charity Main: A Coalfield Chronicle.* London, Allen and Unwin, 1946. 176 pp.

Chalmers, J. M., Ian Mikardo, and G. D. H. Cole, *Consultation or Joint Management? A Contribution to the Discussion of Industrial Democracy.* Fabian Tract No. 277. London, Fabian Publications, Ltd., 1949. 28 pp.

Chester, D. N., *The Nationalised Industries: A Statutory Analysis.* London, Institute of Public Administration, 1951 ed. 48 pp.

Clegg, Hugh, *Labour in Nationalised Industry.* Interim Report of a Fabian Research Group. Research Series No. 141. London, Fabian Publications, Ltd., 1950. 40 pp.

Cole, G. D. H., *A History of the Labour Party from 1914.* London, Routledge, 1948. 517 pp.

———, *The National Coal Board: Its Tasks, Its Organisation, and Its Prospects.* Research Series No. 129. London, Fabian Publications, Ltd., 1948. 45 pp.

Cole, Margaret, *Miners and the Board.* Based on the Report of a Fabian Research Group. Research Series No. 134. London, Fabian Publications, Ltd., 1949. 25 pp.

*Court, W. H. B., *Coal.* London, HMSO and Longmans, Green, 1951. 422 pp. One volume in the series, *History of the Second World War, United Kingdom Civil Service,* W. K. Hancock, ed.

Crenon, Henri, *Le Regime des Mines en Angleterre et la Question de la Nationalisation des Mines.* Paris, Edouard Duchemin, 1921. 157 pp. (Ph.D. Thesis.)

de Man, Henry, *The Psychology of Socialism.* Trans. by Eden and Cedar Paul. London, Allen and Unwin, 1928. 509 pp.

*Dickie, J. P., *The Coal Problem, A Survey: 1910–1936.* London, Methuen, 1936. 368 pp.

Ditz, Gerhard W., *British Coal Nationalized*. New Haven, Conn., The Edward W. Hazen Foundation, 1951. 92 pp.

Edwards, Ness, *The History of the South Wales Miners*. London, Labour Publishing Co., 1926. 122 pp.

Griffiths, James (M.P.), *Coal*. London, The Labour Party, 1942. 24 pp.

Hall, W. S., *A Historical Survey of the Durham Colliery Mechanics' Association: 1879–1929*. Durham, J. H. Veitsch and Sons, 1929. 136 pp.

* Haynes, W. H., *Nationalization in Practice: The British Coal Industry*. Harvard University, Graduate School of Business Administration. Cambridge, 1953. 413 pp.

Heinemann, Margot, *Britain's Coal: A Study of the Mining Crisis*. Foreword by Will Lawther. Labour Research Department. London, Victor Gollancz, 1944. 195 pp.

———, *Coal Must Come First*. Labour Research Department. London, Frederick Muller, 1948. 72 pp.

Jevons, H. Stanley, *The British Coal Trade*. London, Kegan Paul, 1915. 876 pp.

Jones, J. H., G. Cartwright, P. H. Guenault, *The Coal Mining Industry*. London, Pitman, 1939. 394 pp.

Jones, Joseph, *The Coal Scuttle*. London, Faber and Faber, 1936. 168 pp.

*Lancaster, C. G. (M.P.), Sir Charles Reid, and Sir Eric Young, *The Structure and Control of the Coal Industry*. London, Conservative Political Center, 1951.

Lawson, Jack, *A Man's Life*. London, Hodder and Stoughton, 1944. 191 pp.

Livesey, W., *The Mining Crisis*. London, Simkin, Marshall, Hamilton, Kent and Co., 1921. 89 pp.

Lubin, Isador, and Helen Everett, *The British Coal Dilemma*. New York, Macmillan, 1927. 370 pp. plus xii.

Morgan, L. A., *The Dutch State Coal Mines: What Government Administration Has Achieved*. The Fabian Society. London, Victor Gollancz, 1944. 21 pp.

Nef, J. U., *The Rise of the British Coal Industry*. London, Routledge, 1932. 2 vols., 448 pp. and 490 pp.

Neuman, Andrew Martin, *Economic Organisation of the British Coal Industry*. London, Routledge, 1934. 537 pp. plus xxi.

Newsom, John, *Out of the Pit: A Challenge to the Comfortable*. Oxford, Basil Blackwell, 1936. 118 pp.

Political and Economic Planning, *The British Fuel and Power Industries*. London, PEP, 1947. 406 pp.

———, Industries Group, *Report on the British Coal Industry: A Survey of the Current Problems of the British Coalmining Industry and of the Distribution of Coal, with Proposals for Reorganisation*. London, PEP, 1936. 214 pp.

*Raynes, J. R., *Coal and Its Conflicts: A Brief Record of the Disputes Between Capital and Labour in the Coal Mining Industry of Great Britain*. London, Ernest Benn, 1928. 342 pp.

Redmayne, Sir R. A. S. *Men, Mines, and Memories*. London, Eyre and Spottiswoode, 1942. 326 pp.

———, *The Problem of the Coal Mines*. London, Eyre and Spottiswoode, 1945. 60 pp.

———, and Gilbert Stone, *The Ownership and Valuation of Mineral Property in the United Kingdom*. London, Longmans, Green, 1920. 256 pp.

Reid, Sir Charles, Reprinting of articles from the *London Times,* November 22, 23, 24, 1948.

Rowe, J. W. F., *Wages in the Coal Industry.* London, P. S. King and Son, 1923. 174 pp. plus viii.

Sargent, J. J., *Seaways of the Empire.* London, A. and C. Black, 1930. 145 pp.

Walkerdine, R. H., ed., *Guide to the Coalfields* (annual). London, Colliery Guardian Co., 1948 ed. 395 pp. and cix.

Watkins, Harold M., *Coal and Men: An Economic and Social Study of the British and American Coalfields.* London, Allen and Unwin, 1934. 460 pp.

West Yorkshire Coal Owners' Association, *Revised Volume of Agreements and Arrangements with the Yorkshire Mineworkers' Association and Other Trade Unions Having Members Employed at West Yorkshire Collieries.* Leeds, 1935. 106 pp.

White, Eirene, *Workers' Control?* "Challenge" Series No. 4. London, Fabian Publications, Ltd., 1949. 29 pp.

Williams, Francis, *Fifty Years' March: The Rise of the Labour Party.* London, Odhams Press, n.d., 384 pp.

*Wilson, Harold, *New Deal for Coal.* London, Contact Publications, 1945. 264 pp. plus xii.

*Zweig, F., *Men in the Pits.* Foreword by Ronald H. Smith. London, Victor Gollancz, 1949. 177 pp.

2. Publications of the Government of Great Britain

Central Office of Information, *Men and Mining,* by Geoffrey Thomas. London, The Social Survey, N.S. 113, 1948. 23 pp.

————, *The Recruitment of Boys to the Mining Industry: An Inquiry Carried Out in Six Coalfields for the Directorate of Recruitment, Ministry of Fuel and Power, and the National Coal Board, August–October, 1946.* London, the Social Survey, N.S. 77/2/3, 1946. 77 pp., mimeo.

————, *Scottish Mining Communities.* London, The Social Survey, N.S. 61, 1946. 115 pp., mimeo.

Coal Industry Nationalisation Act, 1946, 9 & 10 Geo. 6. Ch. 59. London, HMSO, 1946. 59 pp.

*Fuel and Power, Ministry of, *Coal Mining: Report of the Technical Advisory Committee,* Cmd. 6610. (The Reid Report.) London, HMSO, 1945. 149 pp.

————, *Regional Survey Reports.* London, HMSO, as follows:
Bristol and Somerset Coalfield, 1946. 84 pp.
Durham Coalfield, 1945. 48 pp.
Forest of Dean Coalfield, 1946. 59 pp.
Kent Coalfield, 1945. 4 pp.
Coalfields of the Midland Region, 1945. 67 pp.
North Eastern Coalfield, 1945. 59 pp.
North Midland Coalfield, 1945. 42 pp.
Northumberland and Cumberland Coalfields, 1945. 59 pp.
North Western Coalfields, 1945. 80 pp.
South Wales Coalfield (including Pembrokeshire), 1946. 218 pp.

*National Coal Board, *Annual Report and Statement of Accounts for the Year Ended 31st December . . .* (annually, 1946–1953). London, HMSO. Various pp.

————, *Guide to Consultation.* London, Hobart House, 1948. 48 pp.

——, *Memorandum of Agreements, Arbitration Awards and Decisions, Recommendations, and Interpretations Relating to National Questions Concerning Wages and Conditions of Employment in the Coalmining Industry of Great Britain. Part I: Period 20th March 1940 to 31st July 1946*. Parts II–VIII (annually, 1947–1953). London, Hobart House. Various pp.

——, *Notes on Miners' Welfare*. Published on behalf of the Miners' Welfare Joint Council. London, Hobart House, 1949. 67 pp.

*——, *Plan for Coal: The National Coal Board's Proposals*. London, 1950. 76 pp.

*——, Report of the Advisory Committee on Organisation (The Fleck Report). London, Hobart House, February 1955. 105 pp.

——, East Midlands Division, Area No. 1, Bolsover, Derbyshire. *The Bolsover Story: An Experiment to Reduce the Cost of Coal Production and to Alleviate the Difficulties Arising from the Decline in Manpower in the Coal Mining Industry*. n.p. 1950. 71 pp.

Royal Commission on the Coal Industry, *Report of the Royal Commission on the Coal Industry (1925)*. (With Minutes of Evidence and Appendices.) Vol. I. Report. Cmd. 2600. London, HMSO, 1926. 295 pp. (The Samuel Report.)

Scottish Home Department, *Scottish Coalfields: The Report of the Scottish Coalfields Committee*. Cmd. 6576. Edinburgh, HMSO, 1944. 184 pp.

3. Publications of Trade Unions

Durham Miners' Association, *Information Pamphlet* (Various numbers). Durham, 1943–1951. Various pp.

——, *Rules (Including Political Fund Rules)*. Durham. Revised October 1937. 46 pp.

*National Union of Mineworkers, *Annual Conference. Proceedings*. London.

——, *Information Bulletin*. Vols. I–V (1946–1950). London.

——, *Model Rules: As Adopted at a Special Conference of the Union Held on October 10–12, 1945*. London, 1945. 8 pp.

——, *Rules*. London, 1947. 32 pp.

——, Durham Area. *General Secretary's Annual Report for the Years 1946–1950*. Durham. Various pp.

——, Durham Area. *National and District Agreements to January 1948*. Durham. 283 pp.

——, Yorkshire Area. *Agreements, Etc. and Wage Rates: 1926–1946*. Barnsley. 205 pp.

——, Yorkshire Area. *Findings of Joint Sub-Committees Set Up Under the Conciliation Machinery for the Settlement of Disputes: 19 March 1947 to 29 April 1948*. Barnsley. 174 pp.

Northumberland Miners' Mutual Confident Association, *Agreements Between the Northumberland Coal Owners' Association and the Northumberland Miners' Mutual Confident Association*. Newcastle-Upon-Tyne, 1920. 48 pp.

Trades Union Congress, General Council, *Coal: The Labour Plan*. London, n.d. 31 pp.

Yorkshire Mine Workers' Association, *Price Lists and Agreements of Collieries in the Association*. Barnsley, 1937. 328 pp.

4. Articles in Trade and Professional Journals

Beacham, A., "The Present Position of the Coal Industry in Great Britain." *The Economic Journal*, Vol. 60, No. 237. March 1950, pp. 9–18.

Cole, G. D. H. "The National Coal Board." *The Political Quarterly*, October–December 1946.

* Dahl, Robert A., "Workers' Control of Industry and the British Labor Party." *The American Political Science Review*, October 1947, Vol. 41, pp. 875–900.

George, R. F., "Statistics Relating to the Coal Mining Industry." *Journal of the Royal Statistical Society*. Series A (General), Vol. CXII, Part III, 1949, pp. 331–337.

Henderson, H. D., "The Reports of the Coal Industry Commission." *The Economic Journal*, Vol. 29, September 1919, pp. 265–279.

Iron and Coal Trades Review. Diamond Jubilee Issue. London, 1927. Vol. CXIV. 224 pp.

*Koenig, Robert P., "An American Engineer Looks at British Coal." *Foreign Affairs*, January 1948, pp. 3–16.

Mack, John A., "More Education for Industrial Democracy," *The Highway*, Workers' Educational Association. Vol. 42, December 1950, pp. 53–54.

Moos, S., "The Statistics of Absenteeism in Coal Mining," *The Manchester School for Economic and Social Studies*, January 1951, pp. 89–108.

Patterson, T. T., and F. J. Willett, "Unofficial Strike," *The Sociological Review*, Vol. XLIII, Sec. 4, 1951, pp. 57–94.

The Political Quarterly. "Special Number: Nationalised Industries." April–June 1950. Vol. XXI, No. 2.

Rhodes, E. C., "Output, Labour, and Machines in the Coal Mining Industry in Great Britain." *Economica*, May 1945, pp. 101–110.

Tawney, R. H., "The British Coal Industry and the Question of Nationalisation." *Quarterly Journal of Economics*, Vol. 35, November 1920, pp. 61–107.

*Trist, E. L., and K. W. Bamforth, "Some Social and Psychological Consequences of the Longwall Method of Coal-Getting." *Human Relations*, Vol. IV, No. 1, 1951, pp. 3–38.

5. Miscellaneous Official Publications

International Labour Organisation. Coal Mines Committee, Fourth Session, Geneva, May 1951, *General Report*. Geneva, International Labour Office, 1951. 108 pp.

——, Coal Mines Committee, Fourth Session, Geneva, May 1951, *Productivity in Coal Mines*. Geneva, International Labour Office, 1951. 177 pp.

INDEX

INDEX

DATE DUE

FEB 2 3 1979			
MAY 1 6 1980			
OCT 1 9 1990			
MAY 0 7 2002			
7/24/2002			
GAYLORD			PRINTED IN U.S.A.

Managerial
Cost
Accounting

Harold Bierman, Jr.

The Nicholas H. Noyes Professor of Business Administration,
Graduate School of Business and Public Administration,
Cornell University

Thomas R. Dyckman

Professor of Accounting and Quantitative Analysis,
Graduate School of Business and Public Administration,
Cornell University